Seed Science and Technology

Seed Science and Technology

Edited by **Shirley Doy**

SYRAWOOD
PUBLISHING HOUSE

New York

Published by Syrawood Publishing House,
750 Third Avenue, 9th Floor,
New York, NY 10017, USA
www.syrawoodpublishinghouse.com

Seed Science and Technology
Edited by Shirley Doy

International Standard Book Number: 978-1-68286-154-7 (Hardback)

The publisher's policy is to use permanent paper from mills that operate a sustainable forestry policy. Furthermore, the publisher ensures that the text paper and cover boards used have met acceptable environmental accreditation standards.

Trademark Notice: Registered trademark of products or corporate names are used only for explanation and identification without intent to infringe.

Printed in the United States of America.

Contents

Preface

The recent researches performed in seed science have transformed agricultural practices and have affected agricultural production as well. This book is meant for students who are looking for an elaborate reference text on seed science. It delves into significant topics, such as seed development, seed structure, germination, etc. With state-of-the-art inputs by acclaimed experts of this field, this book targets students and professionals alike.

The information contained in this book is the result of intensive hard work done by researchers in this field. All due efforts have been made to make this book serve as a complete guiding source for students and researchers. The topics in this book have been comprehensively explained to help readers understand the growing trends in the field.

I would like to thank the entire group of writers who made sincere efforts in this book and my family who supported me in my efforts of working on this book. I take this opportunity to thank all those who have been a guiding force throughout my life.

Editor

Adoption of climbing beans in the central highlands of Kenya: An empirical analysis of farmers' adoption decisions

Lara Ramaekers[1], Alfred Micheni[3], Paul Mbogo[3], Jozef Vanderleyden[1]* and Miet Maertens[2]

[1]Centre of Microbial and Plant Genetics, Department of Microbial and Molecular Systems, KU Leuven, Leuven, Belgium.
[2]Division of Bio-economics, Department of Earth and Environmental Sciences, KU Leuven, Leuven, Belgium.
[3]Kenyan Agricultural Research Institute (KARI), Nairobi, Kenya.

Common bean represents the second staple crop in the Kenyan highlands. Decreasing yields and overpopulation in this area demand an intensification of food production. The Kenyan Agricultural Research Institute (KARI Embu) has been promoting improved climbing bean varieties in order to boost yields. To quantify the impact of this work, a survey was done in order to (i) assess awareness, trial and adoption rates of climbing beans by local farmers and (ii) acquire insight in this adoption process. This more detailed analysis of the various adoption stages is the paper's unique contribution to the adoption literature. At the time of the survey, 90% of the farmers were aware of climbing beans, about 40% had grown climbing beans on their farms at least for one season (test/trial) and only about 11% had maintained its production (adoption). Climbing beans seem to be more popular at higher altitudes where they are grown on small areas. Increasing age of the household head, contact with extension services and farmer-to-farmer transmission are important for awareness and testing climbing beans. A serious limitation for both trial and adoption is the poor availability of seed. From this study, recommendations were made on how to improve climbing bean awareness, trial and adoption.

Key words: Climbing bean, technology diffusion, technology adoption, farm-household survey, Heckman two-stage procedure, Kenyan highlands.

INTRODUCTION

The central Kenyan highlands are characterized by a rapidly growing population with over 1000 people per km^2 thereby causing land fragmentation, over-cultivation and declining soil fertility which eventually leads to decreasing yields and decreasing food security (Government of Kenya, 2001; Mugwe et al., 2008). There is a pressing need to intensify food production in this region. One possible way to achieve higher yields and increase food availability is the promotion of climbing bean cultivation.

Common bean (*Phaseolus vulgaris* L.) represents the second most important staple crop in the Kenyan highlands, after maize. Beans are grown by over 95% of farmers in the region, providing over 65% of the protein and 35% of the caloric intake (CIAT, 2004). Mainly bush beans are grown, using varieties introduced by the Kenyan Grain Legume Project (GLP) in 1984. The introduction of climbing beans in this area entails a high potential as substitute of or complement to bush bean production due to its specific characteristics.

The most outstanding characteristic of climbing beans is their high yield potential: up to 4 to 5 tons/ha versus 3 tons/ha for bush beans under optimal conditions (CIAT, 2004). This allows significant climbing bean harvests,

*Corresponding author. E-mail: jozef.vanderleyden@biw.kuleuven.be.

even on very small plots such as backyard gardens (Sperling and Muyaneza, 1995). Typical for climbing beans is their staggered harvesting which is more labor intensive but increases and smoothens the availability of food throughout the growing season as green leaves, pods and fresh or dry grain can be consumed (Sperling et al., 1992). Moreover, their wealthy biomass can be used as fodder for animals or may provide ground cover, control weeds and contribute to soil organic matter when not all leaves are harvested. Climbing bean cropping systems are classified as monoculture or intercropping (Woolley and Davis, 1991). Inter-cropping involves growth in association with other crops (especially maize and banana), either in relay or simultaneous planting, with the other crop providing support for the climbing beans. In monoculture, climbing beans are planted with the support of wood or bamboo stakes (mostly in Africa) or trellis systems (widespread in Andean region). Due to its labour intensity and high yield potential, climbing beans entail a high potential in land-scarce but labour-abundant regions, such as the Kenyan highlands. Some disadvantages of climbing beans perceived by farmers are (i) the longer growing period (4 months compared to 3 months for bush beans), (ii) sensitivity to drought and (iii) increased labour requirements for staking and bird scaring as climbing beans are easily attacked by birds (Sperling et al., 1992; CIAT, 2004; personal communication, KARI Embu).

Since the introduction of climbing beans in the central Kenyan highlands, no quantitative research in this area has been done to assess current diffusion and adoption rates[1] and analyse the determinants of farmers' decision to adopt –or not. In this paper, we present a quantitative adoption analysis and address three sets of central research questions: (i) what are the diffusion and adoption rates of the introduced climbing bean varieties in the central Kenyan highlands; (ii) why do some farmers try growing climbers while others do not and (iii) why do some adopt climbing beans while others stop growing climbers? Identifying the answers on these research questions is important to formulate policy recommendations to increase the diffusion and adoption of climbing beans in this region.

To address these questions, we implemented a quantitative survey among a representative sample of 595 farm-households in the region and collected detailed information on the adoption and cultivation of climbing beans. We use the survey data to estimate a statistical model of technology adoption. Most of the empirical models in the technology adoption literature estimate a dichotomous adoption decision, adoption or non-adoption, using logit or probit models (Feder et al., 1985;

Kaliba et al., 2000; Sunding and Zilberman, 2001; Doss, 2006; Chianu et al., 2007; Ojiako et al., 2007; Agwu et al., 2008; Udoh and Omonona, 2008; Saka and Lawal, 2009). Such an approach ignores the multi-stage procedure of farmers' decision-making process (Dimara and Skuras, 2003). Alternative approaches have been developed by Diagne (2006) and Diagne and Demont (2007) who consider a two-stage decision process and analyse farmers' 'exposure to' and 'adoption of' new rice varieties in Western Africa, correcting for selection bias due to non-exposure in the adoption model by using an average treatment effects method. In this paper, we consider three stages in farmers' decision-making on technology adoption: (1) awareness of climbing beans; (2) trial of climbing bean production on the own farm, and (3) adoption of climbing beans. We do so because a substantial proportion of the examined population does not continue growing climbing beans after a trial period of one season. We estimate the probability at each stage and correct for selection bias at the 'trial' and 'adoption' stage using a Heckman two-stage procedure. Using this approach, it is possible to acquire a more realistic idea about the entire adoption process and to identify the main factors of interest at each level of adoption.

MATERIALS AND METHODS

Historical background

For centuries, climbing beans have been part of traditional agricultural systems in medium to high altitude regions (2000 to 2800 m above sea level) of the Andes and Central America (Voysest, 2000). In 1984, improved climbing bean varieties from CIAT (International Centre of Tropical Agriculture, Colombia) were officially released and promoted in Rwanda (Sperling and Muyaneza, 1995) and gradually spread to neighboring countries, including Kenya. In 1995, six climbing bean varieties (Table 1) were introduced to KARI Embu, one out of the 25 Kenyan governmental institutes for agricultural research which is located in the town Embu, the provincial headquarters of Eastern province in Kenya. The mandate region of KARI Embu covers eight districts in the central highlands of Kenya. Between 1995 and 2002, the varieties were exposed to participatory testing on-farm and on-station and sold by KARI Embu and local extension services to more than 7000 farmers. Due to marketability problems of these six varieties as reported by farmers, a new collection of mid-altitude climbers (MAC) was introduced from CIAT to KARI Embu in 2002. These climbers were designed by CIAT to be early maturing and more heat tolerant (Blair et al., 2007). Participatory testing led to the selection of five varieties which were sent to the Kenyan National Performance Trials (NPTs). This resulted in official release of 3 varieties (MAC13, MAC34 and MAC64; Table 1) by the Kenyan ministry of agriculture in 2008. Currently, KARI-Embu has embarked on seed bulking of these 3 climbing varieties. The bulked seed is constantly disseminated to farmers and other partners through centre visits on a limited scale, field days, on-farm trials or demonstrations, agricultural shows and farmer-to-farmer exchanges.

Description of study area

The study area includes two districts, Embu and Kirinyaga, on the

---

[1]We use the concepts of technology diffusion and adoption as in Diagne and Demont (2007). Technology diffusion refers to the extent of knowledge and awareness of the technology among farmers in the population. Technology adoption refers to the actual use of the technology at the individual farm level as well as at the aggregate population level.

Table 1. List of climbing bean varieties disseminated by KARI Embu in the central Kenyan highlands.

Variety	Farmers' name	Seed size and shape	Seed colour	Origin period
G2333	Umubano	Small, round	Dark red	
G685	Vuninkingi	Small, round	Bright red	
Flor de Mayo	Flora	Intermediate, round	Purplish/pink, cream	Rwanda
G20797	Gisenyi	Intermediate, kidney	Pink/cream with black stripes	1995-2002
/	Ngwinurare	Large, kidney	Red	
G3323	Puebla	Small, round	Yellow	
MAC13	/	Large, kidney	Mottled cream and red	CIAT[a]
MAC34	/	Large, wedge	Mottled bright red and cream	from 2002
MAC64	/	Medium, wedge	Mottled dark red and cream	

[a]CIAT: International Centre of Tropical Agriculture, Cali, Colombia.

eastern slopes of Mount Kenya (Kenya). The districts are part of the mandate region of KARI Embu covering a total of about 88,000 households. In these two districts, only the upper midland zones (UM1 – UM2 – UM3 in Figure 1) are considered as they are the main agro-ecological zones suited for common bean production. The altitude of the study area ranges from about 1200 to 1900 m above sea level. The region has an annual mean temperature of 20 °C and average annual rainfall ranging from 800 to 1200 mm with two main rainy seasons, the 'long rainy season' from March to May and the 'short rainy season' from October to November (Veldkamp et al., 2009). Rainfall increases and temperature decreases with altitude (Gachimbi, 2002; Veldkamp et al., 2009). On the slopes of Mount Kenya, a dense network of small rivers flows down, incising the surface of the slopes and shaping the zone into a system of alternating ridges and depressions. The dominant soil type consists of humic nitisols which are highly weathered and well drained tropical soils with high clay contents. Other minor soil types in the region are andosols, cambisols, acrisols and regosols. pH decreases with increasing altitude and ranges from pH 4.4 in UM1 to pH 5.6 in UM3 (Veldkamp et al., 2009).

Both districts are characterized by a high and increasing population pressure with more than 700 people per km^2 and annual population growth rates of 2.57% (Veldkamp et al., 2009). The inhabitants of Embu district, called Embu, represent a minor tribe in Kenya while Kirinyaga district is inhabited mainly by the ethnic majority Kikuyu tribe. Because of the relatively fertile nitisols, both districts have a high agricultural potential. Mixed farming systems are common with tea (UM1; *Camellia sinensis*), coffee (UM1 and UM2; *Coffeaarabica*) and macadamia nuts (*Macadamia ternifolia*) as the main cash crops. Main food crops are maize (*Zea mays*) and beans (*Phaseolus vulgaris*) and to a lesser extent bananas (*Musa sp.*), irish potatoes (*Solanum tuberosum*), sweet potatoes (*Ipomoea batatas*), cassava (*Manihotesculenta*) and a variety of fruits and vegetables mainly grown for subsistence consumption. Also zero-grazing dairy farming is important, especially in the higher, cooler and humid areas. Despite a high agricultural potential, poverty rates in this region are high, more than 30%, and increasing (Kathuthu et al., 2005).

Sampling frame and survey data collection

The target population for this study consists of all farm-households living in Upper Midland 1, 2 and 3 zones (principal bean growing area) of Embu and Kirinyaga districts. Sampling was done for each district separately using a four-stage sampling design with divisions as primary sampling unit, sub-locations as secondary sampling unit, villages as tertiary sampling and households as ultimate sampling unit.

Stratified random sampling was applied at each stage, except for the last stage, to make sure that locations where climbing beans were promoted are included in the sample. Divisions, sub-locations and villages (primary, secondary and tertiary sampling units respectively) were stratified according to whether they were previously targeted for promotion of climbing beans or not. At each of three initial stages and in each stratum, random sampling was performed with the number of selected units per stratum proportional to its population size. In the last stage, households were randomly selected for each village using pre-established household lists wherever available. When lists were not available, households were selected as dispersed as possible and making sure to cover the entire area of the village. In this way, a self-weighting sample of 550 households in 36 villages was obtained with the number of households selected per village proportional to the village population size and ranging from 12 to 18. This sampling frame ensures our data are representative for the entire study area and allows making population inferences.

The survey was implemented using a structured questionnaire including sections on (i) household and farm characteristics, (ii) bush bean production (if any) and iii) climbing bean production (if any). The interviews were carried out in the period March-May 2010 by four technical assistants from KARI Embu, who managed the local dialects. Field assistants (village elders) residing in the area helped in locating the households and introducing the enumerators. To ensure quality and comparability of the data gathered by the four enumerators, a training was organized and the questionnaire was pre-tested before the actual start of the interviews. Additionally, afterwards, a total of 45 climbing bean adopters located in various villages in Embu district were selected and interviewed in order to increase the total number of this relatively small group for descriptive analyses of climbing bean production.

Statistical data analysis

The survey data were digitized into an SPSS database and cleaned in order to detect and eliminate entry errors in the database. A first descriptive statistical analysis was performed to classify and characterize the adoption levels of climbing beans and to analyse the climbing bean production system. Analysis of variance (ANOVA) was applied to test for differences in the characteristics across various adoption levels. In a subsequent causal statistical analysis using STATA, a probit model and a Heckman probability model are estimated to reveal the factors that influence the likelihood of being aware of, trying out and adopting climbing beans. The extra group of 45 adopters who were selected and

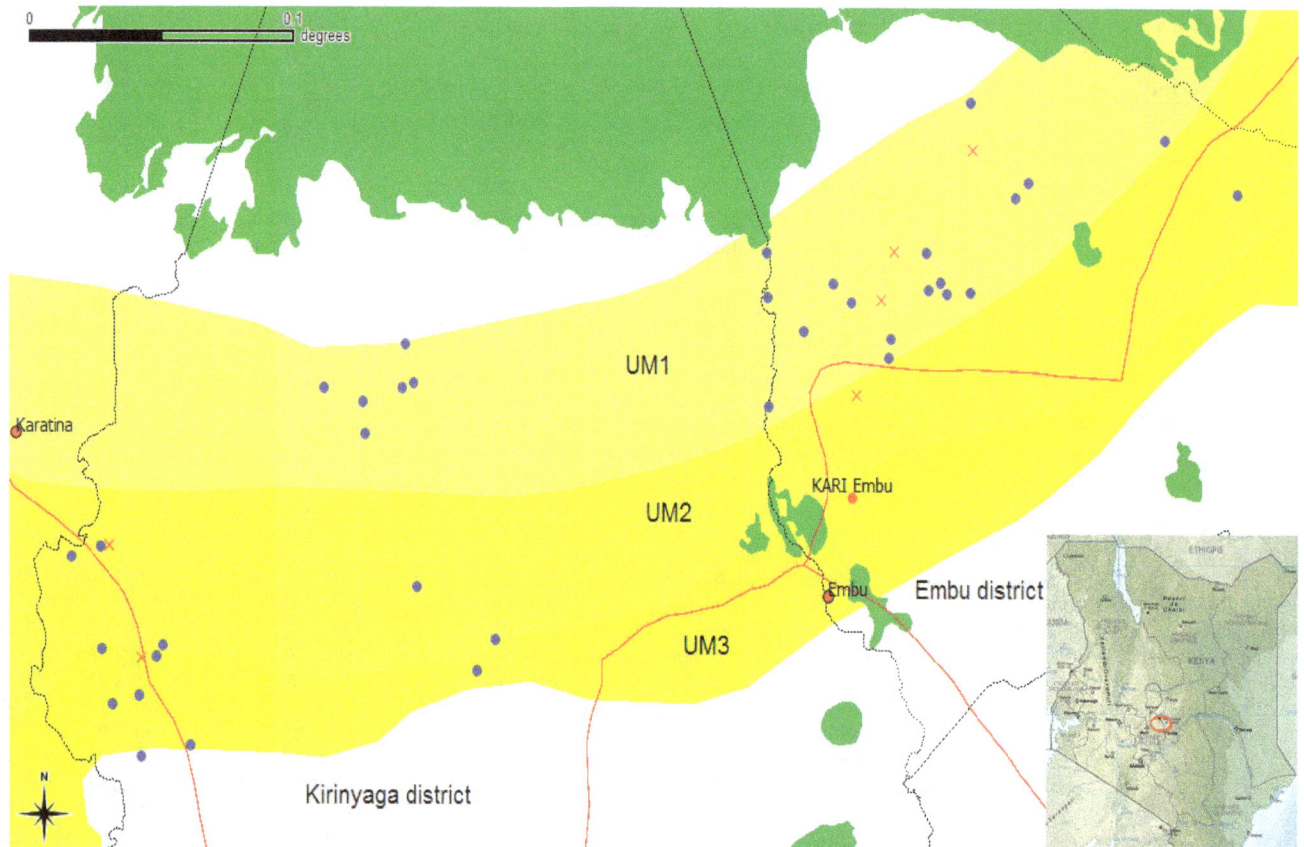

Figure 1. Map of Kirinyaga and Embu districts showing the location of the villages (blue dots) included in the survey. Black dotted line: District boundaries; red line: Main roads; red dots: Main towns; red crosses: Main market areas; green area: Forest; yellow areas: Agro-ecological zones with Upper Midland 1 (UM1), Upper midland 2 (UM2) and Upper Midland 3 (UM3) (qGIS, 2011).

interviewed to increase the number of adopters, were only used for descriptive analyses of climbing bean production.

RESULTS

Here, we discuss the results of the descriptive and causal statistical analysis on climbing bean diffusion and adoption in the study area. First, we unravel farmers' technology adoption decision process in three stages and categorize farm-households in four classes according to the decision outcomes in these three stages. We then insert a short description of climbing bean production by trial and adopter farmers. Subsequently, we describe and compare farm and household characteristics across the four household classes. Finally, multivariate probit and Heckman probability models were developed and estimated to identify the factors determining each of the three stages in the adoption decision-making process of farmers.

Describing the adoption decision process

To examine farmers' decision to adopt climbing beans we

build on and adapt the innovation adoption theory of Rogers (2003) who describes the process of innovation adoption in five stages: (1) knowledge, (2) persuasion, (3) decision, (4) implementation and (5) confirmation. In the first stage, farmers become aware of a new technology. We call this the stage of 'awareness'[2]. In the two subsequent stages of Rogers, interested farmers look for more information on the technology, weigh the costs and benefits, and decide whether or not to try implementing the technology. These stages are empirically hard to observe and our survey data are completely uninformative about these stages. We only empirically observe whether farmers have implemented the new technologies on their field. Our second stage is therefore 'trial'. This corresponds to Rogers' implementation stage in which the decision-maker actually uses the innovation and considers its usefulness. We call this a 'trial' stage, rather than 'implementation' stage. In a final stage, farmers decide, after having tried climbing bean varieties on their field, whether or not to continue growing them and really adopt the technology. We call this the 'adoption' stage, rather than the 'confirmation' stage. This three-stage decision process, including (1) awareness, (2) trial and

---

[2] Others have called this 'exposure' (Diagne, 2006; Diagne and Demont, 2007).

Table 2. Decision tree used to assign each household to one of the four types: Non-aware, aware, trial and adopter households.

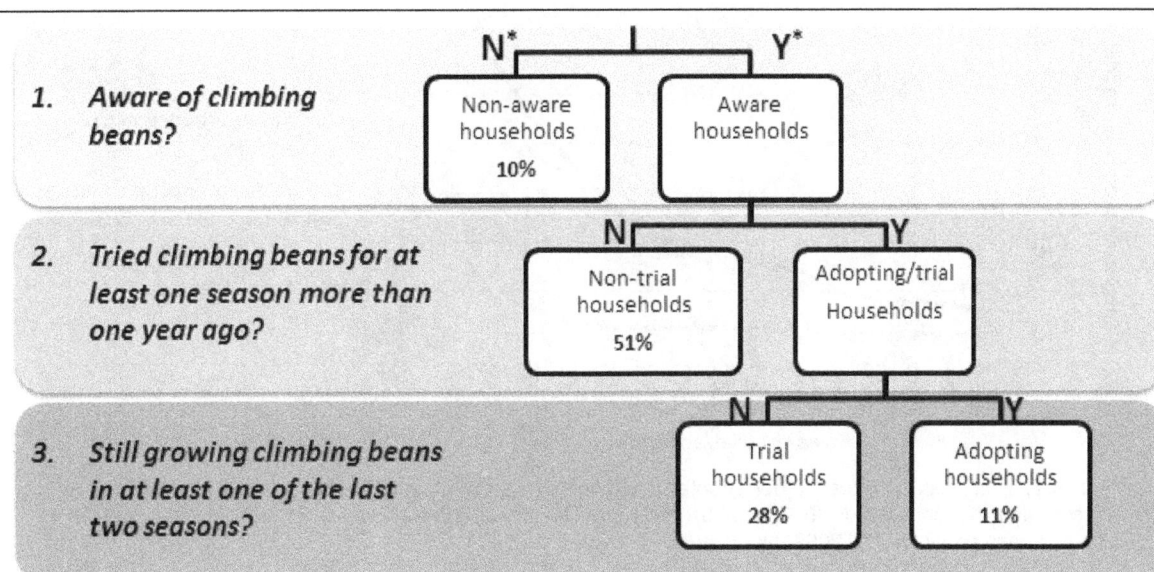

	1.	*Aware of climbing beans?*	N* — Non-aware households 10% / Aware households — Y*
	2.	*Tried climbing beans for at least one season more than one year ago?*	N — Non-trial households 51% / Adopting/trial Households — Y
	3.	*Still growing climbing beans in at least one of the last two seasons?*	N — Trial households 28% / Adopting households 11% — Y

*Y = Yes, N = No

Adopter household:	Household that planted climbing beans on the farm at least for one season in the past (more than two seasons ago) AND has been planting climbing beans during at least one of the last two seasons at the time of the survey.
Trial household:	Household that planted climbing beans on the farm at least for one season in the past, but has not been planting climbing beans anymore during the last two seasons at the time of the survey.
Non-trial household:	Household that has heard of climbing beans but has never planted them on the farm.
Non-aware household:	Household that has never heard of climbing beans.

For each type, the % of farmers is indicated. Below this tree, definitions of the four household types are given.

(3) adoption, is graphically depicted in Table 2 and results in the identification of four types of farm-households. First, non-aware households are households that have never heard about climbing beans. Second, non-trial households are aware of the technology but have never tried cultivating climbing beans on their farm. Third, trial households have planted climbing beans on their farm at least during one season in the past (more than two seasons ago) but did not continue growing them in any of the last two seasons. Fourth, adopter households have grown climbing beans at least during one season in the past (more than two seasons ago) and have continued growing them in at least one out of the two last seasons at the time of the survey.

From our sample, we estimate that the total awareness among selected farmers is 90%, the trial rate is 39% and the adoption rate is 11% (Table 2). As our sample is representative for the population in the two surveyed districts, these results can be extrapolated to the entire population of the two districts. This means that the awareness among selected farmers (90%) can be understood as the technology diffusion rate in the area. The population trial rate is between 39% (if non-aware

households have a zero probability to try out climbing beans) and 43% (=215/495 if non-aware households would have the same probability to try out climbing beans). Finally, the population adoption rate is between 11% (if non-aware and non-trial households have a zero probability to adopt) and 28% (=62/215 if non-aware and non-trial households would have the same probability to adopt as trial households). It can therefore be concluded that climbing beans are widely known throughout the study area, that on average 40% of all farmers have tried this crop at least during one season and that comparatively few farmers (11 to 28%) actually adopted climbing beans.

Trial and adopter farmers were asked in which year they started growing climbing beans. In Figure 2, the adoption rates (%) for trial farmers (blue curve), adopter farmers (black curve) and the sum of these two groups (red curve) are shown per year. It can be observed that in the period 1995 to 2004, there is a significant number of trial farmers, but actual adoption is relatively low. From the year 2004, both trial and adoption rates show a steep increase. Most farmers who were adopters at the time of the survey adopted climbing beans from the year 2006,

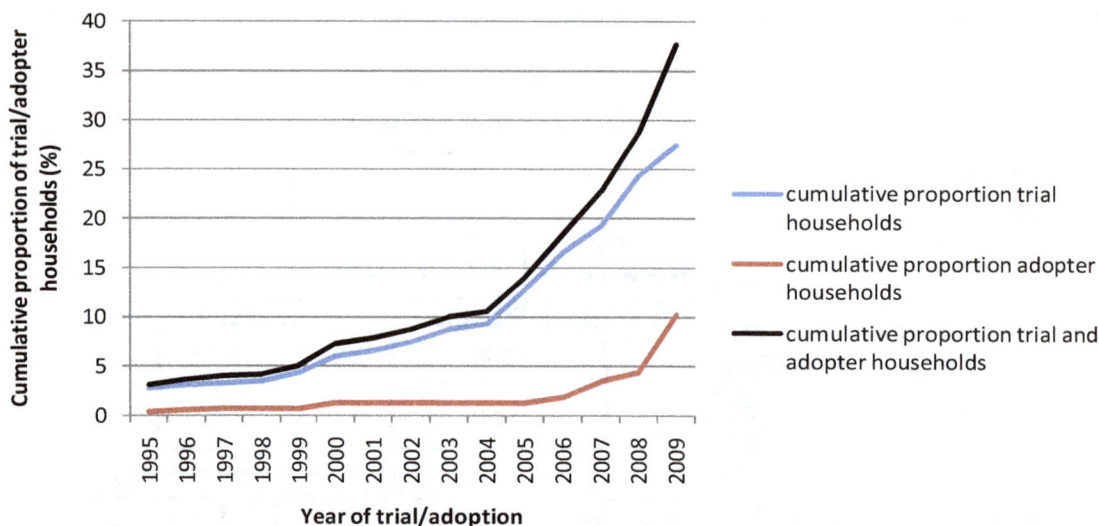

Figure 2. Trial/adoption rates (%) for adopter households (red curve), trial households (blue curve) and both adopter and trial households (black curve) per year. The time period shown is from 1995 (starting year of climbing bean promotion) till 2009 (one year before the survey).

especially in 2008 and 2009. These farmers who were categorized as being adopters at the time of the survey might still disadopt climbing beans after the survey. Hence, both trial and adoption are on the increase since 2004 but it is not known whether adoption which is increasing since 2006, will actually continue its increase.

Most farmers first became aware of the existence of climbing beans through other farmers (69%). Other cited sources of information on climbing beans are the extension staff (14%), the research centre KARI Embu (6.5%), relatives (5.5%) and radio (4%). In Embu district, about 10% of the farmers were informed by KARI Embu while in Kirinyaga district, this was only 3%.

The main reason for farmers to never having grown climbing beans is the unavailabilty of seed (42%). Other minor reasons are: (i) too labour intensive (10%) and (ii) no interest (12%), lack of knowledge about how to grow climbing beans (8%), lack of staking material (6%) and lack of space (4%). The main reason for farmers to abandon climbers after having tried is also the unavailability of seed (39%). Other minor reasons are: (i) too labour-intensive (22%), (ii) lack of staking material (7%), (iii) attack by birds (6.5%) and sensitivity to drought (3%). Labour intensity and attack by birds have also been reported in other studies to limit adoption of climbing beans (Sperling et al., 1992; CIAT, 2004).

Statistical description of improved climbing bean production

In this paragraph, climbing bean production by the groups of trial and adopter farmers is described. About 65% of the persons managing the climbing bean crop on the farm was female. Compared to management of bush beans

(80% female farmers), female contribution in climbing bean production was lower. Climbing bean seed for first time planting was obtained mainly as a gift or exchange with other farmers (60%) or with relatives (6%). Part of the farmers received seed directly from KARI Embu (13%) or through extension workers (11%). A few farmers bought climbing bean seed on the market (4%) or from other farmers (4%).

Most farmers planted only one climbing variety. A minority planted two (16%), three (15%), four (1.5% - four farmers) or even five varieties (0.4% - 1 farmer). The majority of the group of adopter farmers (86%) planted climbing beans during October rains 2009 while during April rains, only 52% of this group planted climbing beans. When looking at popularity of the several climbing bean varieties (Figure 3), Vuninkingi is by far the most popular variety (39%) followed by Umubano and Gisenyi (about 20% each). The climber Ngwinurare was not used by adopters. The more recently introduced varieties which are the Mid-Altitude Climbers (MAC) are only limitedly known and adopted by farmers in the central highlands of Kenya (2 to 5% adoption rate).

The total area of the plots planted with climbing beans were relatively small with 106 ± 31 m^2 on average for October rains (median is 25 m^2) and 105 ± 25 m^2 for April rains (median is 31 m^2). Total land areas ranged from 0.5 to 2400 m^2 and 0.5 to 1000 m^2 for October and April rains respectively. Most farmers (94%) used stakes as support for their climbing beans on the field. Very few farmers did not use any support (3%) and some farmers used permanent fences (2%). Stakes had lengths of 1.83 meters on average with about 63% of the farmers who used stakes that were not sufficiently long (shorter than 2 m). About 21% of the farmers even used stakes shorter than 1.50 m. Stakes were mainly obtained from the farm

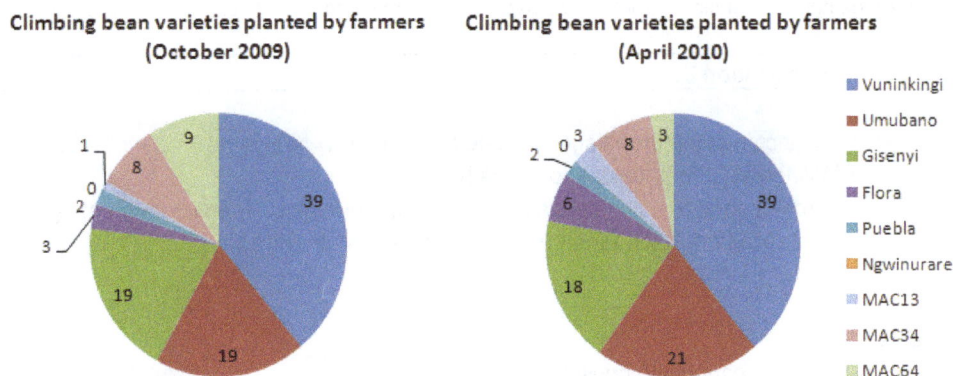

Figure 3. Climbing bean varieties planted by the group of adopters during October rains 2009 (left) and April rains 2010 (right).

(92%) or were bought (5%). Climbing beans were staked on average at 2 weeks with a range from 0 to 4 weeks as is recommended for climbing bean production. Most climbing beans were planted as a single crop on the field (88%) while some farmers intercropped climbing beans with maize (8%), coffee or bush beans (1% each). For October rains, the median of yields for climbing beans according to farmers was 1333 kg/ha.

Farmers were asked to list the several uses of climbing beans on their farm and to order these uses according to their importance. The primary use of climbing beans is home consumption of the dry seed (75%). Other main uses are consumption of green pods (18%) and selling the dry seed (3%). The second function of climbing beans according to farmers is consumption of green pods (35%) and the use as fodder for their animals (22%). About 10% of the farmers also consume the leaves of climbing beans. The third use as indicated by the farmers is again the use as animal fodder (47%), consumption of green pods (19%), exchanging the seed (16%) and selling it (9%). The few farmers who ever sold the dry seed, reported relatively high market prices and good appreciation of the seed by the buyers.

The main advantages of climbing beans according to the farmers are their high yielding capacity (53%), their sweet taste (18%), the need of only a small land area (3%), fast cooking (5%) and more tolerant to heavy rains and wet soils (2%). A total of 41% of the farmers could not think of any disadvantages related to climbing beans. The other farmers considered the need for staking (12%) and susceptibility to bird attack (11%), due to the sweet taste of leaves, pods and seed, as the main drawbacks of climbing beans.

Farmers were asked to give a value judgment (on a 1 to 5 scale with 1 being bad and 5 very good) for several characteristics related to climbing bean production. The yield, seed colour and taste of climbing beans were considered as good to very good. Storage capacities of the seed were considered as fair. Improvement of soil fertility by the climbing bean crop was noticed by about 60% of the farmers while 27% did not see any effect on soil fertility. In a participatory evaluation study by KARI Embu (Muthamia et al., 2003), farmers also observed better performance of their maize crop grown where climbing beans were previously planted. Tolerance to diseases and drought were rated as being average to good. In general, climbing beans have a growing season which is extended by 2 to 4 weeks compared to bush beans. Most farmers however did not consider this longer growing season as an important disadvantage. The labour requirements in the field were rated as intensive to very intensive by 19% of the farmers. The remainder estimated that the amount of labour was not especially higher for climbing beans. This result might indicate that adopter farmers do not perceive climbing bean production to be labour intensive which is contradictory to the initial hypothesis that climbing beans are labour intensive and to earlier reports on farmers' appraisal of climbing beans (Sperling et al., 1992; CIAT, 2004). Most farmers agreed that climbing beans cannot be intercropped. About 25% of the farmers did consider intercropping with climbing beans as a realistic option.

Descriptive analysis of factors influencing farmers' technology adoption decision

The technology adoption literature, especially studies focussing on the diffusion and adoption of agricultural technologies in developing countries, have stressed the importance of socio-economic and demographic attributes of the farmer or farm-household in explaining adoption decisions (Feder et al., 1985; Rogers, 2003; Sunding and Zilberman, 2001).In this literature, farmers are assumed to adopt a new technology if it increases their profits or utility. Factors that lower the cost or risk of adopting a certain technology or factors that increase the benefits of it, will likely enhance adoption. Also farmers'

Table 3. Description of village and household characteristics hypothesized to influence farmers' adoption technology decisions.

Independent variable	Description
Village level	
District	District in which the village is located (0=Embu, 1=Kirinyaga)
Altitude	Village altitude (meters above sea level)
Promotion	Promotion of climbing beans by KARI Embu through on-farm experiments (no=0, yes=1)
Distance	Distance from KARI Embu by car (in minutes when driving)
Household level	
Demographic variable	
Age_hh	Age in years of the household head
Gender_hh	Gender of the household head (male=0, female=1)
Dependency_ratio	Total number of persons from the household younger than 15 proportional to the total number of persons from the household
Human capital	
years_school	Total number of years the household head went to school
labour_endowment	Total number of persons from the household older than 16 years
Physical capital	
Income	Main source of cash income for the household (farm produce sales=0, other=1)
Land size	Land size owned by the household (acres)
TLU	Livestock size (Tropical Livestock Unit*)
non_mot_vehicle	Household owns at least one non-motorized vehicle (bicycle, cart or wheelbarrow; no=0, yes=1)
mot_vehicle	Household owns at least one motorized vehicle (motorcycle or motorcar; no=0, yes=1)
Social capital	
Farmer_group	Membership of a farmer group (no=0, yes=1)
Visit_extension	Total number of visits by extension staff on the farm in the past year
Visit_events	Total number of visits to agricultural events in the past year (show, training, field day, baraza)
Farming_info	Sources of information on farming (radio, extension, other farmers, other; no=0, yes=1)
nr_farming_info	Number of sources used to obtain information on farming

*Sum of livestock units using 1 per cow; 0.8 per donkey and 0.2 per pig, sheep or goat.

preferences towards consumption or leisure, which might be strongly related to the demographic structure of the farm-household, may play a role. In addition, farmers may face constraints, most importantly capital constraints that may hinder needed investment to adopt the new technology. Previous empirical studies on technology adoption in general (Godoy et al., 1998; McDonald et al., 2003) and on climbing bean adoption in Africa in particular (Sperling et al., 1992; Grisley, 1994; Sperling and Muyaneza, 1995; Musoni et al., 2001; CIAT, 2004) have shown that the following factors may play a role in explaining farmers' technology adoption decisions: (1) age was shown to be a factor with a positive impact on adoption meaning that older, more experienced farmers were more likely to adopt new technologies; (2) promotion of new technology through farmer-managed, on-farm field experiments has proven to be an effective but limited method for diffusing new varieties; (3) higher labour endowments would facilitate the decision to test and adopt a new technology; (4) a higher level of education

has shown to stimulate amenability to innovation and provide better access to information; (5) small land sizes could give incentives to test and adopt climbing beans and (6) membership of farmers' associations was shown to positively influence adoption of new technology.

In our analysis, we focus on (1) village level variables that represent the institutional and agro-ecological setting; (2) demographic variables, including the age and gender of the household head and the household dependency ratio; (3) human capital variables, including labour endowments and education; (4) physical capital variables, including households' land and non-land assets and access to off-farm income; and (5) social capital variables, including membership of a farmers' group and access to information. These variables are described in Table 3. Average values among the four types of farmer-households were calculated for each characteristic separately and subsequently, it was tested whether there were significant differences among these types (Table 4).

Table 4. Village and farm/household characteristics (%) of the 4 household types of improved climbing bean varieties in the central highlands of Kenya.

Parameter		Adopter households	Trial households	Non-trial households	Non-aware households	
		n=62	n=153	n=280	n=55	
Village level	Total mean		Means per household type			ANOVA or Welch[a]
District	0.46±0.02	0.15±0.04[c]	0.33±0.04[b]	0.60±0.03[a]	0.69±0.06[a]	0.000***
Altitude (m)	1568±7	1630±22[a]	1622±12[a]	1540±9[b]	1490±21[b]	0.000***
Promotion	0.31±0.02	0.35±0.06[ab]	0.40±0.04[a]	0.28±0.03[ab]	0.18±0.05[b]	0.007**
Distance (min)	41±0.9	37±2.4[ab]	37±1.6[b]	43±1.3[a]	44±2.7[ab]	0.009**
Household level						
Demographic characteristics						
Age_hh (years)	47.4±0.4	48.3±1.1[ab]	49.3±0.7[a]	46.8±0.6[b]	44.2±1.3[b]	0.002**
Gender_hh	0.17±0.02	0.15±0.05	0.15±0.03	0.19±0.02	0.15±0.05	0.600
Dependency_ratio	0.29±0.01	0.25±0.03	0.28±0.02	0.29±0.01	0.29±0.03	0.598
Human capital						
Years_school	7.6±0.2	7.9±0.6	7.7±0.3	7.5±0.2	7.4±0.6	0.861
Labour_endowment	3.05±0.07	2.84±0.18	3.19±0.13	3.05±0.11	2.85±0.24	0.431
Physical capital						
Income	0.22±0.02	0.16±0.05	0.22±0.03	0.24±0.03	0.22±0.06	0.487
Land size (acres)	1.96±0.08	2.01±0.25	2.19±0.16	1.86±0.10	1.76±0.26	0.255
TLU	1.77±0.06	1.40±0.15[b]	2.04±0.12[a]	1.75±0.09[ab]	1.46±0.19[ab]	0.011*
Non_mot_vehicle	0.69±0.02	0.68±0.06	0.76±0.04	0.67±0.03	0.60±0.07	0.104
Mot_vehicle	0.08±0.01	0.05±0.03	0.12±0.03	0.05±0.01	0.15±0.05	0.071
Social capital						
Farmer_group	0.25±0.02	0.24±0.06[ab]	0.35±0.04[a]	0.22±0.03[b]	0.16±0.05[b]	0.016*
Visit_extension	0.30±0.06	0.35±0.10[a]	0.50±0.17[a]	0.23±0.06[a]	0.02±0.02[b]	0.000***
Visit_events	1.13±0.09	1.02±0.16	1.32±0.18	1.06±0.12	1.06±0.34	0.613
Farming_information						
Radio	0.76±0.02	0.68±0.05	0.76±0.04	0.78±0.03	0.80±0.05	0.233
Extension	0.18±0.02	0.36±0.05[a]	0.15±0.03[b]	0.14±0.02[b]	0.11±0.04[b]	0.000***
Other farmers	0.27±0.02	0.29±0.04	0.28±0.04	0.28±0.03	0.16±0.05	0.314
Other	0.06±0.01	0.07±0.02[b]	0.03±0.01[b]	0.05±0.01[b]	0.16±0.05[a]	0.004**
nr_farming_info	1.38±0.02	1.59±0.07[a]	1.35±0.04[b]	1.33±0.03[b]	1.35±0.07[ab]	0.001**

Values represent mean±mean standard error of each adoption level. [a]Levene's test was used to check equality of variances among household types. For equal variances, ANOVA and Tukey as post-hoc test for differences among each pair of groups were used. For unequal variances, Welch test and Tamhane as post-hoc test were used. Esterics indicate significance level: *0.05 > p ≥ 0.01; **0.01 > p ≥ 0.001; ***p < 0.001.

Characteristics at village level

Generally, 46% of the interviewed households are from Kirinyaga district (Table 4). The proportion of trial and especially adopter households living in Embu district appears to be higher compared to Kirinyaga district (Table 4). It seems therefore that trial and particularly adoption of climbing beans are most spread in Embu compared to Kirinyaga.

For every village, altitude was recorded. There is a difference in altitude across the villages in the sample and the average village altitude for the groups of trial and adopter households is significantly higher than for the group of non-aware and non-trial households. This implies that trial and adoption are more frequent at higher altitudes.

One of the promotion and evaluation strategies for climbing beans by KARI Embu in the field are farmer-managed, on-farm field experiments. This strategy has proven to be an effective but limited method for diffusing new varieties (Grisley, 1994; Kaliba et al., 2000). From Table 4, it can be concluded that a higher proportion of trial and adopter households live in villages where promotion was done, although adopter and non-trial households were not significantly different. The group of non-aware households was least represented in villages with promotion activities by KARI Embu. This might indicate that promotion through on-farm field experiments contributes to increased awareness and trial.

The distance by car from KARI Embu to each village in minutes drive was also recorded as KARI Embu has limited financial resources and therefore mainly chooses to carry out the field activities close to the station. The results from Table 4 indicate that trial and adopter households are located closer to KARI Embu on average than the non-trial and non-aware households, although the difference between adopter households and non-trial households was not significantly different due to higher mean standard errors of these two smaller groups.

Demographic characteristics

Mean age among all household heads (age_hh) was 47 years with trial and adopter households being somewhat older on average than non-trial and non-aware households. However, the difference in age between adopters and both non-trial and non-aware households was not significant.

Of all the households surveyed, 83% was male-headed. Equal proportions of female- and male-headed households can be found among the various household types implying that female-headed households are not running behind at the level of climbing bean awareness, trial and adoption.

Based on information on the number of household members and their age, the dependency ratio was calculated which estimates the relative number of dependent individuals (children younger than 16) in the household. The higher the dependency ratio in a household, the higher the need for food consumption. However, dependency ratios were not significantly different among the various household types.

Human capital

Education of the household head was recorded in the total number of years this person had gone to school (years_school). In general, a higher level of education is expected to stimulate amenability to innovation and provide better access to information, that is, leaflets, brochures (Grisley and Shamambo, 1993; Udoh and Omonona, 2008). Trial and adopter households showed a slightly higher education level compared to non-trial and non-aware households although differences were not significant.

A first source of labour is the household members who are (potentially) able to work on the farm (labour_endowment). These are all the household members older than 15 years. As climbing bean adoption requires more labour, it is expected that higher labour endowments facilitate the decision to test and adopt (Sperling et al., 1992; CIAT, 2004; Sperling and Muyaneza, 1995). However, there were no significant

differences for labour endowment among the various household types.

Physical capital

The main cash income of most households in the regions surveyed is generated through farm produce sales (78%). Only about one in five households (22%) gets its main source of income off-farm, mainly in the private sector. It could be expected that the latter have less incentives to know and eventually to test and adopt climbing beans. No significant differences for the main source of income among the four household types are found although the group of adopters includes less farmers with off-farm incomes than the 3 other groups.

The increasing population pressure in the central highlands of Kenya causes land fragmentation and small land sizes per farm. The mean land size owned by the household was about 1.96 acres or 0.79 ha. Small land sizes could give incentives to test and adopt climbing beans as this high-yielding crop is able to generate the same yield as bush beans on a smaller land space (Sperling et al., 1992; CIAT, 2004; Sperling and Muyaneza, 1995). However, land sizes did not vary significantly among the 4 household types although trial and adopter households had higher land sizes compared to non-trial and non-aware households. Also, trial households have higher land sizes compared to adopter households although the difference is not significant.

As the climbing bean biomass can and is actually used as fodder for animals (Sperling et al., 1992; CIAT, 2004), it is expected that households with more animals (expressed in TLU or Tropical Livestock Units) have higher chances to test climbing beans and to adopt them. However, significant differences were only found among adopter and trial households with the latter group having more livestock which is against expectations.

The ownership of non-motorized vehicles and motorized vehicles by the household are both indicators for wealth. Only 8% of the households owned at least one motorized vehicle. The possession of non-motorized vehicles including bicycles, carts or wheelbarrows were more common with almost 70% of the population owning at least one non-motorized vehicle. However, these wealth indicators did not vary significantly among the four household types.

Social capital

Only 25% of the farmers were member of a farmer group. Farmer groups gather regularly and these gatherings are potential sources of information on climbing beans and possibly also of seed of climbing varieties. Several adoption studies found that membership of farmers' associations positively influenced adoption of improved varieties (Ojiako et al., 2007; Chianu et al., 2007). Of the group of non-aware households, only 16% were members of farmer groups. Trial and adopter households were more frequently member of a farmer group compared to non-trial and non-aware households although only trial households differed significantly from both non-trial and non-aware households.

Contact with extension services was measured by the number of visits of an extension worker to the farm in the year before the survey. Non-aware households had significantly lower number of visits by extension compared to the other three household types implying that extension services played a role in making farmers aware of the existence of climbing beans.

Another way of possible access to climbing bean technology could be by visits to agricultural events. However, among the four household types, there was no difference in number of visits to agricultural events.

Farmers were also asked about the main sources they use to obtain information on farming. About 76% of the farmers received farming information through the radio, 27% from other farmers and 18% from extension services. The radio did not appear to play a role in spreading information on climbing beans. Remarkably, among the adopters, 35% used extension services as farming information source versus around 11 to 15% only among the other groups. This implies that adopters making use of extension services were more likely to adopt climbing beans after trial. Farmers using other sources of information, such as newspaper or television (other), were less likely to know about climbing beans.

Finally, it also mattered how many different sources on farming information were used (nr_farming_info). Adopter households used more sources compared to the other household types (although the difference with non-aware households was not significant but this is probably due to the higher mean standard errors of both groups).

Causal analysis of farmers' technology adoption decisions

Description of estimated models

For each decision level, we estimate the probability of a household to be aware of, to have tried and to have adopted climbing beans as follows:

$$P(\text{awareness} = 1 \mid X_i) = \Phi(\beta_0 + \beta_i X_i + \varepsilon) \quad (1)$$

$$P(\text{trial} = 1 \mid X_j, \text{awareness} = 1) = \Phi(\beta_0' + \beta_j X_j + \varepsilon') \quad (2)$$

$$P(\text{adoption} = 1 \mid X_k, \text{trial} = 1) = \Phi(\beta_0'' + \beta_k X_k + \varepsilon'') \quad (3)$$

where awareness, trial and adoption are the dichotomous dependent variables (0 for no and 1 for yes). The vectors of independent variables $X_i X_j$ and X_k include the variables identified in Table 3, apart from nr_farming_info as this is collinear with and captured by the farming_info dummy variables. The error terms are denoted by ε, ε' and ε'' and Φ denotes the cumulative distribution function of the standard normal distribution (Wooldridge, 2001). Equations 1 to 3 describe probit models that can be estimated using maximum likelihood estimation (Wooldridge, 2001).

The decision of trial and adoption are conditional on being aware of and having tried climbing beans respectively. A problem of selection bias arises as awareness and trial are not randomly distributed across the population. This might lead to biased estimates in the probit models. We use a Heckman two-stage procedure to test and correct for possible selection bias in model 2 and 3 (Heckman, 1976, 1979; Wooldridge, 2001). A first Heckman probability model (Heckprob) is estimated, explaining trial of climbing beans among aware farm-households in the second stage and selecting aware households in the first stage. The variables defining how the farmer receives information on farming (radio, extension, other_farmers, other) are used as selection variables in the first stage selection equation. The test for possible selection bias for this Heckprob model was not significant (Likelihood-Ratio test of independent equations had a value of 2.66 with p-value 0.1027), indicating that sample selection does not bias the estimates in the adoption model. This result is likely related to the fact that awareness is quite high in the study region. As a result, we can use two independent probit regressions to estimate the probability of awareness and trial of climbing beans independently (Tables 5 and 6 respectively). In order to verify robustness of each probit regression, we estimate two slightly different specifications of the models. Model specification A includes a set of village level variables identified in Table 3 as explanatory variables while model specification B includes village dummies to correct for village level variation in institutional and agro-ecological conditions (estimated coefficients for village dummies are not shown).

A second Heckman probability model (Heckprob) is estimated, explaining adoption of climbing beans among trial farmers in the second stage and selecting trial households in the first stage. We use the variables promotion and distance to the research centre KARI Embu as selection variables in the first stage. Promotion of a new technology is important in creating awareness among farmers and in convincing farmers to try out the new technology. However, once farmers have tested the technology, they are less influenced by promotion activities and likely base their judgement and decision on whether to continue using the new technology or not on their own experience. Therefore, promotion is a good selection variable. In addition, promotion has a significant effect on trial (Table 6) but is not correlated with adoption.

Table 5. Parameter estimates of the probit model for the factors potentially influencing awareness of improved climbing bean varieties in the central highlands of Kenya.

Parameter	Model A (village level variable)				Model B (village dummies)			
Number of observations	540				389			
Log-likelihood value	-145.275				-119.247			
LR Chi-squared	60.54***				79.89***			
Pseudo R-squared	0.172				0.239			
Predictor variable	Coef[a]	Std. Err.	dy/dx[b]	Std. Err.[c]	Coef.[a]	Std. Err.	dy/dx[b]	Std. Err.[c]
Village level								
District	-0.286	0.260	-0.031	0.029	-	-	-	-
Altitude	0.002***	0.001	0.0002	0.000	-	-	-	-
Promotion	0.306	0.200	0.297	0.183	-	-	-	-
Distance	0.005	0.005	0.001	0.001	-	-	-	-
Household level								
Demographic variable								
age_hh	0.020**	0.010	0.002	0.001	0.034***	0.012	0.004	0.002
gender_hh	0.393	0.278	0.021	0.020	0.452	0.319	0.046	0.027
dependency_ratio	0.075	0.116	0.080	0.012	0.151	0.130	0.019	0.016
Human capital								
years_school	0.018	0.024	0.002	0.003	0.015	0.028	0.002	0.003
labour_endowment	0.260	0.135	0.028	0.015	0.109	0.151	0.014	0.019
Physical capital								
Income	0.290	0.221	0.027	0.0187	0.350	0.261	0.038	0.026
Land size	-0.055	0.057	-0.006	0.006	-0.093	0.071	-0.012	0.009
TLU	0.087	0.069	0.009	0.007	0.095	0.084	0.012	0.011
non_mot_vehicle	0.236	0.193	0.027	0.024	0.390*	0.226	0.055	0.037
mot_vehicle	-0.619**	0.280	-0.099	0.062	-0.698**	0.309	-0.132	0.081
Social capital								
farmer_group	0.259	0.225	0.028	0.020	0.297	0.267	0.034	0.027
visit_extension	0.693*	0.364	0.074	0.030	0.884**	0.438	-0.117	0.040
visit_events	-0.019	0.043	-0.002	0.005	-0.020	0.064	0.002	0.008
Radio	0.004	0.235	0.038	0.025	0.008	0.275	0.001	0.035
Extension	0.198	0.269	0.047	0.023	0.360**	0.324	0.037	0.028
other_farmers	0.458**	0.226	0.070	0.019	0.589**	0.264	0.061	0.026
Constant	-3.949**	1.135	-		-2.364**	0.884	-	-

[a]Stars indicate significance level: *$0.10 > p \geq 0.05$; **$0.05 > p \geq 0.01$; ***$0.01 > p \geq 0.001$; ****$p < 0.001$. [b]dy/dx is for discrete change of dummy variables from 0 to 1. [c]Standard errors for these discrete changes.

The second selection variable which is the distance to the research centre KARI Embu, is a parameter for the probability of a farmer-household to be involved in any kind of promotional activity with climbing beans and is therefore also a proper selection variable. The Heckman selection model is difficult to estimate with a large number of dummy explanatory variables; this results in problems of convergence. We therefore estimate the model only using specification A, without village dummies, and using the variable nr_farming_info instead of separate farming_info dummy variables. These results are shown in Table 7. The test for possible selection bias (Likelihood Ratio test of independent equations) is significant (value of 2.77 with p-value 0.0960), indicating that sample selection needs to be corrected by using the Heckman two-stage procedure.

Causal analysis of farmers' technology adoption decisions

The first independent probit regression model which estimates the probability of awareness of climbing beans is shown in Table 5. The Log-likelihood value is used in the Likelihood Ratio Chi-Square test (LR Chi-squared) which tests significance of the model. The test points out

Table 6. Parameter estimates of the probit model for the factors potentially influencing trial of improved climbing bean varieties in the central highlands of Kenya.

Parameter	Model A (village level variables)				Model B (village dummies)			
Number of observations	540				525			
Log likelihood value	-309.501				-279.984			
LR Chi-squared	104.48****				134.80****			
Pseudo R-squared	0.144				0.194			
Predictor variable	Coef.[a]	Std. Err.	dy/dx[b]	Std. Err.[c]	Coef.[a]	Std. Err.	dy/dx[b]	Std. Err.[c]
Village level								
District	-0.808****	0.187	0.299	0.066	-	-	-	-
Altitude	0.002****	0.000	0.001	0.000	-	-	-	-
Promotion	0.275**	0.132	0.105	0.051	-	-	-	-
distance	0.011**	0.004	0.004	0.002	-	-	-	-
Household level								
Demographic variable								
Age_hh	0.020***	0.008	0.008	0.002	0.029***	0.009	0.011	0.003
Gender_hh	-0.202	0.188	-0.074	0.067	-0.224	0.200	-0.080	0.069
Dependency_ratio	-0.019	0.081	-0.007	0.031	-0.075	0.089	-0.028	0.033
Human capital								
Years_school	0.005	0.017	0.002	0.006	0.017	0.018	0.006	0.007
Labour_endowment	0.031	0.077	0.012	0.029	0.020	0.085	0.007	0.032
Physical capital								
Income	-0.086	0.156	-0.032	0.058	-0.133	0.173	-0.048	0.062
Land size	-0.001	0.040	-0.000	0.015	-0.022	0.047	-0.008	0.018
TLU	0.001	0.046	0.000	0.017	-0.011	0.050	-0.004	0.018
Non_mot_vehicle	0.258*	0.141	0.096	0.051	0.256*	0.151	0.092	0.053
Mot_vehicle	0.333	0.232	0.130	0.092	0.390	0.250	0.151	0.099
Social capital								
Farmer_group	0.308**	0.145	0.119	0.057	0.283*	0.160	0.107	0.061
Visit_extension	0.103	0.063	0.039	0.024	0.117*	0.061	0.043	0.023
Visit_events	-0.001	0.031	-0.000	0.012	0.011	0.032	0.004	0.012
Radio	-0.225	0.160	-0.086	0.062	-0.231	0.173	-0.087	0.066
Extension	0.020	0.175	0.008	0.066	-0.071	0.198	-0.026	0.072
Other_farmers	0.054	0.153	0.021	0.059	0.002	0.166	0.001	0.061
Constant	-4.958****	0.872	-	-	-1.890***	0.637	-	-

[a]Esterics indicate significance level: *0.10 > p ≥ 0.05; **0.05 > p ≥ 0.01; ***0.01 > p ≥ 0.001; ****p < 0.001. [b]dy/dx is for discrete change of dummy variables from 0 to 1. [c]Standard errors for these discrete changes.

that both models are significant. The pseudo R-squared value (McFadden's pseudo R-squared) is a measure for goodness-of-fit of the model and indicates that model B fits the data better than model A. For each predictor variable in each model, the model coefficients, standard errors, marginal effects (dy/dx) and standard errors of marginal effects are shown.

Altitude appears to be highly significant and positively influences awareness indicating that at higher altitudes, there are more farmers who have heard of climbing beans.

The age of the household head positively influences awareness in both model A and B, implying that older households have a higher probability of being aware of climbing beans than younger households.

Also in both models, wealth of households as measured by the number of motorized vehicles owned, negatively influences awareness. Therefore, wealthy households appear to be less aware of climbing beans. In model B only, the number of non-motorized vehicles plays a role in awareness of climbing beans (at 10% probability level). This might indicate that the poorest households are less aware of the existence of climbing beans.

The number of visits by extension services made to the household appears to play a role in awareness of climbing beans. With every visit paid to a household, the chances on awareness of climbing beans increases with 7.4 or 11.7% (marginal value of 0.074 or 0.117 in model A or B respectively). Also receiving information on farming through extension services influences awareness significantly.

The second probit regression model built to explain the decision by farmers to test climbing beans or not is shown in Table 6. Both model A and B are significant. Model B fits the data better than model A as the pseudo R-squared value is higher for model B.

Both probit models A and B are very similar with the same predictor variables being significant. The variable district is highly significant indicating that in Embu, there are significantly more households who have tested climbing beans on their farm compared to Kirinyaga. Altitude is again highly significant indicating that at higher altitudes, there are more farmers who tested climbers. Also promotion by KARI Embu appears to have a significant positive influence on climbing bean trial. The distance from the village to the KARI Embu center also influences climbing bean testing in a positive way. This might be because KARI Embu center is located at low altitudes (Figure 1) and adoption appears to be higher at higher altitudes which are further away from the research center.

Age of the household head has a positive influence on climbing bean trial suggesting that older households have higher chances to test climbing beans compared to younger households. The number of non-motorized vehicles a household owns which is a measure for wealth does play a significant, positive role in the decision to test climbing beans or not. In other words, it seems that the poorest households are the ones having a smaller probability of testing climbing beans. Being member of a farmers' group plays an important role in the decision to whether or not test climbing beans. Farmers who are member of a farmers' group are 11.9 or 10.7% (model A or B) more probable of testing climbing beans as compared to farmers not being member of a farmers 'group. Also visits made by extension officers to the households play a significant, positive role in the decision to test climbing beans (the variable visit_extension is not significant in model A, but close to significance with a p-value of 0.101, while in model B, this variable is significant at 10% probability level).

According to the Likelihood Ratio Chi-Square test (LR Chi-squared), the model explaining adoption of climbing beans is significant (Table 7). The variable district is significant at 1% probability level pointing out that in Kirinyaga district there is less adoption compared to Embu district. The variable land size has a positive effect on adoption indicating that farmer households with larger land sizes are more likely to adopt climbing beans after having tested them. Another significant explanatory variable showing a negative effect is the amount of livestock owned by the household. Households with more animals are less likely to adopt climbing beans. Finally, the number of sources through which farming information is obtained influences adoption positively at 10% probability level.

DISCUSSION

In this study, the adoption process of climbing bean varieties by farmers in the central Kenyan highlands, promoted by KARI Embu and extension services, was evaluated. The main objective is to come to a better understanding of this adoption process and to formulate useful recommendations on how to enlarge the group of farmers who are adopters of climbing beans in view of the intrinsic advantages of climbing beans.

In the central highlands of Kenya, awareness of climbing beans by farmers is high but adoption is quite low (11 to 28%). For comparison, assessment studies on adoption of climbing beans by farmers in Rwanda (Sperling et al., 1992; Sperling and Muyaneza, 1995) report overall rates of adoption increasing from under 5% in the early 1980s to 42% in 1992 and 47% in 1995. Adoption ranged between 47 and 90% in six of ten provinces that have conducive environments, or in those that were deliberately targeted by research and development projects. The main differences between the central highlands of Kenya and the success story in Rwanda might be: (i) a higher availability of climbing bean seed through research and extension (Musoni et al., 2001) and (ii) high incidence of root rot (*Fusarium oxysporum*), which severely attacked and damaged local and improved bush germplasm in farmers' fields (David et al., 2002).

The most remarkable variable playing a highly significant role for both awareness and trial is the altitude of the village. This result indicates that climbing beans are clearly more popular at higher altitudes. A possible reason could be the difference in soil and climate conditions as pH and temperature decrease and rainfall increases with increasing altitude. Therefore, at higher altitudes (mainly UM1), soils are more acidic (also observable because main zone of tea growth is UM1) and the climate is cooler with more rainfall mainly as heavy showers. Compared to bush beans, climbing beans are more suitable for these conditions and therefore, the difference in performance between bush and climbing beans will be explicitly larger at high altitudes. Additionally, as global warming makes temperatures rise, the lower regions become less suitable every day for the production of climbing bean varieties, even though the MAC-lines are considered to be heat-tolerant lines. Another potential reason for the popularity of climbing beans at higher altitudes could be the vicinity of the forest which could provide staking material. However, according to the study, 92% of the farmers obtain their stakes from their own farm and 5% buy stakes. Still, it

Table 7. Parameter estimates of the Heckman probability model for the factors potentially influencing adoption of improved climbing bean varieties in the central highlands of Kenya.

Total number of observation			540	
Censored observations			328	
Uncensored observations			212	
Log likelihood value			-426.435	
LR Chi-squared			29.28**	
Likelihood ratio test of independent equation			2.77*	
Predictor variable	**Coef.**[a]	**Std. Err.**	**dy/dx**[b]	**Std. Err.**[c]
Village level				
District	-0.519***	0.189	0.088	0.032
Altitude	0.0005	0.0006	0.0001	0.0001
Household level				
Demographic variable				
Age_hh	0.008	0.009	0.001	0.002
Gender_hh	-0.191	0.244	-0.029	0.034
Dependency_ratio	-0.124	0.114	-0.021	0.019
Human capital				
Years_school	0.020	0.021	0.003	0.0034
Labour_endowment	-0.051	0.128	-0.009	0.021
Physical capital				
Income	-0.251	0.214	-0.039	0.030
Land size	0.090*	0.051	0.015	0.009
TLU	-0.173*	0.069	-0.029	0.012
Non_mot_vehicle	0.031	0.176	0.005	0.029
Mot_vehicle	-0.323	0.359	-0.045	0.040
Social capital				
Farmer_group	-0.094	0.188	-0.015	0.030
Visit_extension	-0.015	0.084	-0.002	0.014
Visit_events	-0.005	0.049	-0.001	0.008
nr_farming_info	0.087	0.151	0.015	0.026
constant	-2.388**	1.053	-	-

In order to test and correct for sample selection bias, due to unobservability of the decision to adopt or not for non-trial and non-aware households, the probit model for trial of climbing beans (Table 6) was used as a selection model with promotion and distance to the KARI center as selection variables (selection model not shown).[a]Stars indicate significance level: *$0.10 > p \geq 0.05$; **$0.05 > p \geq 0.01$; ***$0.01 > p \geq 0.001$; ****$p < 0.001$; [b]dy/dx is for discrete change of dummy variables from 0 to 1. [c]Standard errors for these discrete changes

could be possible that also forest material is used, but as the forest is protected by Kenyan law, farmers will not mention it. Very few reported studies on agricultural technology adoption take altitude into account. A study on production constraints of banana based cropping systems in the Great Lakes region (Bouwmeester et al., 2009) revealed that biotic constraints and drought are more severe at lower altitudes. However, the altitude range in this study was much broader (430 to 1900 masl) and biotic constraints were only significantly higher at very low altitudes.

Direct promotion by KARI Embu is done on field days and agricultural shows, by organizing farmer visits to the centre, selling limited amounts of seed at the centre and by setting up experimental fields on-farm which are also used as demonstration plots. A significant effect of these promotional activities on trial of climbing beans could be observed in this study. Farmers living in villages where promotional activities were carried out, have a 10.4% higher chance of testing climbing beans. Also indirect promotion of climbers by KARI Embu was done through collaboration with extension services which have an elaborate and dense, rural network. The effect of extension services is observable both at the level of creating awareness and of climber testing. The approach of collaboration with extension services therefore seems to be a very fruitful strategy. Many other adoption studies report a significant positive effect of contact with extension services on adoption of new technologies (Kaliba et al., 2000; Ojiako et al., 2007; Udoh and Omonona, 2008;

Saka and Lawal, 2009).

The role of farmer-to-farmer transmission seems to be important both at the level of creating awareness and of testing climbing beans. Most of the farmers (69%) first heard of climbing beans through other farmers and also the probit model for awareness points out that receiving farming information through other farmers effectively contributes to an enhanced awareness of climbing beans. Also membership of a farmers' group significantly contributes to the probability of testing climbing beans. This is also shown by descriptive statistics as most farmers (64%) obtained their first seed from climbing beans through other farmers. In a previous small-scale study (60 farmers) by J.J. Ouma (unpublished), similar results were found for farmer-to-farmer transmission of information and seed. Also several other adoption studies in soybean and maize found that membership of farmers' associations positively influenced adoption of improved varieties (Ojiako et al., 2007; Chianu et al., 2007). Hence, it can be concluded that farmer-to-farmer transmission is important in both creating awareness of climbing beans and in testing the climbers. Therefore, a promotion campaign of climbing beans through farmers' groups might succeed although it has to be taken into account that only a minority of farmers would be reached (only 25% is member of a farmers' group). However, it could be argued that by further farmer-to-farmer transmission, awareness would be created beyond farmers' groups and climbing bean seed would be further dispersed beyond members of farmers' groups.

Another variable that appears to be important both at awareness- and trial level is the age of the household head. Older households have higher probabilities on knowing climbing beans and on testing them (Grisley and Shamambo, 1993; Godoy et al., 1998; Agwu et al., 2008). In both cases, a more elaborate social network and a broader farmer experience possibly leading to differences in attitude towards risk, could explain this tendency. Also wealth of the household measured as the number of non-motorized vehicles and motorized vehicles owned, plays a role in awareness as well as trial of climbing beans (Godoy et al., 2008). First, the wealthiest households are less aware of climbing beans probably because their need for improving livelihoods is not particularly important. Secondly, testing climbing beans might be more difficult for the poorest households in the central Kenyan highlands. Possible reasons could be that the poor are particularly dependent on seed purchases and additionally cannot afford to buy seed (David and Sperling, 1999; Almekinders et al., 1994). Also in general, poorer households are less keen on taking risks. Therefore, making climbing bean seed available also to the poorest households might be important in future climbing bean promotion campaigns.

A first variable which exclusively plays a role in the decision to adopt climbing beans is the amount of livestock. In the central Kenyan highlands, zero-grazing

dairy farming is an important land use, especially in the higher, cooler and humid areas. As the wealthy biomass of climbing beans can be used as fodder for animals, it would be expected that having many animals would be an incentive to continue growing climbing beans. However, the opposite result was obtained which indicates that livestock tends to act more as a competitor instead of an ally. Probably, competition arises due to limited labour and land availability. When looking at use of the climbing beans by farmers, it appears that use as fodder is only of minor importance. Future promotion campaigns should therefore stress the use of climbing beans as healthy and diversified fodder for animals.

A second variable with an exclusive role in the adoption of climbing beans is land size. As explained in the introduction, climbing beans are often represented as the solution to decreasing land sizes as the same yield can be obtained on a smaller piece of land compared to bush beans. Therefore, the expected result would be that smaller land sizes favour the adoption of climbing beans. Surprisingly however, in this study, land size positively influences the decision to adopt climbing beans meaning that larger land sizes favour adoption. When farmers adopt climbing beans, they actually add this crop to their existing farming system. So often climbing beans are not substituting another crop or part of bush bean cultivation and so this 'extra' crop demands more land. Therefore, researchers and extensionists should realize that small land sizes can actually be a limiting factor for adoption rather than a stimulating one. The addition of novel climbing bean varieties to the agricultural system can actually be considered as advantageous as it brings more biodiversity into the existing farming system and lowers the risk of harvest failure. A serious limitation, possibly the most important for attaining widespread trial and adoption of climbers as indicated by the farmers themselves, is the poor availability of seed. Many aware farmer-households want to test climbing beans but lack the access to seed (42%) and many trial households stopped growing climbing beans because they ran out of seed (39%) and were not able to find it anymore at local markets. Farmers in the central Kenyan highlands obtain seed for planting from their own produce of previous season (42%) or by buying seed at the local market (53%). However, at these markets, climbing bean seed is not available. Farmers who grow climbing beans in the region use it mainly for home consumption and very few of them have ever sold the seed. These results indicate that the missing link in climbing bean trial and adoption is the availability of climbing bean seed at local markets. This problem seems to be specific for Africa (Grisley, 1990; David et al., 2002) and is often an ignored factor in crop varietal adoption studies (David et al., 2002). The few farmers who ever sold climbing bean seed, claim that marketability and prices for the seed are high, especially for the MAC-lines as they represent the locally preferred grain type very well. As a consequence, there is high potential for selling climbing bean seeds at markets. In this sense, the official release of the MAC-varieties by the Kenyan Government in 2008 could be very important as now, the possibility arises to collaborate with seed companies who can produce and sell certified climbing bean seed at large scale. However, it has to be taken into account that very few farmers (0.5%) buy certified seed. This finding is in concordance with previous studies pointing out that the share of the formal seed sector in the total seed supply rarely exceeds 10% in most staple crops in several African and Latin-American countries (Almekinders et al., 1994; David and Sperling, 1999; David, 2004). Main reasons are (i) very high market prices for certified seed; (ii) competition from farmer-saved seed, (iii) strong, region-specific varietal preferences and (iv) low adaptation of certified seed to regional biotic and abiotic stresses (Almekinders et al., 1994; David, 2004). This is why there are no private seed companies who want to produce and market climbing bean seed at large scale (KARI Embu, personal communication). Therefore, an alternative approach could be the set-up of so-called farmer-seed enterprises including local production, selection and selling of seed of improved varieties by farmers themselves. This approach makes use of the existing informal local seed systems in which farmer-to-farmer transmission of farming know-ledge and seed is crucial and most farmers (53%) buy seed for planting at local markets. Additional advantages of this approach are: (i) also the poorest households will have better access to seed of improved varieties as it will be cheaper than certified seed, (ii) seed is locally produced and selected making it highly adapted to both local agro-ecological conditions and market preferences, (iii) seed quality and cleanness is equally good as certified seed as proven by various comparative studies (Janssen et al., 1992), (iv) it is sustainable as it is market-driven and demand for climbing bean seed is high in the region, (v) a greater varietal diversity can be offered to farmers, promoting enhanced genetic diversity in the region and (vi) there is a possibility to establish linkages to formal institutions including extension services and commercial seed companies. When setting up farmer-seed enterprises, the main challenge that has to be carefully considered is a proper selection of farmers' groups including farmers who preferably have business experience or who have the reputation to be reliable seed suppliers in the local community. Selected farmers should be trained and supported in the beginning without creating a dependency mentality. Together with seed distribution, these farmers should also provide information and technical assistance to other farmers on climbing bean management. A nice description on the establishment of farmer-seed enterprises in rural Uganda is provided by David (2004).

An interesting opportunity in order to increase awareness of climbing beans could be the use of the radio. According to the results of this study, about 70% of the farmers retrieve farming information through the radio. However, there was no significant effect on awareness nor on climbing bean trial possibly because climbing beans were never promoted through the radio. Promotion of climbing beans through the radio is relatively easy and would reach many farmers. This would have an immediate effect on increasing awareness of climbing beans. Additionally, if appropriate information on how to obtain seed and how to grow climbing beans is included, there could also be an effect on trial and even on adoption of this crop. Agwu et al. (2008) report that radio farmer programmes actually enhanced the extent of adoption of agricultural technologies in Nigeria mainly through the spreading of information. This study also gives some recommendations to maximize the positive effect of a radio programme on agricultural technology adoption.

It is claimed that one of the largest advantages of climbing beans is their high yield potential. In this study, for October rains, the median of yields for climbing beans according to farmers was 1333 kg/ha. This figure is not even close to scientific reports of 4 to 5 tons/ha (which is under optimal conditions with use of fertilizers and pesticides), but is still considerably higher than reported yields for bush beans in farmers' fields in the region (averages of 300 to 800 kg/ha depending on variety, management, zone and weather condition). Also, in this study, adopters indicate high yields of climbing beans as their main advantage. Therefore, when promoting climbing beans, the argument of higher yields can certainly be used but it is advisable to use the actual figures obtained in farmers' fields, not the ones obtained under optimal conditions in experimental fields as this would only deceive farmers when they would test or adopt climbing beans.

Management of climbing beans by trial and adopter households was also verified in this study. All farmers staked the climbing beans at an appropriate time (that is, within the first four weeks after planting). However, about 1/5 of the farmers used stakes that were actually far too short (shorter than 1.50 m). Proper staking is an important requirement for good production of climbing beans. However, it might not be easy for farmers to find sufficient appropriate stakes and to fix these long stakes in the soil. Future promotional activities should again stress the importance of using sufficiently long stakes.

In contrast to other countries such as Rwanda where planting mixtures of up to 30 varieties is very common (Sperling and Muyaneza, 1995), the majority of farmers in the central Kenyan highlands choose to plant few varieties (maximum 3) and pure stands of bush beans. Additionally, since the introduction of the bush GLP-lines by the Kenyan Grain Legume Project in the 80's, no new bean varieties have been introduced. This exceptionally low diversity in bean varieties reflects a very vulnerable system with little adaptability and at high risk if the region would be challenged by severe biotic or abiotic stresses.

In Rwanda, increased incidence of root rot (*Fusarium oxysporum*) and fear of reduced genetic variability on-farm have resulted in the release of many new cultivars simultaneously by the agricultural research institute (ISAR) in Rwanda (Sperling and Muyaneza, 1995). In this sense, introduction of new climbing bean varieties is the first step toward a broadening of genetic diversity. It would also be worthwhile to select more MAC lines in order to improve genetic diversity in climbing beans and beans in general in farmers' fields of the central Kenyan highlands.

Conclusion

In conclusion, in the central Kenyan highlands, the current situation is one in which many farmers are aware of climbing beans (90%), about 39% of the farmers have tested climbing beans on their farms at least for one season and about 11% of all farmers have adopted climbing beans. Since 2004, trial and adoption rates show a steep increase.

Awareness and trial of climbing beans are significantly higher at higher altitudes. Climbing beans are grown on relatively small areas. From this study, some recommendations can be made on how to improve awareness, trial and most importantly, adoption of climbing beans:

1. As climbing beans seem to be more successful at higher altitudes, promotional strategies should first be focused there.
2. The official release of the MAC-varieties provides the logical step of collaboration with seed companies for certified seed production. However, the establishment of farmer-seed enterprises might have higher local impacts as the existing informal, local seed system would be used to provide farmers with improved climbing bean seed. Careful selection of existing farmers' groups to set up these kinds of enterprises is key in order to obtain success.
3. It could be worthwhile to select more MAC lines in order to improve genetic diversity in climbing beans and beans in general in the central Kenyan highlands. This can also be done by participatory selection of MAC-lines through the farmer-seed enterprises.
4. Future promotion campaigns should stress: (i) the use of climbing beans as healthy and diversified fodder for animals and (ii) the importance of using sufficiently long stakes. In particular this information should be provided to farmers together with seed distribution through farmer-seed enterprises.
5. An interesting opportunity in order to increase awareness of climbing beans could be the spreading of appropriate information on climbing beans and their advantages, on how to obtain seed and how to grow climbing beans through a radio programme.

ACKNOWLEDGEMENTS

We thank J. Muthamia, J. J. Ouma and B. Rono (KARI Embu) for their advisory contributions to this work. We also thank the enumerators Paul Mbogo, Lucy Ireri, Susan Wangari and Madrin Nthiga (KARI Embu). We acknowledge the bean breeding work of the MAC lines by Dr. M. W. Blair (CIAT) and Professor P. M. Kimani (University of Nairobi). We also thank J. Muthamia for leading promotional activities of climbing beans with farmers in the mandate region of KARI Embu. This work is supported by a grant from the Flemish Interuniversity Council (VLIR) to L.R. and by an Own Initiative Project 'Nutribean' (ZEIN2006PR331) supported financially by VLIR.

REFERENCES

Agwu AE, Ekwueme JN, Anyanwu AC (2008). Adoption of improved agricultural technologies disseminated via radio farmer programme by farmers in Enugu state, Nigeria. Afr. J. Biotech. 7(9):1277-1286.

Almekinders CJM, Louwaars NP, de Bruijn GH(1994). Local seed systems and their importance for an improved seed supply in developing countries. Euphytica 78:207-216.

Blair MW, Hoyos A, Cajiao C, Kornegay J (2007). Registration of two Mid-Altitude Climbing bean germplasm lines with yellow grain color, MAC56 and MAC57. J. Plant Reg. 1:143-144.

Bouwmeester H, van Asten PJA, Ouma EA (2009). Mapping key variables of banana based cropping systems in the Great Lakes Region, Partial outcomes of the baseline and diagnostic surveys, International Institute of Tropical Agriculture, Ibadan, Nigeria. p. 51.

Chianu JN, Tsujii H, Mbanasor J (2007). Determinants of the decision to adopt improved maize variety by smallholder farmers in the savannas of northern Nigeria. J. Food Agric. Environ. 5(2):318-324.

CIAT (2004). Impact of Improved bean varieties in Western Kenya, Highlights. No.18 December. 2004.

David S (2004). Farmer seed enterprises,a sustainable approach to seed delivery? Agric Human Values 21:387-397.

David S, Mukandala L, Mafuru J (2002). Seed availability, an ignored factor in crop varietal adoption studies: A case study of beans in Tanzania. J. Sust. Agric. 21(2):5-20.

David S, Sperling L (1999). Improving technology delivery mechanisms, Lessons from bean seed systems research in eastern and central Africa. Agric. Human Values 16:381-388.

Diagne A (2006). The diffusion and adoption of NERICA rice varieties in Côte d'Ivoire. Dev. Econ. 44, 2 June 2006.

Diagne A,Demont M (2007). "Taking a New Look at Empirical Models of Adoption: Average Treatment Effect Estimation of Adoption Rates and Their Determinants. Agric. Econ. 37(2-3):201-210.

Dimara E, Skuras D (2003). Adoption of agricultural innovations as a two stage partial observability process. Agric. Econ. 28:187-196.

Doss CR (2006). Analyzing technology adoption using microstudies: limitations, challenges, and opportunities for improvement. Agric. Econ. 34(3):207-219.

Feder G, Just RE, Zilberman D (1985). Adoption of agricultural innovations in developing countries, A survey. Econ. Dev. Cult. Change 33:255-299.

Gachimbi LN (2002). Technical report of soil survey and sampling results, Embu-Mbeere Districts, Kenya. LUCID Working Paper Series No.9, ILRI.

Godoy R, Morduch J, Bravo D (1998). Technological adoption in rural Cochabamba, Bolivia. J. Anthrop. Res. 54(3):351-372.

Government of Kenya (2001). The 1999 Kenya National Census results. Ministry of Home Affairs, Nairobi, Kenya.

Grisley W (1990). Seed for bean production in Sub-Sahran Africa, issues, problems and possible solutions. Paper presented at the second regional workshop on Bean Research in Eastern Africa,

Nairobi, Kenya, 5-9 March 1990.

Grisley W (1994). Farmer-to-farmer transfer of new crop varieties, an empirical analysis on small farms in Uganda. Agric. Econ. 11:43-49.

Grisley W, Shamambo M (1993). An analysis of the adoption and diffusion of carioca beans in Zambia resulting from an experimental distribution of seed. Exp. Agric. 29:379-386.

Heckman JJ (1976). The Common Structure of Statistical Models of Truncation, Sample Selection and Limited Dependent Variables and a Simple Estimator for Such Models. Ann. Econ. Soc. Meas. 5(4):475-492.

Heckman JJ (1979). Sample Selection Bias as a Specification Error. Econometrica 47(I):53-161.

Janssen W, Luna CA, Duque MC (1992). Small-farmer behaviour towards bean seed, evidence from Colombia. J. Appl. Seed Prod. 10:43-51.

Kaliba ARM, Verkuijl H, Mwangi W (2000). Factors affecting adoption of improved maize seeds and use of inorganic fertilizer for maize production in the intermediate and lowland zones of Tanzania. J. Agric. Appl. Econ. 32(1):35-47.

Kathuthu , Muchoki T, Nyaga C (2005). KIRINYAGA District Strategic Plan 2005-2010 for Implementation of the National Population Policy for Sustainable Development. Ministry of Planning and National Development,   Kenya. <http://www.ncapdke.org/UserFiles/File/District%20Strategic%20Plans/ Kirinyaga%20FINAL%20Modified.pdf>

McDonald H, Corkindale D, Sharp B (2003). Behavioral versus demographic predictors of early adoption, A critical analysis and comparative test. J. Market. Theor. Pract. 11(3):84-94.

Mugwe J, Mugendi D, Mucheru-Muna M, Merckx R, Chianu J, Vanlauwe B (2008). Determinants of the decision to adopt integrated soil fertility management practices by smallholder farmers in the central highlands of Kenya. Exp. Agric. 45:61-75.

Musoni A, Buruchura R, Kimani PM (2001). Climbing beans in Rwanda, development, impact and challenges. PABRA Millenium Workshop, Arusha, Tanzania.

Muthamia JGN, Nyaga C, Mbogo P (2003). Participatory evaluation of the influence of climbing beans on soil productivity. KARI Embu Reports 2003.

Ojiako IA, Manyong VM, Ikpi AE (2007). Determinants of rural farmers' improved soybean adoption decisions in northern Nigeria. J. Food Agric. Environ. 5(2):215-223.

Rogers EM (2003). Diffusion of Innovations 5th edition New York, Free Press.

Saka JO, Lawal BO (2009). Determinants of adoption and productivity of improved rice varieties in southwestern Nigeria. Afri. J. Biotech. 8(19):4923-4932.

Sperling L, Muyaneza S (1995). Intensifying production among smallholder farmers: the impact of improved climbing beans in Rwanda. Afr. Crop Sci. J. 3(1):117-125.

Sperling L, Scheidegger R, Burachara P, Nyabyenda P, Muyaneza S (1992). Intensifying production among smallholder farmers, the impact of improved climbing beans. CIAT Africa Occasional Publication Series No. 12. CIAT, Cali, Colombia.

Sunding D, Zilberman D (2001). The agricultural innovation process: research and technology adoption in a changing agricultural sector. In: Gardner B,Rausser,G. (eds) Handbook of Agricultural and Resource Economics. Amsterdam: Elsevier Science. pp. 207-261.

Udoh EJ, Omonona BT (2008). Improved rice variety adoption and its welfare impact on rural farming households in Akwalbom state of Nigeria. J. New Seeds 9(2):156-173.

Veldkamp T, Stoorvogel J, Claessens L, Hebinck P, Slingerland M, Van De Vijver C (2009). Getting to the bottom of Mount Kenya, Analysis of Land Dynamics and Sustainable Development in an Interdisciplinary Perspective. Interdisciplinary course organized by PE&RC and supported by CERES; MG3S and Sense. December 7-18. 2009.

Voysest O (2000). Mejoramiento genético del fríjol Phaseolus vulgaris L.: llegado de variedades de América Latina. Cali, Colombia: CIAT. pp. 1930–1999.

Wooldridge JM (2001). Introductory econometrics, a modern approach. Fourth edition, Macmillan, Portland, USA.

Woolley J, Davis JHC (1991). The agronomy of intercropping with beans. In: Common beans: Research for crop improvement (van Schoonhoven, A. and Voysest, O., eds). C.A.B. Intl. Wallingford UK and CIAT Cali Colombia. pp. 707-735.

2

Inheritance of various yield contributing traits in maize (*Zea mays* L.) at low moisture condition

Muhammad Ahsan*, Amjad Farooq, Ihsan Khaliq, Qurban Ali, Muhammad Aslam and Muhammad Kashif

Department of Plant Breeding and Genetics, University of Agriculture Faisalabad, Pakistan.

Six maize inbred lines were selected on the basis of their overall performances in preliminary screening. They were subjected to complete diallel analysis for various physio-genetic parameters under water deficit conditions. There were highly significant (P<0.01) differences for all parameters, except leaf venation, among the inbred lines and their possible crosses. All traits showed additive gene action, except stomatal frequency and stomata size, which showed complete dominance and over-dominance respectively. The strength of dominant and recessive genes for each trait was found different in each inbred. The narrow and broad sense heritabilities for the traits under study ranged between 35 to 80 and 32 to 78% respectively. However, stomata frequency and size might be useful while selecting maize inbred lines for hybrid seed production under limited water conditions and all other traits might be used while selecting inbred lines for synthetic genotype development.

Key words: Drought, corn, stress breeding, gene action, physiology.

INTRODUCTION

Maize plays a pivotal role in more food production program to cope with ever increasing population of Pakistan. It is used in the human diet in both fresh and processed forms. Besides providing food for human beings and feed for animals, it is used extensively as raw material in industry that benefits a large proportion of the world. The value addition has been an economic driver in the specialty corn markets (Hallauer and Miranda, 1988). The global food supply demand model predicts that global demand for maize will increase from 526 M tons to 784 M tons from 1993 to 2020, with most of the increased demand coming from developing countries (Rosegrant et al., 1999). Plants may experience many distinct abiotic stresses in the fields like water stress, salinity and tempe-rature extreme either continuously or discontinuously at different times during the growing season (Tester and Basic, 2005). Abiotic stresses limit crop productivity (Araus et al., 2002). According to Jamieson et al. (1995) water requirement of maize at the time of tesseling is 135 mm/month (4.5 mm/day) and this requirement may increase up to 195 mm/month (6.5 mm/day) during hot windy conditions. The rainfall in Pakistan is low and irregular (less than 100 mm) in maize growing areas and 70% of total rainfall occurs in July to September (Anonymous, 2003).

Drought results in reduced crop yield, pasture deterioration and livestock death. It strongly affects the production of cereals and poses a serious threat to the food security of households, countries and even entire subcontinents. In future, the destructive impact of drought may grow, as the climate change becomes a reality. The maize crops may experience reductions in grain yields when subjected to water deficit during the critical period of crop cycle from tesseling stage to beginning of grain filling. During 1998/1999 long drought period, 48.8 mm rain only allowed grain yield of 8 t/ha while during the year 2002/03 a short duration drought at critical period reduced grain yield less than to 2 t/ha, resulting from the effects on the ear per plant and kernel per ear (Bergamaschi et al., 2004). The use of genetics to improve drought tolerance and provide yield stability is an important part of the solution to this problem. Agronomical interventions also have their importance, since

*Corresponding author. E-mail: ahsanpbg@uaf.edu.pkm or saim_1692@yahoo.com.

genetic solutions are unlikely to close more than 30% of the gap between potential and realized yield under water stress (Edmeades et al., 2004). However, improved genetics can be packed in a seed and easily be adapted than improved agricultural practices that depend more heavily on input availability, infrastructure, and access to markets and skill in crop and soil management (Campos et al., 2004).

There are two pre-requisites of evolving new lines, one is the existence of variations for the character and the second is the heritability of that character. The genetically controlled variations for drought tolerance can only enable the plant breeders to evolve drought tolerant maize lines. The previous research is evident that variation for drought tolerance exists in various crop species like wheat (Guttieri et al., 2001) and in maize (Frova et al., 1999). Because of this rich genetic diversity for drought, new methodologies can help to evolve drought tolerant varieties. Hybrids of the parents have more diversity yield than of similar parents (Troyer et al., 1998). This research program was initiated to examine the existence of variability and its genetic basis and to evaluate a set of inbred lines of maize to obtain information on the relative importance of general and specific combining ability for grain yield and its gene action for various physio-genetic traits under drought conditions.

MATERIALS AND METHODS

Preliminary screening

The experimental material was made of 25 maize inbred lines, collected from different research institutions of Pakistan and was evaluated against drought stress of below 50% field capacity (FC) moisture deficit level. The moisture level was maintained using moisture meter (∆T-NH2, Cambridge, England). An experimental unit was a Polythene bag (18 × 9 cm) filled with sand containing one seedling. Ten seedlings of each inbred line were grown in each of the established four replications. Two replications were harvested after 15 days and the rest after 20 days from date of sowing. On the basis of their survival rate and better performance under water deficit condition, six inbred lines, including 20P2-1, L5-1, L7-2, 70NO2-2, 150P1 and 150P2-1 were selected as drought tolerant and further evaluated on the field under 50% water deficit conditions as recommended irrigation.

Final assessment phase

The inbred lines selected at the preliminary screening were crossed in full diallel fashion in spring season of 2007 to generate genetic material for final assessment and hybrid (F_1) seeds were collected. The hybrid seeds (F_1) along with the parental inbred lines were planted in triplicate in a randomized complete block design during autumn 2007. Sowing was done in rows at two seeds per hill and later thinned to one plant per stand at 4 to 5 leaf stage. The intra-row and inter-row spacings were 15 and 75 cm, respectively. 50% water stress was given by reducing half number of irrigations alternatively. Five tagged plants were selected from each entry at random and data were recorded for the following physio-genetic traits.

Leaf venation

Leaf venation was examined with the help of microscope (NIKON-H3, JAPAN). The low power of microscope (10X) was used to examine the leaf venation. These observations were recorded from different places of each leaf slide.

Stomatal frequency

The stomata frequency was observed from the upper surface of the middle part of the leaf blades. Two leaves from each of ten plants of each entry were taken. These leaf samples were placed in Carnoy's solution (Absolute alcohol 100 parts, Chloroform pure 50 parts, Glacial acetic acid 16 parts) for 24 h to arrest stomatal movement and to remove chlorophyll. The leaves were washed in acetone and stored in formalin solution. The cleaned leaves were examined under (10X) objective of the microscope and numbers of stomata were counted.

Stomata size

Stomata size was measured with the help of ocular micrometer using microscope (NIKON-H3, JAPAN). The medium power (40X) was used to examine the stomata size. Ocular micrometer was calibrated with the help of stage micrometer. The length and width of the stomata was measured in microns. These observations were recorded from three different places of each leaf and area was calculated by using the following formula:

Area = Length × Width

Epidermal cell size

Epidermal cell size was also measured with the help of ocular micrometer using microscope (NIKON-H3, JAPAN). The medium power (40X) was used to examine the epidermal cell size. Ocular micrometer was calibrated with the help of stage micrometer. The length and width of the epidermal cell size was measured in microns. These observations were recorded from three different places of each leaf and area was calculated by using the following formula:

Area = Length × Width

Silk elongation rate

Silk elongation rate was recorded following Anderson et al. (2004). Five cobs from five tagged plants of each entry were selected to record the data under 50% stress condition. On the day of first silk appearance, the silk length was marked as zero. Length of silk was recorded daily in the morning between 7.00 to 8.00 am. Silks were kept covered with butter paper bag to rule out any chance of pollination during this period.

Elongation rate = Total length of silk/Total number of days after first silk appearance

Anthesis to silking interval

The anthesis to silking interval is the difference between the first day of anthesis to the day of silk appearance. The five tagged plants were observed for the date of anthesis and then the date of silking. Then, the days between anthesis to silking of each plant

Table 1. Mean squares of analysis of variance for six inbred lines along with all their direct and indirect crosses for various parameters in maize.

S.O.V	df	Leaf venation	Stomata frequency	Stomata size	Epidermal cell size	Silk elongation rate	Anthesis to silking interval	Grain yield per plant
Replication	2	1.0278^{ns}	0.5832^{ns}	5147^{ns}	47093^{ns}	0.6025^{ns}	0.1111^{ns}	4.22^{ns}
Genotype	35	0.5976^{ns}	1.0303^{**}	15071^{**}	146088^{**}	0.8987^{**}	0.5402^{**}	193.62^{**}
Error	70	0.3992	0.3207	4249	31532	0.4242	0.1378	66.43

*, Significance at 5%; **, significance at 1% probability level; ns, non significant; SOV, source of variation.

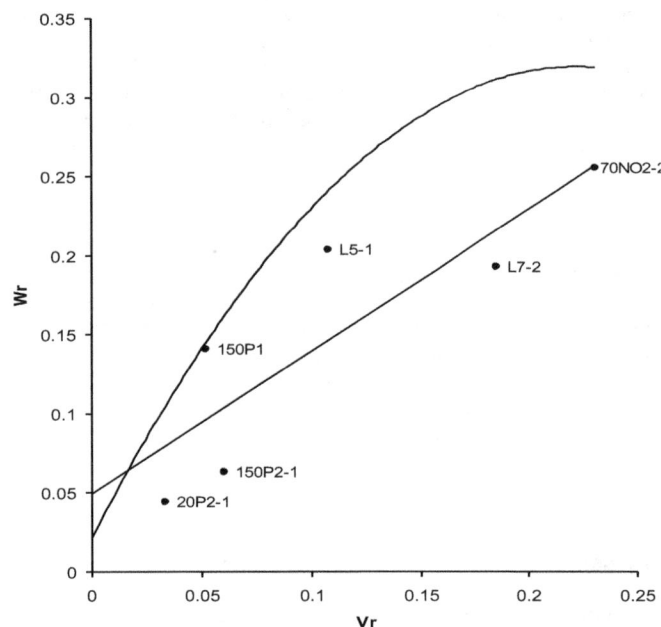

Figure 1. Leaf venation.

were counted and the mean of five plants was calculated.

Grain yield per plant

The cobs from five selected plants of each entry were harvested and threshed separately. The grains from each plant were weighed in grams using an electronic balance (OHAUS-GT4000, USA) and their mean was calculated.

Statistical analysis

The recorded data of all the parameters were analyzed by using analysis of variance (ANOVA) (Steel et al., 1997) in ordered to determine the variability among the crosses. The parameters showing significant genotypic differences among the thirty hybrids and six parents were further analyzed to determine gene action by following additive-dominance model developed by Hayman (1954) and Jinks (1954, 1955, 1956).

RESULTS

Results indicate significant differences at 0.01%

probability level for all the parameters among the genotypes (parents and crosses) except leaf venation (Table 1). In order to check the adequacy of the data for simple additive-dominance model, after analysis of variance, the data were subjected to joint regression analysis. For this purpose, the mean values of each cross and six parents were arranged (Table 3). Variance (Vr) and covariance (Wr) values were used to construct Vr/Wr graph (Figures 1 to 6) to observe the mode of gene action and distribution of dominant and recessive genes.

Leaf venation

Results of joint regression analysis (Table 2) suggested that data for leaf venation were fit for simple additive-dominance model. The results of Figure 1 exhibits that the regression line intercepted the Wr-axis above the point of origin, which indicated partial dominance with additive type of gene action. The regression line did not deviate significantly from unit slope, which suggested the absence of non-allelic interactions. The position of array points on regression line (Figure 1) showed that inbred line 20P2-1 had maximum dominant genes for leaf venation, being the closest to the point of origin while inbred line 70NO2-2 possessed maximum recessive genes being the farthest from origin. Table 3 shows that inbred line 70NO2-2 possessed maximum array mean (4.47) showing its best general combining ability and L7-2 secured minimum mean value (3.93) from all the inbred line, while within arrays of crosses, the cross 70NO2-2×L7-2 having the highest value (5.0) showed the best specific combining whereas 150PI×L7-2 secured minimum mean value (3.33) effects for the trait of leaf venation. The presence of additive genetic effects described moderate estimates (Table 3) of narrow sense heritability (0.36) and broad sense heritability with value of (0.32) indicative of additive with partial dominance type of gene action suggesting the feasibility of selection in early generations.

Stomatal frequency

In case of stomata frequency, the values of Wr and Vr plotted a graph (Figure 2) and the results of joint

Table 2. Scaling test (Joint regression analysis) of plant parameters in maize.

Parameter	Regression coefficient			Remark	Conclusion
	B	b = 0	b = 1		
Leaf venation	0.90±0.26	3.44*	0.37	b value deviate significantly from zero but not from unity	The data were adequate for simple additive-dominance model
Stomata frequency	1.09±0.28	3.79*	-0.32	-do-	-do-
Stomata size	0.91±0.25	3.56*	0.32	-do-	-do-
Epidermal cell size	1.05±0.20	5.27*	-0.29	-do-	-do-
Silk elongation rate	0.91±0.24	3.70*	0.36	-do-	-do-
Anthesis to silking interval	0.95±0.15	6.02*	0.30	-do-	-do-
Grain yield per plant	0.93±0.15	5.90*	0.42	-do-	-do-

-t of b_0 should be >2.7764; -t of b_1 should be < 2.7764.

Table 3. Mean values of parents and crosses.

Inbred line	Leaf venation	Stomata frequency	Stomata size	Epidermal cell size	Silk elongation rate	Anthesis to silking interval	Grain yield per plant
20P2-1	4.00	11.67	1078.82	3207.05	8.20	5.93	126.13
L5-1	4.33	11.93	1127.00	2628.03	7.39	5.80	111.40
L7-2	3.00	12.07	1127.82	3086.18	8.18	5.80	109.10
70NO2-2	5.00	12.13	1081.27	2773.40	7.18	5.53	101.13
150P1	4.33	10.20	1161.30	3549.23	8.29	7.00	103.27
150P2-1	4.67	12.53	1011.03	3163.77	6.62	5.80	112.56
Crosses							
20P2-1×L5-1	4.33	12.20	1171.10	3278.10	7.49	6.87	114.29
20P2-1×L7-2	4.67	13.07	981.63	3303.42	6.85	5.87	110.64
20P2-1×70NO2-2	4.67	12.27	1136.80	3336.90	7.35	5.20	119.18
20P2-1×150P1	4.00	11.27	1324.63	3314.03	8.16	4.81	112.01
20P2-1×150P2-1	4.00	11.27	1033.90	3306.68	7.29	4.80	116.68
Mean	4.33	12.02	1129.61	3307.83	7.43	5.51	114.56
L5-1×20P2-1	4.00	11.13	1266.65	3221.75	6.25	6.13	109.13
L5-1×L7-2	3.67	10.67	1220.92	3398.15	6.81	5.93	110.47
L5-1×70NO2-2	4.67	11.27	1327.90	3670.10	7.42	5.27	106.29
L5-1×150P1	4.00	11.73	1141.70	3232.37	9.12	5.73	112.42
L5-1×150P2-1	4.00	11.93	1233.98	3118.03	7.54	5.40	139.24
Mean	4.07	11.35	1238.23	3328.08	7.43	5.69	115.51
L7-2×20P2-1	3.67	13.93	1039.62	2654.17	7.45	4.87	119.19
L7-2×L5-1	4.00	10.93	1292.78	3164.58	7.67	5.73	119.89
L7-2×70NO2-2	3.33	12.80	1229.08	2944.08	5.50	4.53	153.14
L7-2×150P1	4.33	12.87	1144.15	3278.92	7.10	5.73	91.65
L7-2×150P2-1	4.33	12.60	1233.98	3515.75	7.43	5.93	90.36
Mean	3.93	12.63	1187.92	3111.5	7.03	5.36	114.85
70NO2-2×20P2-1	4.33	11.67	1148.23	3354.05	7.14	4.73	102.43
70NO2-2×L5-1	5.00	9.93	1046.15	3801.58	7.98	6.00	112.65
70NO2-2×L7-2	4.00	13.53	1128.63	2833.83	6.68	5.90	115.51
70NO2-2×150P1	4.67	12.47	1208.67	2707.25	5.87	5.33	108.63
70NO2-2×150P2-1	4.33	11.93	1108.22	2719.50	7.24	5.33	96.57
Mean	4.47	11.91	1127.98	3083.24	6.98	5.46	107.16

Table 3. Contd.

150P1×20P2-1	4.33	10.67	1262.40	3920.00	6.97	5.80	94.38
150P1×L5-1	4.33	13.87	1046.97	3259.32	8.00	4.67	132.10
150P1×L7-2	3.33	10.93	1027.37	3334.45	6.40	5.60	90.19
150P1×70NO2-2	4.00	11.80	1073.92	3327.92	6.71	5.93	141.91
150P1×150P2-1	4.67	11.47	1143.33	3056.78	8.26	5.27	119.80
Mean	4.13	11.75	1110.80	3379.69	7.27	5.45	115.68
150P2-1×20P2-1	4.00	11.93	1064.93	3527.18	8.16	5.47	95.11
150P2-1×L5-1	4.67	9.60	1247.05	3521.47	7.08	4.49	140.63
150P2-1×L7-2	4.00	9.93	989.80	3477.37	6.95	4.53	100.83
150P2-1×70NO2-2	4.00	10.60	1192.33	4847.73	7.34	4.53	113.11
150P2-1×150P1	4.33	12.13	1091.07	2850.98	7.09	6.67	107.50
Mean	4.20	10.83	1117.04	3644.95	7.32	5.13	111.44
$(h^2{}_{ns})$	0.36	0.56	0.52	0.80	0.62	0.68	0.58
$(h^2{}_{bs})$	0.32	0.35	0.51	0.78	0.55	0.67	0.50

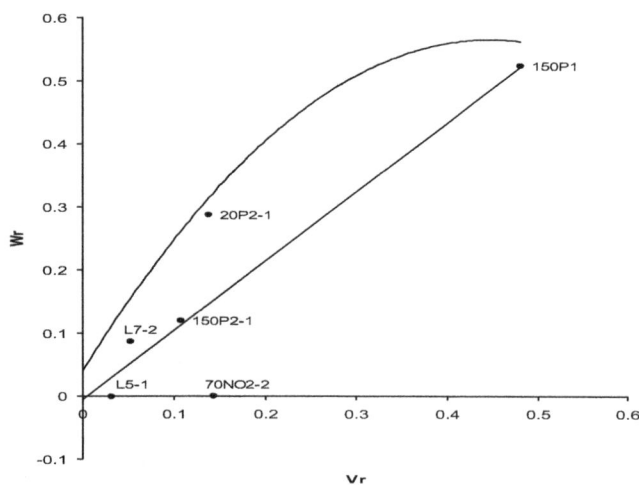

Figure 2. Stomatal frequency

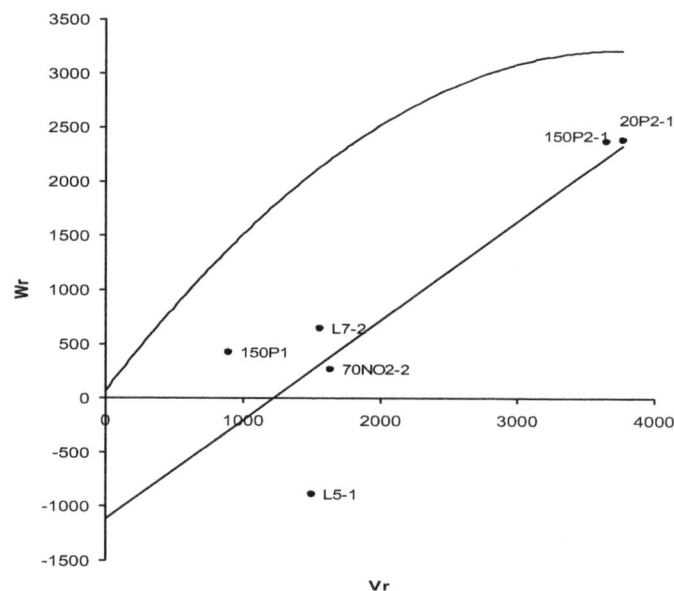

Figure 3. Stomatal size

regression analysis (Table 2) suggested the fitness of data for additive-dominance model. The graphical representation showed that the regression line (b) passed through the origin indicating the complete dominant gene action for stomatal frequency. The position of the array points (Figure 2) on regression line exhibited that the inbred line L5-1 being closer to the origin received maximum dominant genes and 150P1secured maximum recessive genes being farther from the origin. Table 3 reveals that inbred line L7-2 with higher array mean (12.63) proved to be the best general combiner while the cross L7-2×20P2-1 showed the highest specific combining ability as it secured maximum value (13.93). The stomatal frequency (Table 3) showed narrow sense heritability (0.56) but broad sense heritability (0.35) estimates, suggesting the presence of additive genetic

effects.

Stomata size (µm²)

Results of the joint regression analysis (Table 2) proved the fitness of data for genetic analysis. The inheritance of stomata size was determined as over dominance gene action because the regression line intercepted the Wr-axis below the origin. As the position of array points on regression line was concerned, inbred line 150P1 secured maximum dominant genes being the nearest to origin than the 20P2-1 being the farthest from origin

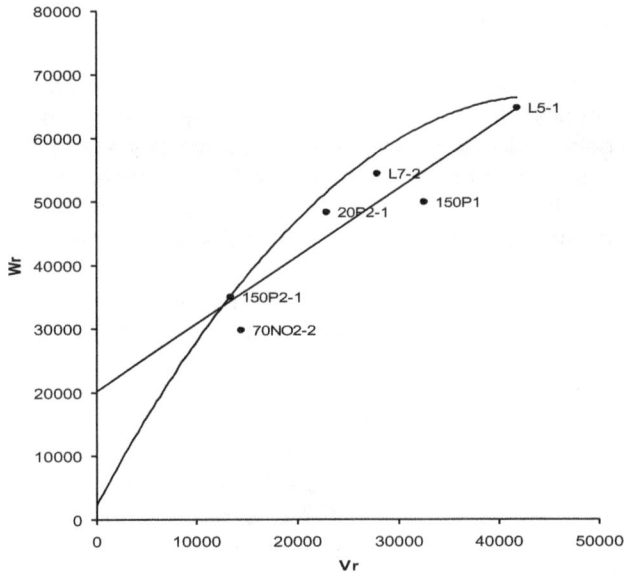

Figure 4. Epidermal cell size

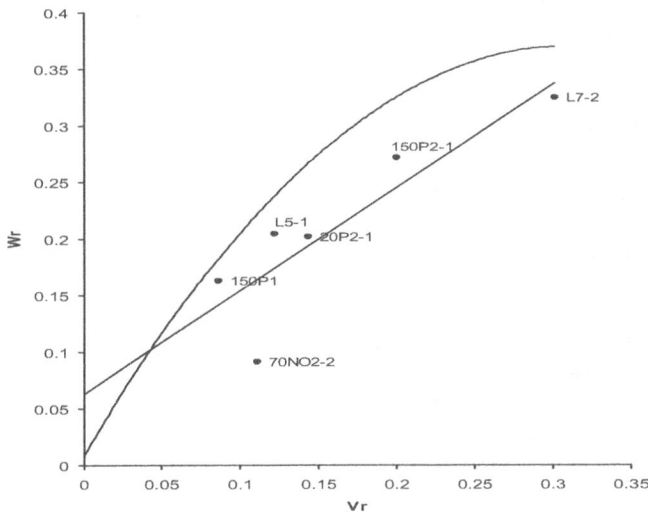

Figure 5. Silk elongation rate

(Figure 3). The data of mean values (Table 3) revealed that inbred line L5-1 proved to be the best general combiner with mean value of 1238.23 μm^2. The cross L5-1×70NO2-2 exhibited the maximum specific combining ability with mean value of 1327.90 μm^2. The estimates of heritability in Table 3 showed narrow and broad sense heritability 0.52 and 0.51 respectively that was an indication for the presence of complete dominance gene action.

Epidermal cell size

The results of the joint regression analysis (Table 2) proved that the data was fit for genetic analysis and fully adequate for additive-dominance model. The graphical representation (Figure 4) revealed that regression line (b) intercepted the Wr-axis above the origin indicating additive type of gene action involved in the expression of this character. The relative position of array point on the regression line determined that inbred line 70NO2-2 received maximum dominant genes and inbred line L5-1 secured maximum recessive genes because of their closer and farther positions from the origin respectively. The mean values (Table 3) revealed that inbred line 150P1 proved to be the best general combiner with maximum mean value of 3379.69 μm^2 whereas inbred line L5-1 proved to be the poor general combiner because of the minimum mean value of (3083.24 μm^2). The cross 20P2-1×150P1 secured the highest specific combining ability mean value of 3801.58 μm^2 and L7-2×20P2-1 proved to be the poor specific combiner with mean value of 2654.17 μm^2. There was high narrow sense heritability (0.80) and broad sense heritability (0.78) respectively.

Silk elongation rate

The results of joint regression analysis (Table 2) suggested that data were quite valid for additive-dominance model. Figure 5 illustrates the additive type of gene action controlling the inheritance of this trait as the regression line intercepted the Wr-axis above the point of origin. The position of array points on regression line concluded that inbred line 70NO2-2 received maximum dominant gene and inbred line L7-2 secured maximum recessive genes being the closest and the farthest relative to the origin, respectively. The inbred lines 20P2-1 AND L5-1 proved to be the best general combiner with highest mean value of 7.43 cm while inbred line 70NO2-2 possessed minimum mean value of (6.98). The cross L5-1×150P1 also proved to be the best specific combiner with the highest cross value of 9.12 cm while L5-1×20P2-1 proved to be the poorest specific combiner with mean value of 6.25 cm (Table 3). As indicated by Table 3, that narrows sense heritability (0.62) as well as broad sense heritability (0.55) was moderate due to the additive type of inheritance pattern in this trait. Moderate narrow sense heritability suggested that improvement could be made through individual plant selection in the later generations.

Anthesis to silking interval

The joint regression analysis (Table 2) suggested that data were suitable for additive-dominance model. As the regression line passed through Wr-axis above the point of origin, it signified that the inheritance of this trait was controlled by additive type of gene action. The position of array points on regression line (Figure 6) illustrated that inbred line 70NO2-2 received maximum dominant gene

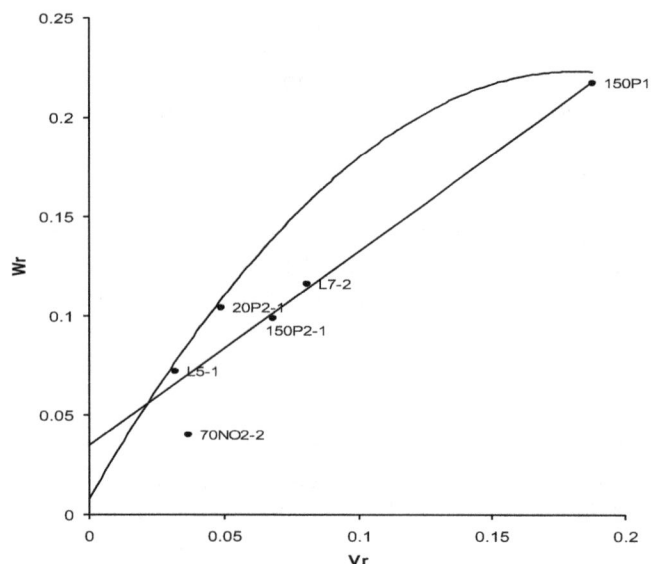

Figure 6. Anthesis to silking interval.

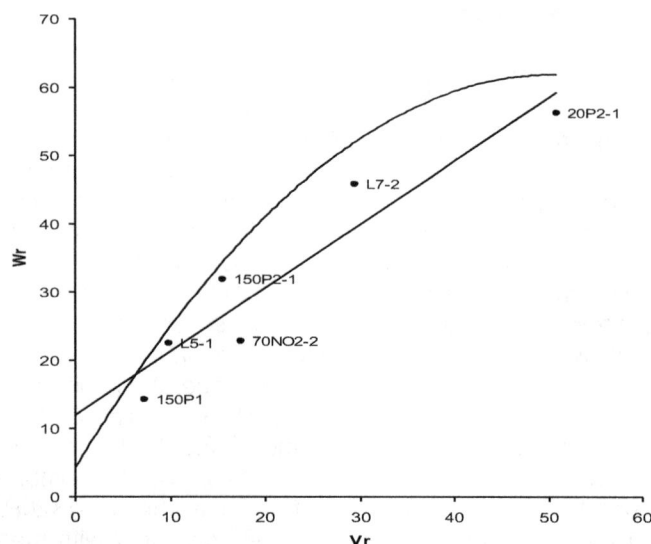

Figure 7. Grain yield per plant

and inbred line 150P1 secured maximum recessive genes being the closest and the farthest relative to the origin, respectively. The inbred line L5-1 proved to be the best general combiner with highest mean value of 5.69 whereas inbred line 150P2-1 proved to be the poorest general combiner with mean value of 5.13 for anthesis to silking interval. The cross 20P2-1×L5-1 also proved to be the best specific combiner with the highest cross value of 6.87 and cross L7-2×70NO2-2 exhibited poorest mean value of 4.53 for specific combining ability. The narrow sense heritability (0.68) as well as broad sense heritability (0.67) was due to the additive type of inheritance pattern in this parameter (Table 3).

Grain yield per plant

The results of regression analysis (Table 2) indicated the adequacy of the data for genetic analysis. Additive type of gene action was observed in the inheritance of this trait as the regression line (b) passed through the Wr-axis above the point of origin (Figure 7). The distribution of array points along with the regression line (b) indicated that inbred line 150P1 received most of the dominant genes because of its presence nearest to the origin while inbred line 20P2-1 secured maximum recessive genes being the farthest from origin respectively. Table 3 suggests that inbred line L5-1 and 150P1 had good general combining ability as it secured maximum array mean value (115.51 and 115.68 g) whereas 70NO2-2 showed minimum mean value107.16 g as within arrays. The cross L7-2 × 70NO2-2 showed good specific combining ability effects and secured the highest value153.14 g whereas the cross 150P1 × L5-2 showed the lowest specific combining ability effects and secured the value of 90.19 g. Table 3 shows the values of narrow sense heritability (0.58) and broad sense heritability (0.50) that was an indication of the presence of additive gene action.

DISCUSSION

For general combining ability (GCA), 20P2-1 was the best general combiner for the epidermal cell size and grain yield and L7-2 for stomata size. The inbred line 70NO2-2 exhibited best GCA for leaf venation and 150P1 proved a good general combiner for stomata size, silk elongation rate and anthesis to silking interval. The crosses 20P2-1×L7-2 and L5×70NO2-2 had highest specific combining ability (SCA) values for grain yield per plant and leaf venation respectively whereas 20P2-1×150P1 exhibited best SCA for parameters like stomata size, epidermal cell size and anthesis to silking interval. Regarding inheritance patterns, over dominance was found to control stomata size and these results are in accordance with the results of Tabassum et al. (2005) under normal and water stress conditions. Negative combining ability results for stomata size were found which were according to the findings of Saeed et al. (2001). Regarding epidermal cell size, additive effects were found which were supported by Mahmood et al. (2003). In this present study, silk elongation rate increased at value which was supported by Carcova et al. (2003). For anthesis to silking interval, additive gene action with partial dominance was found in contrast to the results of Afarinesh et al. (2005) and diverse anthesis to silking interval which was supported by the results of Sari et al. (1999) and Ribaut et al. (1996). For grain yield per plant, low specific combining ability effects were observed and these were supported by the report of Gissa et al. (2007). Water stress reduces grain yield

(Kebede et al., 2001; Ge et al., 2005). Additive effects for this trait were reported by Farshadfar et al. (2002).

REFERENCES

Afarinesh A, Farshadfar E, Choukan R (2005). Genetic analysis of drought tolerance in maize (*Zea mays* L.) using diallel method. Seed Plant 20:457-473.

Anderson SR. Lauer MJ. Schoper JB. Richard MS (2004). Pollination timing effects on kernel set and silk receptivity in four maize hybrids. Crop Sci. 44:464-473.

Anonymous (2003). Strategic plan of the Pakistan Council of Research in Water Resources. *Pakistan Council of Research in Water Resources* Islamabad.

Araus JI. Slafer GA. Reynolds MP, Toyo C (2002). Plant breeding and drought in C_3 cereals, what should we breed for. *Ann. Bot.* 89:925-940.

Bergamaschi H, Dalmago AG, Bergonci IJ, Bianchi MAC. Muller GA, Comiran F, Heckler MMB (2004). Water supply in the critical period of maize and the grain production. *Pesquisa Agropecuaria Brasileira.* 39:831-839.

Campos H, Cooper M, Habben JE, Edmeades GO, Schussler JR (2004). Improving drought tolerance in maize: A view from industry. Field Crops Res. 90:19-34.

Carcova J, Andrieu B, Otegui ME (2003). Silk elongation in maize relationship with flower development and pollination. Crop Sci. 43:914-920.

Edmeades GO, Banziger M, Schussler JR, Campos H (2004). Improving abiotic stress tolerance in maize: a random or planned process? In: Proceedings of the Arnel R. Hllauer International Symposium on Plant Breeding. Mexico City, 17-22 August 2003, Iowa State University Press, in press.

Farshadfar E, Afarinesh A, Sutka J (2002). inheritance of drought tolerance in maize. Cereal Res. Commun. 30:285-291.

Frova C, Villa M, Sari GM, Krajewski P, Fonzo ND (1999). Genetic analysis of drought tolerance in maize by molecular markers. I. Yield components. Theor. Appl. Genet. 99:280-288.

Ge T, Sui F, Bai L, Lu Y, Zhou G (2005). Effect of different soil water content obn photosynthetic character and pod yields of summer maize. J.Shanghai. Jiaotong Univ. Sci. 23:143-147.

Gissa DW, Zelleke H, Labuschagne MT. Hussien T, Singh H (2007). Heterosis and combining ability for yield and its components in selected inbred lines. South Afr.J. Plant Soil. 24:133-137.

Guttieri MJ, Jeffrey C, O'Brien SK, Souza E (2001). Relative sensitivity of spring wheat grain yield and quality parameters to moisture deficit. Crop Sci. 41:327-335.

Hallauer AR. Miranda JB (1988). Quantitative genetics in maize breeding. 2nd ed. Iowa State Univ. Press, Ames, IA. p. 468.

Hayman BJ (1954). The analysis of variance of diallel tables. Biometrics 10:235-244.

Jamieson PD, Martin RJ, Francis GS (1995). Drought influence on grain yield of Barley, wheat and maize. NZ J. Crop. Hortic. Sci. 23:55-66.

Jinks JL (1954). The analysis of continuous variation in diallel crosses of *Nicotiana* rustica L. Varieties. Genetics 39:767-788.

Jinks JL (1955). A survey of genentical bases of heterosis in variety of diallel crosses. Heredity 9:223-238.

Jinks JL (1956). The F_2 and back cross generation from set of diallel crosses. Heredity 1:1-30.

Kebede H, Subudhi PK, Rosenow TD, Nguyen HT (2001). Quantitative trait loci influencing drought tolerance in grain sorghum (*Sorghum bicolor* L.Moench). Theor. Appl. Genet. 103:266-276.

Mahmood N, Chowdhry MA, Kashif M (2003). Genetic analysis of some physio-morphic traits of wheat under drought conditions. J. Genet. Breed. 57:385-391.

Ribaut JM, Hoisington DA, Deutsch JA, Jiang C, Gonzalez DLD (1996). Identification of quantitative trait loci under drought conditions in tropical maize. Flowering parameters and the anthesis to silking interval. Theoret. Appl. Genet. 92:905-914.

Rosegrant MW, Leach N, Gerpacio RV (1999). Alternative future for world cereal and meat consumption. Summer meeting of the Nutrition Society. Guildford, UK. 29June- 2July 1998. Proc. Nutr. Soc. 58:1-16.

Saeed A, Chowdhry MA. Saeed N, Khaliq I, Johar MZ (2001). Line × tester analysis for some morpho-physiological traits in bread wheat. Int. J. Agric. Biol. 3:444-447.

Sari GM, Villa M, Frova C, Krajewski P, Fonzo ND (1999). Genetic analysis of drought tolerance in maize by molecular markers. II. Plant height and flowering. Theor. Appl. Genet. 99:289-295.

Steel RGD, Torrie JH, Dickey DA (1997). Principles and procedure of statistics. A biometric approach. Mc Graw Hill Book Co., New York, USA. pp. 400-428.

Tabassum MI. Saleem M, Ali A, Malik MA (2005). Genetic mechanism of leaf characteristics and grain yield in maize under normal and moisture stress conditions. Biotechnology 4:243-254.

Tester M, Bacic A (2005). Abiotic stress tolerance in grasses. From model plants to crop plants. Plant Physiol. 137:791-793.

Troyer AF. Openshwa SJ, Knittle KH (1998). Measurement of genetic diversity among popular commercial corn hybrids. Crop Sci. 28:481-485.

3

Overview of pepper (*Capsicum* spp.) breeding in West Africa

Sokona Dagnoko[1,2], Niamoye Yaro-Diarisso [3] Paul Nadou Sanogo [1] Olagorite Adetula [4]
Aminata Dolo-Nantoumé[3], Kadidiatou Gamby-Touré[3], Aissata Traoré-Théra[3], Sériba Katilé[3]
and Daoulé Diallo-Ba[2]

[1]Rural Polytechnic Institute for Training and Applied Research (IPR/IFRA), Katibougou, Koulikoro, BP 06, Mali Republic.
[2]Seneso Limited, BPE 5459, Bamako, Mali Republic.
[3]Institute of Rural Economy (IER), Rue Mohamed V, BP 258, Bamako, Mali Republic.
[4]National Horticultural Research Institute (NIHORT), Jericho Reservation Area - Idi-Ishin, PMB 5432, Ibadan, Oyo State, Nigeria.

The genus *Capsicum* (sweet and hot pepper) harbors an incredible intra and inter-specific diversity in fruit type, color, shape, taste, and biochemical content. Its potential uses and benefits to mankind cover many areas such as food and nutrition, medicine, cosmetic, plant based insecticides (PBI), and income. The cash income potential combined with the fact that peppers are easy to grow, harvest, and process makes them suitable for use in poverty reduction and food security improvement programs. Efforts were made in West Africa to improve peppers in terms of germplasm collection and conservation, variety introduction and testing, but due to the wide genetic diversity within and between species and inter-specific crossings, there is still room for further improvement of yield and fruit quality. However, to better exploit the various potentials of peppers, there is need to promote improved varieties and improve seed systems through enhanced public/private partnership.

Key words: *Capsicum*, pepper varieties, fruit morphotypes, capsaicin, seed systems.

INTRODUCTION

Peppers (hot and sweet) belong to the Solanaceae family, genus *Capsicum* (Greenleaf, 1986). This genus originated from Central and South America (Grubben and El Tahir, 2004) and comprises about 30 species, of which, five domesticated that comprise *Capsicum annuum* L. (hot and sweet peppers), *Capsicum frutescens* L. or bird pepper, *Capsicum chinense* Jacq. or aromatic chili pepper, *Capsicum baccatum* L. (aji), and *Capsicum pubescens* Ruiz and Pav. (rocoto). The first three species are the most cultivated in both tropical and temperate zones. *C. annuum* often forms a complex with *C. frutescens* and *C. chinense*. In Africa, they are generally considered together as *C. annuum* L (Grubben and El Tahir, 2004).

In this review paper, the morphologic and genetic diversity of *Capsicum* sp. peppers are highlighted as well as their potential uses. Also highlighted are the efforts deployed by the private sector, national and international agricultural research institutions in breeding, germplasm collection and management, and the perspectives for further improvement in West Africa.

GERMPLASM COLLECTION AND CHARACTERIZATION

Gene banks of *Capsicum* species are available throughout the world. The USDA has established a pepper gene bank from accessions collected worldwide. AVRDC-The World Vegetable Center has collections in Bamako (Mali), Arusha (Tanzania), and in Shanhua (Taiwan). Gene bank collections available in West Africa also include that of the Nigerian National Horticultural Research Institute in Ibadan. The Brazilian *Capsicum* sp gene bank maintained at the "Universidade Estadual do Norte Fluminense" is a good source of *C. chinense and C. frutescens* accessions. Private seed companies in Asia and North America have developed sets of pepper varieties and may provide small samples for testing. West African agro-dealers and seed companies offer adapted pepper varieties obtained from national and/or international research institutions and from private sector agricultural organizations located in the sub region and abroad. Activities conducted with West African collections include characterization of morphologic diversity and/or genetic diversity using molecular markers and field evaluations to identify gene sources for yield improvement (Adetula, 2006; Adetula and Olakojo, 2006).

USEFULNESS AND DIVERSITY OF MORPHOLOGIC DESCRIPTORS

Numerous identifiable and measurable morphologic descriptors are used for germplasm characterization and/or evaluation (IPGRI, 1995). They are related to the seedling, plant and stem, the inflorescence, the fruits, or to the seeds and are more or less species specific. Some descriptors are often considered as useful for assignment of accessions to species of the genus. Hence, Baral and Bosland (2004) used morphologic descriptors to verify species assignment at the germplasm repository for the *C. chinense* and *C. frutescens* accessions they studied. Using discriminant analysis, their study confirmed the usefulness of morphological descriptors in species classification, though not for all descriptors, but only two (calyx constriction and flower position). Fonseca et al. (2008) also assigned successfully 100% of the *C. chinense* accessions they studied to their corresponding species at the germplasm repository using mainly inflorescence characters. More recently, these results were confirmed by Ortiz et al. (2010) who reported the efficacy of inflorescence descriptors and seed color to reliably distinguish among *Capsicum* species. Using about 90 chili accessions, these authors reported up to three flowers/node and calyx constriction in *C. chinense* accessions, single white flowers/node and white filaments

in *C. annuum*, greenish corolla and purple filaments in *C. frutescens*, purple flowers and black seeds in *C. pubescens*, and corolla spots only in *C. baccatum*. Ortiz et al. (2010) highlighted also the usefulness of qualitative descriptors for assessing genetic diversity between species and of quantitative descriptors for assessing genetic diversity within species.

Large variation in the morphological descriptors used for germplasm characterization and evaluation, and assignment of accessions to species of the genus has been reported. Fonseca et al. (2008) evaluated 38 *C. chinense* accessions from Brazil for 51 characters and found significant genetic variation for all characters except only for eight traits (leaf shape, corolla color and shape, calyx pigmentation, annular constriction of the calyx, days to fructification, duration of fructification and seed color). Lannes et al. (2007) highlighted also the large variation of *C. chinense* in fruit color and shape. Large variation in qualitative and quantitative descriptors was reported also by Idowu-Agida et al. (2012). Worth noting is the fact that the subjectivity and difficulty of measurement of qualitative descriptors and the influence of environment on quantitative descriptors are impediments to their use for genetic diversity assessment.

CHARACTERISTICS OF PEPPER FARMS AND PRODUCTION SYSTEMS

The majority of pepper and other vegetable fields in most West African countries (Mali, Burkina Faso, Chad, Guinea, Niger and Senegal) are located in rural areas and exploited mainly by male smallholder farmers (AVRDC, 2008). Women are the main marketers, processors, buyers, and users (Yaméogo et al., 2002). As for other vegetables, farm size ranges from one or more plots of a few squared meters to less than 1 ha (USAID/Guinea, 2006; AVRDC, 2008). The crop may be produced in single stands or intercropped with staple crops or other vegetables, sometimes even with other crops of the Solanaceae family. In urban and periurban settings, peppers and other vegetables are produced to provide nearby city dwellers with fresh produces and growers and marketers with cash income. Farm labor is provided mostly by family members in rural areas, and family members plus paid personnel in the urban and periurban settings. Seed supply is through the private sector, National Agricultural Research and Extension Systems (NARES) or farmers' own saved seeds. The public/private sector partnership involving NARES and/or International Agricultural Research Institutions is very limited. Conventional production methods in open fields are commonly practiced. Although there is an increasing awareness of farmers and consumers in the health and environmental consequences related to the abusive use of chemical inputs, farmers have little interest in organic

Table 1. Profile of a preferred chili pepper variety in Mali according to various stakeholders involved in vegetable value-chain.

	Characters	Most preferred attributes
Standing plant	Number of fruits/plant	Large
		Viruses
	Disease resistance	Bacterial spot
		Bacterial wilt
Fresh fruit	Fruit size	Small – big fruits
	Shelf-life	> 3 days
	Firmness	Medium - high
	Wall thickness	Medium - high
Dry fruit	Resistance to breakage	High
	Resistance to transport	High
Fresh and dry fruit	Pungency	Low - high
	Aroma	Low - strong
	Taste	Medium good - very good

Source AVRDC – The World Vegetable Center/Mali, 2010.

farming perhaps due to the lack of resistant varieties to the main biotic constraints, and lack of access to organic markets and regulatory mechanisms.

POPULAR VARIETIES AND CROP PROFILE

There are many local cultivars grown in West Africa. Nigeria alone has more than 200 selections of pepper (Idowu-Agida et al., 2012). The literature regarding other countries is rather limited. However, it is generally recognized that the species *C. annuum sensus stricto* is popular in other West African countries, while almost unknown to Malian consumers who by far preferred the bird pepper *C. frutescens* and the aromatic pepper *C. chinense*. However, since 2008, Malian stakeholders are gaining more and more interest in *C. annuum sensus stricto* as a result of the efforts of the Vegetable Breeding and Seed Systems for the sub-Saharan Africa Project (vBSS) and partners in introducing, testing, promoting, and disseminating this type of pepper (Afari-Sefa et al., 2012). Popular varieties in the sub-region include among others Legon 18, Gbatakin, Bird's eye, Safi, Cayenne, and Jaune du Burkina for hot pepper and Yolo wonder, California wonder, and Morti (Diffa) for sweet pepper.

The pepper crop profile (quality requirements) has been investigated in Mali using participatory variety selection and organoleptic tests, where the quality requirements for chili in that country have been related not only to field appearance of the standing plants and post-harvest quality (Table 1), but also organoleptic quality (data not shown). Further, varieties that have scored high for field appearance may score low for

organoleptic quality, indicating the need to select simultaneously for both traits (Dagnoko and Gniffke, unpublished). Another important quality requirement often neglected and none verified is the absence or low content of aflatoxin in the dried whole products and/or powder. Aflatoxin is a pathogenic compound that occurs in the fruits as a result of contamination by the fungus *Aspergillus flavus* Link. However, the extent of pepper contamination in most countries is poorly known and needs exploring for West African pepper products to be competitive in both local and external markets.

MAIN BIOTIC CONSTRAINTS

Peppers are usually considered a robust crop compared with tomatoes and chili pepper is more robust than sweet pepper. However, both peppers are susceptible to a number of pests and pathogens causing considerable economic losses.

The most economically important pests to peppers in West Africa are thrips (*Frankliniella* sp) feeding on the leaves, flowers or fruits; aphids (*Myzus persicae* Sulzer, *Aphis* ssp, and *Macrosiphum euphorbiae* Thomas) feeding on young leaves and shootss; whitefly (*Bemisia tabaci* Gennadius) feeding on the leaves, root knot nematodes (*Meloidogyne* spp) feeding on the roots, the Mediterranean fruit fly [*Ceratitis capitata* (Wiedemann)] feeding on the fruit flesh, red spider mites (*Tetranychus* spp) feeding on the leaves, and fruit borers (*Lepidopterae* spp). In addition to damages caused to the plants by direct feeding, some pests such as nematodes, whiteflies, aphids and thrips are vectors of viruses.

Peppers are susceptible to pathogens such as bacteria, fungi, and viruses. The most economically important pepper bacterial diseases are bacterial wilt [*Ralstonia solanacearum (Smith)*], bacterial leaf spot [*Xanthomonas campestris,* var vesicatoria (Doidge)], and bacterial soft rot [*Erwinia carotovora* (Smith)]. Most threatening fungi for peppers in West Africa are *Phytophthora* root rot, Southern blight [*Sclerotium rolfsii* (Sacc)], and anthracnose [*Colletotrichum capsici* (Syd) E.J. Butler and Bisby)]. About 35 viruses are known to infect peppers worldwide (Green and Kim, 1994). Among viruses, 11 were reported in Africa out of which Pepper Veinal Mottle Virus (PVMV) and Tomato Yellow Leaf Curl Virus (TYLCV) are the most economically important and the most widespread in the Western Africa sub-region (Dafalla, 2001). The latter is restricted to semi-arid and arid zones. Detailed information on economically important biotic stresses of peppers and other vegetables in West Africa, their biological control methods and integrated pest management strategies are provided by James et al. (2010).

ABIOTIC CONSTRAINTS

Abiotic constraints pertaining to the climate (drought, flooding, strong winds, extreme temperature and sun light) and to the soil (moisture and nutrients content) may add up to biotic constraints and lead plants to stress and undergo anatomical and physiological disorders that reduce yield (Jackson, 1986). One of the most common physiological disorders of pepper is blossom-end rot, a calcium deficiency disorder that appears only at the blossom end of the fruit (Hochmuth and Hochmuth, 2009). The threat of abiotic constraints is getting increasingly alarming as a result of population growth and climate change with expected greater adverse effects in vulnerable regions such as semi-arid West and Central Africa. The use of adapted varieties combined with careful crop management practices notably the control of root damaging factors and proper irrigation and nitrogen fertilization help control the effects of abiotic constraints (Hochmuth and Hochmuth, 2009).

YIELD POTENTIAL OF EXOTIC AND LOCAL VARIETIES

Chili pepper may yield up to 18 t/ha and sweet pepper up to 30 t/ha in open-fields (Grubben and El Tahir, 2004).Yields greater than 20 t/ha have been reported for introduced *C. annuum* chili varieties grown on-station in Mali (AVRDC/Mali, unpublished). Identifying varieties with high yield potential is important to breeders, growers, and seed producers in the sub-region. From 2005 through 2010, multi-location trials involving 39 entries of exotic and local varieties were conducted jointly by the National Agricultural Research Systems of six West African

countries (Benin, Burkina Faso, Chad, Gambia, Niger, and Togo) and the AVRDC. The World Vegetable Center's sub-regional office in Mali, hence, bringing at seven the number of test countries. Results indicated country yield range of 0.14 to 21.91 t/ha and a mean sub-regional yield of 6.78 t/ha, the highest yield of 14 t/ha being observed in Niger (Table 2). In the FAOStat (2012) estimates, the highest pepper yield of the sub-region was also observed in Niger (9.25 t/ha) followed by Nigeria (8.33 t/ha). Compared to the mean chili and green pepper yield of the world (14.82 t/ha) (FAOStat, 2012) and the yield potentials reported by Grubben and El Tahir (2004) and AVRDC/Mali (unpublished), this regional mean yield is low, suggesting that further improvement of pepper yield in West Africa is needed. Improvement of pepper yield in West Africa can be achieved by introducing and evaluating more exotic varieties in multi-location trials. Since yield is dependent on the genetic background of the plant and the environment, focus should be on varieties with moderate to high tolerance to the prevailing environmental conditions such as high temperatures, viruses, bacterial wilt and *Phytophthora* spp., etc. Also required for reducing the pepper yield gap in West Africa is promotion and adoption of best cultural practices.

CAPSAICIN AND ITS POTENTIAL IN NUTRITION AND HEALTH

Capsaicin ($C18H27NO3$) is an alkaloid compound believed to be found only in peppers. It is responsible of their characteristic hot taste or pungency. The level of hotness depends on the concentration of capsaicin in the fruit and is variable between species, among varieties within species, among plants within varieties, among fruits of the same plant, and among different parts of the same fruit. *C. chinense* species is traditionally reported to contain the hottest cultivars (Canton-Flick et al., 2008). The placenta tissues and seeds of Habanero pepper (*C. chinense*) are reported to contain most of the capsaicin with 62 and 37%, respectively (González et al., 2004). Sweet pepper has no hot taste as capsaicin is controlled by a single dominant gene and this pepper is recessive for this gene. Capsaicin content is traditionally measured by organoleptic tests involving preferably a panel of non-addicted consumers. Today, high-performance liquid chromatography (HPLC) and enzyme imunoassay (EIA) tests are used to more precisely measure capsaicin content (Guthrie et al., 2004; Canton-Flick et al., 2008). Capsaicin benefits include anti-carcinogenic (American Association for Cancer Research, 2009), anti-oxidant, anti-mutagenic, immunosuppressive, hypocholesterolaemic, and bacterial growth inhibition effects (Grubben and El Tahir, 2004). In traditional medicine, chili pepper is used to ease digestion, stimulate the gut, combat constipation, and relieve pain. Capsaicin may have also a potential role in the development of pain-killing agents (Patwardhan et al., 2010).

Table 2. Minimum, maximum and average yield [t/ha] of chili pepper evaluated in seven countries of West Africa from 2005 through 2010.

Country	Trial period RS (rainy season) DS (dry season)	# Entries	Chili pepper total yield [t/ha]			
			Minimum	Maximum	Average	Coefficient of variation %
Benin	RS 2005	8	1.08	20	10.1	20.6
Burkina Faso	RS 2005	9	5.08	14.61	9.59	15.9
	DS 2005-2006	11	2.13	11.67	5.49	31.1
Chad	RS 2005	6	1.75	3.87	2.95	31.2
Gambia	RS 2005	7	1.46	6.52	3.67	56.8
Niger	DS 2005 - 2006	3	9.17	19.53	14	34.2
Togo	RS 2005	8	1.08	11.8	5.35	41
	DS 2005 -2006	8	3.81	9.93	6.8	27.4
Mali	RS 2005 DS 2006	15	2.43	11.36	5.3	32
	RS 2008	20	6.9	9.88	7.9	39.4
	RS 2009	8	2.42	5.02	4.12	19.7
	DS 2009-2010	10	8.69	21.91	12.43	13.01
	DS 2009-2010	10	0.14	0.73	0.38	38.2
Across all countries [t/ha]						
Minimum			0.14	0.73	0.38	
Maximum			9.17	21.91	14	
Mean			3.55	11.29	6.78	

Data source: AVRDC – The World Vegetable Center.

Peppers are also good sources of nutrients - Vitamins A, C, K, and B6, calcium, iron, zinc, and fiber. The nutrient and other phytochemical composition of peppers vary with color and/or maturity stage (Deepa et al., 2007). Given the wide variability in capsaicin and nutrient content of peppers, more breeding can be undertaken for improvement. Since peppers are easy to grow, harvest, process and utilize, efforts should be undertaken by extension workers, nutrition and health promoting specialists to disseminate and promote improved varieties.

BIOPESTICIDE POTENTIAL OF *CAPSICUM* SPECIES

Although the use of plant based insecticides (PBI) to control insect pests prior to and after harvest has been practiced for many centuries (Jacobson, 1958, 1975), it was rather limited in potential and ignored (Oparaeke et al., 2005). Nowadays, PBIs are gaining more and more interest in IPM strategies worldwide as a means to promote agricultural production, environment sustainability, and human health. In this regard, the biopesticide potential of *Capsicum* spp. to control insects feeding on diverse parts of various plants has been widely explored.

Bouchelta et al. (2003, 2005) reported a toxic effect of pepper extracts on eggs and adults of the *tabaci* whitefly infesting solanaceous crops. Oparaeke et al. (2005) reported a reduced number of thrips, pod borers, and pod suckers on cowpea after treatment with chili pepper based extracts. Prior to these studies, other authors have reported a repellent effect of pepper extracts on the grain borer *Rhyzopertha dominica* (L.) (El-Lakwah et al., 1997) and the cowpea bruchids *Callosobruchus maculatus* (F.) (Onu and Aliyu, 1995). The toxicity of *Capsicum* spp on insects is thought to be the effects of secondary metabolites including alkaloids, saponins and flavonoid compounds of this plant (Bouchelta et al., 2005). The efficacy of these toxic compounds may be enhanced by combining chili pepper extracts with extracts from other plants such as cashew nutshell and garlic bulb to create a synergistic effect between their respective toxic compounds (Oparaeke et al., 2005). Worth mentioning is that chili extracts have been used by some authors in small quantity (2% compared with 10% for other botanical species) perhaps due to its potential phytotoxicity. Yepsen (1976) suggested a rate of only two teaspoonfuls per 4 L of water. Adding a small level of soap to a solution of PBI may help it adhere to the plant upon spraying. Advantages of using chili pepper extract in pest control in West Africa encompass availability and affordability in most local markets, environmental safety, safety on human health, and ease of use. However, more research is needed before the commercial use of chili extracts as bio-pesticides can be feasible. Such studies should focus on the development of synergistic combinations of PBIs with chili extracts, identification of susceptible insects, and quantification of the toxicity levels, and efficient spray time and frequency, etc.

ECONOMIC IMPORTANCE AND POTENTIAL IN POVERTY REDUCTION

The world fresh chili and sweet pepper production was 27.6 million tons in 2010, to which West Africa contributed 888,400 tons or 3.2%. The biggest West African contributors are Nigeria and Ghana that ranked 8th and 13th, respectively (FAOStat, 2012). The vast majority of West Africa's pepper is sold in local or regional (Senegal, Gambia, Liberia, Sierra Leone and Mali), and international markets (Europe and North America). The crop therefore constitutes a source of income for resource poor households in rural, periurban and urban areas. Women are the main processors, traders, buyers, and users in West African cuisine. They can benefits from the cash income potential of chili pepper. There are some *C. annuum* chili varieties that have high fruit yield, many fruits/plants, and attractive fruit color and shape, and that are easy to grow and harvest. They are suitable for use in poverty reduction programs targeting resource poor households, including women in developing countries.

CONCLUSIONS AND PERSPECTIVES FOR FURTHER IMPROVEMENT

Breeding programs in West Africa have focused on pungent types of the *C. annuum - C. chinense -C. frutescens* complex, perhaps due to their large adaptation as compared with *C. annuum* sweet pepper. Both types of peppers hold tremendous potential in the sub-region in terms of health and nutrition improvement and poverty reduction. Although progress has been made in terms of germplasm collection and characterization, variety introduction and testing, there is still room for further fruit yield and quality improvement of both chili and sweet pepper especially in the face of climate change. The great genetic diversity in the *C. annum - C. frutescens - C. chinense* complex and possibilities of inter-crossing between species of this complex offer potential for genetic improvement of the crop. Such improvement should focus on agronomically important traits such as disease resistance, yield, persistence of the mature fruits on the pedicel (it determines fruit resistance to winds under open field cultivation), and heat tolerance for sweet pepper. Other characters to consider in breeding programs would be fruit market value adding traits such as longer shelf-life and wall thickness (both desirable for transport over long distances and storage of fresh and dried fruits, respectively). Taste and visual appeal are important issues to consider as preferences vary among consumers between and within communities. Fortunately, the great diversity in fruit characters offers opportunity for breeders to collect, characterize, evaluate, and to select *Capsicum* spp. varieties with different fruit characters. Another area for pepper improvement in West Africa is large-scale dissemination and promotion of improved varieties. To that end, efforts are needed from extension workers, National and non-government organizations, nutrition and health promoting specialists and agro-dealers for seed availability to growers through a dynamic public/private sector partnership.

ACKNOWLEDGEMENTS

AVRDC - The World vegetable Center, Dr Virginie Levasseur and the NARES partners are acknowledged for their efforts in providing data from 2005 to 2006, under the USAID funded project on "Promotion of superior vegetable cultivars in West Africa". Data from 2008 to 2010 were obtained through the Bill and Melinda Gates funded project on "Vegetable Breeding and Seed Systems in Sub Saharan Africa" and the WASA seed project.

REFERENCES

Adetula OA (2006). Genetic diversity of *Capsicum* using random amplified polymorphic DNAs. Afr. J. Biotechnol. 5(2):120-122.

Adetula OA, Olakojo SA (2006). Genetic characterization and evaluation of some pepper accessions, *Capsicum frutescens* (L.): The Nigerian 'Shombo' collections. American-Eurasian J. Agric. Environ. Sci. 1 (3):273-281.

Afari-Sefa V, Dagnoko S, Endres T, Tenkouano A, Kumar S, Gniffke PA (2012). Tools and approaches for vegetable cultivar and technology transfer in West Africa: A case study of new hot pepper variety dissemination in Mali. Journal of Agricultural Extension and Rural Development. 4(15):410-416. Review (2012) in: http://academicjournals.org/JAERD

American Association for Cancer Research (2006). Pepper component hot enough to trigger suicide in prostate cancer cells. www.eurekalert.org/pub_releases/2006-03/aafc-pch031306.php.

AVRDC (2008). Baseline study on vegetable production and marketing in Mali and spoke countries. AVRDC - The World Vegetable Center, Bamako, Mali. p. 46.

Baral J, Bosland PW (2004). Unravelling species dilemma between *Capsicum. frutescens* and *C. chinense* accessions (Solanaceae): A multiple evidence approach using morphology, molecular analysis, and sexual compatibility. J. Am. Soc. Hort. Sci. 129(6):826–832.

Bouchelta A, Boughdad A, Blenzar A (2005). Effets biocides des alcaloïdes, des saponines et des flavonoïdes extraits de *Capsicum frutescens* L. (Solanaceae) sur *Bemisia tabaci* (Gennadius) (Homoptera : Aleyrodidae). Biotechnol. Agron. Soc. Environ. 9(4): 259–269.

Bouchelta A, Blenzar A, Beavougui AJP, Lakhlifi T (2003). Étude del'activité insecticide des extraits de *Capsicum frutescens* (Solanacées) sur *Bemisia tabaci* (Gennadius) (Homoptera : Aleyrodidae). Rev. Méd.Pharm. Afr. 17:19–28.

Canton-Flick A, Balam-Uc E, Jabın Bello-Bello J, Lecona-Guzman C,

Solıs-Marroquın D, Aviles-Vinas S, Gomez-Uc E, Lopez-Puc G, Santana-Buzzy (N 2008). Capsaicinoids content in Habanero pepper (*Capsicum chinense* Jacq.): Hottest known cultivars. HortScience 43(5):1344-349.

Dafalla GA (2001). Situation of tomato and pepper viruses in Africa. In: Proceedings of a Conference on "Plant Virology in Sub Saharan Africa". d'A. Hughes J.BO Odu (eds.), International Institute of Tropical Agriculture, June 4-8, Ibadan, Nigeria, pp 18-24.

Deepa N, Kaur C, George B, Singh B, Kapoor HC (2007). Antioxidant constituents in some sweet pepper (*Capsicum annuum* L.) genotypes during maturity. LWT- Food Sci. Technol. 40(1):121 -129.

EL-Lakwah F, Khaled OM, Kattab MM, Abdel-Rahman TA (1997). Effectiveness of some plant extracts and powders against the lesser grain borer *Ryzopertha dominica* (F.). Ann. Agric. Sci. **35**(1):567–578.

FAOSTAT (2012). Food and Agricultural Organization of the United Nations Statistical Database, Rome, Italy.

Fonseca RM, Lopes R, Barros WS, Gomes-Lopes MT, Ferreira FM, (2008). Morphologic characterization and genetic diversity of *Capsicum chinense* Jacq. accessions along the upper Rio Negro – Amazonas. Crop Breed. Appl. Biotech. 8:187-194.

González T, Villanueva L, Cisneros O, Gutiérrez L, Contreras F, Peraza S, Trujillo J, Espadas G. (2004). Analysis of fruit morphology of Habanero pepper (*Capsicum chinense* Jacq.). Book of Abstracts of the 17th International Pepper Conference, November 14-17, Naples, Florida, USA. p. 6.

Green SK, Kim JS (1994). Characteristics and control of viruses infecting peppers: a literature review. Asian Vegetable Research and Development Center, Tec. Bulletin Nº 18:60.

Guthrie K, Perkins B, Fan T, Prince A, Jarret R. (2004). Determination of capsaicinoids in oleoresins and dried hot peppers by enzyme immunoassay. Book of Abstracts of the 17th International Pepper Conference, November 14-17, Naples, Florida, USA. p. 9.

Greenleaf WH (1986). Breeding vegetable crops, Chapter 3. Pepper breeding. Basset MJ (ed.). The AVI Publishing Company Inc. Westport, Connecticut. pp. 67-134.

Grubben GJH, El Tahir IM (2004). *Capsicum annuum* L. In: Grubben, GJH & OA Denton (eds.). PROTA 2: Vegetables/Légumes. [CD-Rom]. PROTA, Wageningen, The Netherlands.

Hochmuth GJ, Hochmuth R (2009). Blossom-end rot in bell pepper: causes and prevention. SL 284, Institute of Food and Agriculture Science, University of Florida. p. 5.

Idowu-Agida OO, Ogunniyan DJ, Ajayi EO (2012). Flowering and fruiting behavior of long cayenne pepper (*Capsicum frutescens* L.). Int. J. Plant Breed. Genet. 6(4):228-237.

IPGRI (1995). Descriptors for *Capsicum* (*Capsicum* spp.). International Plant Genetic Resources Institute, Rome. p. 54.

Jackson RD (1986). Remote sensing of biotic and abiotic plant stress. Ann. Rev. Phytopathol. 24:265-287.

Jacobson M (1958). *Insecticides from plants: a review of the literature, 1941 - 1953*; U. S Dept. Agric. Handbook 154. U.S Department of Agriculture, Washington DC. p. 299.

Jacobson M (1975). *Insecticides from plants: a review of the literature, 1954 - 1961*; U. S. Dept. Agric. Handbook 461. U.S. Department of Agriculture, Washington DC. p. 38.

James B, Atcha-Ahowé C, Godonou I, Baimey H, Goergen G, Sikirou R et Toko M (2010). *Gestion intégrée des nuisibles en production maraîchère : Guide pour les agents de vulgarisation en Afrique de l'Ouest.* Institut International d'Agriculture tropicale (IITA), Ibadan, Nigeria. p. 120.

Lannes SD, Finger FL, Schuelter AR and Casali VWD (2007). Growth and quality of Brazilian accessions of *Capsicum chinense* fruits. Sci. Hort. 112:266-270.

Onu I, Aliyu M (1995). Evaluation of powdered fruits of four peppers (*Capsicum* spp.) for the control of *Callosobruchus maculatus* (F.) on stored cowpea seed. Int. J. Pest Manag. 41(3):143–145.

Oparaeke AM, Dike MC, Amatobi CI (2005). Evaluation of botanical mixtures for insect pests management on cowpea plants. Journal of Agriculture and Rural Development in the Tropics. Subtropics 106(1):41–48.

Ortiz R, de la Flor FD, Alvarado G, Crossa J (2010). Classifying vegetable genetic resources. A case study with domesticated *Capsicum* spp. Sci. Hort. 126(2):186-191.

Patwardhan AM, Armen NA, Ruparel NB, Diogenes A, Weintraub ST, Uhlson C, Murphy RC, Hargreaves KM (2010). Heat generates oxidized linoleic acid metabolites that activate TRPV1 and produce pain in rodents. J. Clin. Invest. 126 (5): 1617–1626.

USAID/Guinea (2006). Plan de développement de produit: le petit piment (*Capsicum frutescens*) en Guinée: Activités liaisons de la Guinée avec les marchés agricoles. Rapport préliminaires, Novembre 2006. p. 55.

Yaméogo CK, Karimou AR, Kabore S, Diasso K (2002). Les pratiques alimentaires à Ouagadougou, Burkina Faso : Céréales, légumineuses, tubercules et légumes. CNRST/CIRAD. pp. 111-132.

Yepsen RB (1976). Organic plant protection. Rodale Press, Emmaus, Pennsylvania.

4

Effects of UV-B radiation and butylated hydroxyanisole(BHA) on the response of antioxidant defense systems in winter wheat (*Triticum aestivum L.*Yildirim) seedlings

E. TASGIN[1]* and H. NADAROGLU[2]

[1]Primary Department, Bayburt University, 69000 Bayburt, Turkey.
[2]Department of Food Technology, Erzurum Vocational Training School, Atatürk University, Erzurum, Turkey.

Effects of UV and butylated hydroxanisol (BHA) on the activities of antioxidant enzymes were studied on the leaves of winter wheat cultivars (*Triticum aestivum L.* Yildirim). Fifteen days old wheat seedlings were treated with UV radiation (240 nm, 3 day) before they are treated with BHA (20°C). Supplementary UV-B radiation and UV-B+BHA significantly decreased chlorophyll and total phenol contents. The activities of enzyme extracts; polyphenol oxidase, catalase, paraoxonase and peroxidase were determined in the leaves under normal, UV-B (<315 nm) and UV-B + BHA conditions for 4 days. The antioxidant enzymes affected and showed enhanced activities in peroxidase, paraoxonase and polyphenoloxidase (except catalase) in UV-B and then BHA irradiated seedling. UV-B and then BHA-treated winter wheat seedling tries to counteract high level of reactive oxygen species that are produced under UV-B stress through the increased activities of antioxidant enzyme. It brings to mind that BHA tries to counteract high concentrations of oxygen species produced under UV-B radiation stress through increase in UV absorbing compound and antioxidant enzymes. In this study, the changes in quantities of polypeptide in wheat was investigated by sodium dodecyl sulfate polyacrylamide gel electrophoresis (SDS-PAGE). While the accumulation and pattern of the polypeptides unchanged in UV-treated leaves, they increased in UV+BHA-treated leaves compared to the control. It is concluded that BHA can be involved in UV radition tolerance by regulating antioxidant enzyme activities.

Key words: Oxidative damage, butylated hydroxanisol (BHA), peroxidase, catalase, polyphenol oxidase, paraoxonase, UV, winter wheat.

INTRODUCTION

Due to thinness of the ozone layer, it absorbs less UV-B in sunlight. Due to thinness of the ozone layer, less UV-B radiation of the sun is adsorbed by ozone. The world's surface is exposed to UV-B radiation which is harmful for the plant growth, development and physiology. Excessive UV-B negatively affects the growth process and development of almost all of the green plants (Frohnmeyer and Staiger, 2003; Brian, 2011). A number of studies have been carried out to study whether this increased UV-B radiation has a significant impact on plants (Dėdelienė and Juknys, 2010). Plants' exposure to UV radiation stress cause production of deleterious free radicals/reactive oxygen species (ROS) such as singlet oxygen, superoxide radical, hydrogen peroxide, hydroxyl ion and free hydroxyl radical ($^1O_2$, $\cdot O^{2-}$, H_2O_2, OH^- and $\cdot OH$) which are produced in mitochondria, endoplasmic reticulum, micro-bodies, plasma membranes and chloroplasts (Escoubas et al., 1995; Olga et al., 2003). Reactive oxygen species presence and induce antioxidant

*Corresponding author. E-mail: esent25@yahoo.com

system to protect plants (Foyer and Noctor, 2005; Foyer and Graham, 2009; Parra-Lobata et al., 2009). Antioxidants can act by scavenging reactive oxygen species, by inhibiting their formation, by binding transition metal ions and preveting formation of OH /O_2^-, $H_2O_2^-$ (Zancan et al., 2008; Hideg et al., 2002; Mackerness et al., 1998; Foyer and Graham, 2009).

Antioxidants occur naturally in plants during stress. The antioxidant system includes enzymes such as peroxidase (POX, EC 1.11.1.7), catalase (CAT, 1.11.1.6), paraoxonase (PON, aryldialkylphosphatase, EC 3.1.8.1) etc...; besides vitamins, phytochemicals, minerals and food additives (Azzedine et al., 2011). SOD accelerates. O^{2-} to H_2O_2 dismutation and is localized in cytosol, chloroplast, mitochondria and peroxisomes. POX, an iron heme protein, catalyses the reduction of H_2O_2 with a concurrent oxidation of a substrate, mostly located in cell wall and involved in oxidation of phenol compounds in terms of lignin synthesis. CAT also catalyses the reduction of H_2O_2 to water and molecular oxygen and is localized in mitochondria and peroxisomes, and absent in chloroplast (Kuk et al., 2003). Paraoxonase 1 (PON1) is a member of the multigene family of paraoxonases (PONs), which is observed in various tissues and cells (Primo-Parmo et al., 1996; La Du et al., 1999) including plants (Demir et al., 2011). There is also evidence in some studies that human serum paraoxonase (PON1)'s antioxidant function begins at the level of lipoprotein protection against oxidative modification by reactive oxygen species (ROS). The enzyme also reduces lipid hydroperoxides to hydroxides and presents a peroxidase-like activity, as PON1 seem to degrade hydrogen peroxide (H_2O_2), one of the ROS that is produced under oxidative stress (Aviram et al., 1998). Previous studies have shown that UV-B radiation and O_3 reduced biomass, yield, photosynthetic rate, chlorophyll, carotenoid and ascorbic acid contents and catalase activity, whereas increased total phenol content and peroxidase activity in wheat plants (*Triticum aestivum* L.) (Ambasht and Agrawal, 2003).

Butylated hydroxyanisole (BHA, 2-tert-Butyl-4-hydroxyanisole and 3-tert-butyl-4-hydroxyanisole) is a food additive and an aromatic organic compound. BHA has been widely used as antioxidants in the food industry including beverages, ice creams, candies, baked goods, instant mashed potatoes, edile fats and oils, breakfast cereals, dry yeast and sausages (Williams et al., 1999; Young, 1997; Verhagen et al., 1991). The structure of Butylated hydroxyanisole (BHA) is as follows:

It is very important to determine and evaluate the effects of UV-B on wheat which is used to produce bread for nutrition. In this study, a type of winter wheat (*T.aestivum L.* Yildirim) which is used for producing first- class bread

was used. It can outlast drought and cold. It is thought that this winter wheat variety can also resist to UV-B effects which is another reason why it is selected as the wheat type.

In addition, it is thought that BHA, which is an antioxidant substance, has a protective role against UV stress on wheat leaves in the antioxidant system of plants. In this study, effects of UV- B radiation and BHA were investigated on the amount of activities of some antioxidant enzymes such as paraoxonase (PON 1), catalase (CAT), peroxidase (POX), polyphenol oxidase (PPO), protein and the phenolic compound in winter wheat (*T. aestivum L.* Yildirim).

MATERIALS AND METHODS

The growth of plants

In this study, *T. aestivum* L. Yildirim was used as a type of winter wheat. Before sowing, the plant seeds' surface were sterilized for 10 min. with 10:1 water/bleach (commercial NaOCl) solution and then washed five times with distilled water. The plants were grown hydroponically in a growth chamber under controlled environmental conditions for 15 days (day/night temperature of 22/20°C, relative humidity of 75%, and a 16-h photoperiod with photosynthetically active radiation at photon flux density of 400 µmol m^{-2} s^{-1}).

Plants were divided into three groups: (1) control, (2) plants that exposed to UV-B, (3) plants exposed to UV- B radiation and then exposed to BHA (10 mM). Three replicates were used in each experiment. Firstly, two groups of plants were exposed to UV-B (312 nm) radiation with a density of 5.8 W/m^2 and measured with a UV sensor model of Leybold Didactic, Germany, during 4 days. UV-B lamp was purchased from Philips. After the plant was exposed to UV-B, BHA solution (10 mM) was sprayed on a group of wheat leaves in plastic pots during 2 days. Distilled water was used to spray control plants. Then, 10 g wheat leaves were harvested from plants exposed to UV-B radiation, UV-B + BHA and harvested from control leaves of plants, respectively. All harvested leaves were used to determine the antioxidant enzymes, amount of protein and phenolic content.

Sampling and laboratory analysis

Harvested leaves (10 g) were carefully cut with a sharp bistoury into 2 cm lengths, and rinsed 6 times in distilled water to remove BHA from the surfaces of plants. Winter wheat (*T. aestivum L.* Yildirim) sample of 10 g fresh weight (FW) was frozen immediately in liquid nitrogen and stored at -80°C until assay. The enzyme extract for POX, PPO, CAT and PON was prepared by grinding frozen tissue with 50 ml extraction buffer [0.1 M potassium phosphate buffer, pH 7.5 containing 0.5 mM ethylene diamine tetra acetic acid (EDTA)]. The extract was centrifuged for 20 min. at 15,000 × g and the supernatant was used for enzymatic assay (Nadaroglu and Demir, 2009).

Determination of catalase activity

CAT (mM H_2O_2 reduced min^{-1} g^{-1} FW) assay was based on the absorbance at 240 nm on UV spectrophotometer, a decrease in absorbance was recorded awhile as described by Aebi (1984). The complete reaction mixture contained 1.5 ml of 100 mM potassium phosphate buffer (pH 7.0), 0.5 ml of 75 mM H_2O_2, 0.2 ml of enzyme extract and 0.8 ml of distilled water.

Determination of peroxidase (POX) activity

POX activity (μmol tetraguaiacol min^{-1} g^{-1} FW) analysis is based on the increase in optical density due to the oxidation of guaiacol to tetraguaiacol (Castillo et al., 1984). The complete reaction mixture contained 1.0 ml of 100 mM potassium phosphate buffer (pH 6.1), 0.5 ml of 96 mM guaiacol, 0.5 ml of 12 mM H_2O_2, 0.1 ml of enzyme extract and 0.4 ml of distilled water. The enzyme activity was measured at 470 nm for 1 min. and calculated using the extinction coefficient (ε) of tetraguaiacol (26.6 mM/cm).

Determination of paraoxonase activity

Paraoxonase activity was determined at 25°C with paraoxon (diethyl *p*-nitrophenyl phosphate) (1 mM) in 50 mM *Tris*/HCl (pH 8.0) containing 1 mM $CaCl_2$. The enzyme analyze was based on the estimation of *p*-nitrophenol at 412 nm. Molar extinction coefficient of *p*-nitrophenol (ε = 18,290 M^{-1}cm^{-1} at pH 10.5) was used to calculate enzyme activity (Gan et al., 1991; Renault et al., 2006; Demir et al., 2011). One enzyme unit was defined as the amount of enzyme that catalyzes the hydrolysis of 1 μmol of substrate at 25°C. Analyses were carried out using a spectrophotometer (PG Instrument T80, USA) (Kuo and La Du, 1995; Stafforini et al., 1990; Reiner and Radic, 1985).

Determination of polyphenol oxidase

PPO activity was determined by measuring the increase in absorbance at 420 nm with a spectrophotometer (UV-Beckman). 50 μl of crude extract was added to a 3 ml substrate mixture containing 0.20 M sodium phosphate buffer (pH: 6.5), 25 mM catechol. Enzyme activity was calculated from the linear portion of the curve. One unit of PPO activity was defined as the amount of enzyme that can cause an increase in absorbance of 0.001/min (Flurkey, 1986).

Determination of total phenolic content

Total phenolic content was determined (Wu et al., 2007) with a concomitant formation of a blue complex by reduction of Folin-Ciocalteu's reagentby phenolic compounds. In this study, 0.50 ml of the extract was mixed with 3.0 ml of distilled water and 0.25 ml of Folin-Ciocalteau reagent. Immediately, 0.75 ml of saturated sodium carbonated and 0.95 ml of distilled water were added. Then, the mixture was incubated for 30 min at 37°C; the absorbance was read at 765 nm using an UV-Vis spectrophotometer (T80 PG Instrument, USA). Total phenolic concentration of samples was calculated on the basis of a standard curve obtained by using gallic acid; and then calculations were expressed as milligrams of gallic acid equivalents (GAE) per gram of sample.

Determination of chlorophyll content

The chlorophyll content was analyzed according to the method described by Wintermans and De Mots (1965). Chlorophyll was extracted by 80% ethanol and measured spectrophotometrically at 654 nm.

Protein concentration

Total soluble protein concentration was determined spectrophotometrically by Bradford's method (Bradford, 1976), using bovine serum albumin (BSA) as the standard.

SDS polyacrylamide gel electrophoresis

SDS polyacrylamide gel electrophoresis was performed control,control+BHA, UV and UV+BHA treated proteins. It was carried out in 3 and 10% acrylamide concentrations for the stacking and running gels, respectively, each of them containing 0.1 % SDS (Laemmli, 1970). The sample (20 μg) was applied to the electrophoresis medium. Brome tymol blue was used as tracking dye. Gels were stained in 0.1% Coomassie Brilliant Blue R-250 in 50% methanol, 10% acetic acid and 40% distilled water for 1.5 h. It was destained by washing with 50% methanol, 10% acetic acid and 40% distilled water several times (Laemmli, 1970). The electrophoretic pattern was photographed (Figure 3).

Statistical analysis

All results were analysed by analysis of variance, and means were compared by Duncan's multiple range test.

RESULTS AND DICUSSION

It was thought that loss of plant species would start with the ozone layer depletion which accordingly starts reduction of global food stocks. Negative effects of UV lights on wheat (*T. aestivum L.* Yildirim), the basic raw material for the bread which is the most important basic food of the communities, will cause a global problem. UV-B enhancement disturbs plant metabolism and causes oxidative injury by enhancing the production of reactive oxygen species. Metabolism of reactive oxygen species depends on low molecular anti-oxidant systems as well as enzymes such as paraoxonase, peroxidase, polyhenoloxidase and catalase. Therefore, in this study, it was determined that antioxidant enzymes activities changed in winter leaves on which firstly UV-B radiation and then BHA which is known as an antioxidative food additive is applied.

Activities of PPO, CAT, PON and POX in winter wheat were decreased very little in BHA treatment. However, activity of PON was significantly increased under UV-B radiation and BHA treatments as 196.83 EU/g leaf and 249.32 EU/g leaf, respectively; with respect to the control value 52.49 EU/g leaf. POX activity was significantly increased at 5.8 W/m^2 UV-B treatments (10.53 and 2.15% over control values). Activity of PPO was increased at both UV-B radiation and BHA treatments as 15 and 27.2 EU/g leaf, respectively, with respect to the control value 10 EU/g leaf. CAT activity significantly was significantly decreased at 5.8 W/m^2 UV-B and then BHA treatments (66.7 and 46.7% over controls, respectively) (Figures 1 and 2).

The results suggest that excess UV-B radiation could promote and stimulate the generation of ROS leading to increase in the activities of antioxidant enzymes as a defense system induced antioxidant defenses protecting plant against major fatal effects of ROS (Mittler et al.,

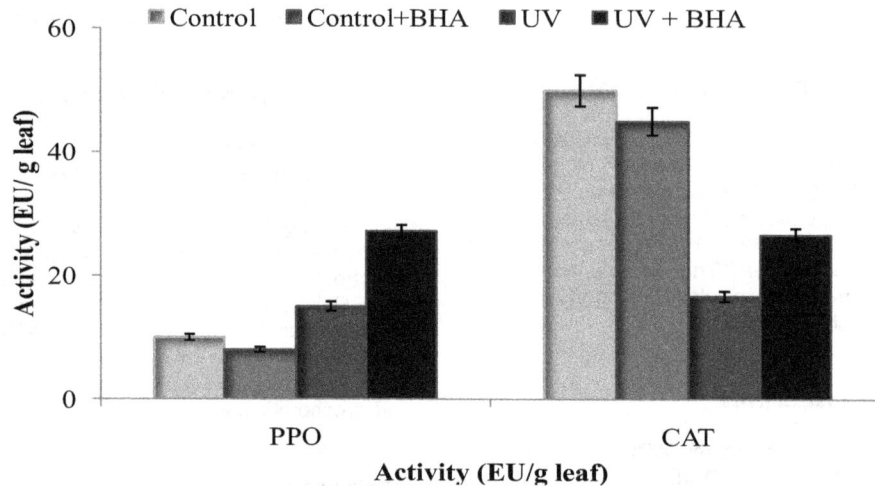

Figure 1. Effect of BHA, UV and UV + BHA on PPO and CAT antioksidant enzymes activity. Means with different letters are significantly different at $p<0.05$ based on Duncan's multiple range test

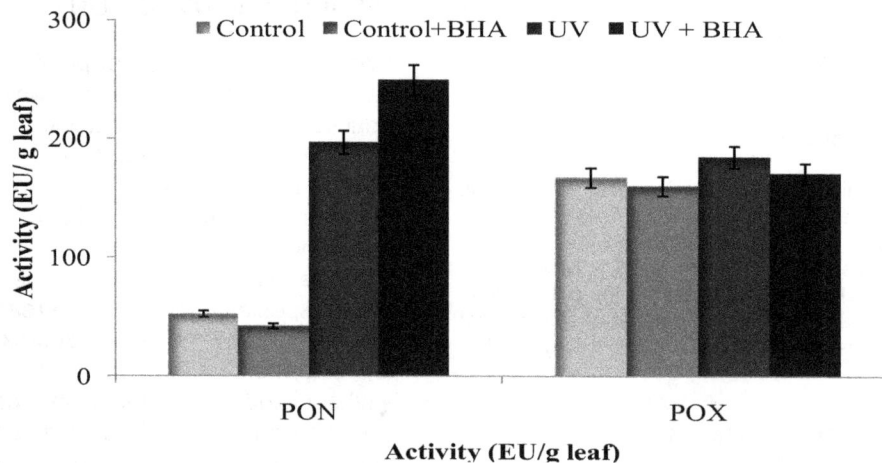

Figure 2. Effect of BHA,UV and UV + BHA on PON and POX antioksidant enzymes activity. Means with different letters are significantly different at $p<0.05$ based on Duncan's multiple range test.

2004; Fariduddin et al., 2009). The damage caused by these radicals which is indicated by the decrease in chlorophyll and total phenol contents was limited. This showed that water soluble non-enzymatic antioxidants and enzymatic antioxidant defense system represented by PON, POX, PPO and CAT are changeable in the winter wheat (*T. aestivum L.* Yildirim). The UV-B enhancement of CAT and POX activities which are both responsible for detoxification of H_2O_2 generated by PON, are probably equally important in the process. BHA has not increased antioxidant enzyme activities (PON, POX and PPO).

Chlorophyll content was greatly reduced in both treatments with UV-B and then BHA in winter wheat leaves (Table 1). It was shown that it was a possible

damage in photosynthetic capacity of chloroplasts (Malanga et al., 1997; Kakani et al., 2004; Santos et al., 2004; Strid, 1993; Mackerness et al., 1998). In a field experiment with *Vigna radiate*, (Pat et al., 1999) and *Phyllanthus amarus* (Indrajith and Ravindran, 2009) an initial increase and subsequent decrease in chlorophyll content was observed, which is also reflected in this present study. Total phenol content also decreased for treated UV-B and then BHA as 22.64 and 22.13% respectively, with respect to control (Table 1).

Amount of total soluble protein in 1 g wheat was determined by using Bradford method and standard graphic of serum albumin. In 1 g wheat, the amount of protein for control, treated with UV-B radiation and treated with BHA was determined as 94.04 ± 1.78 µg of

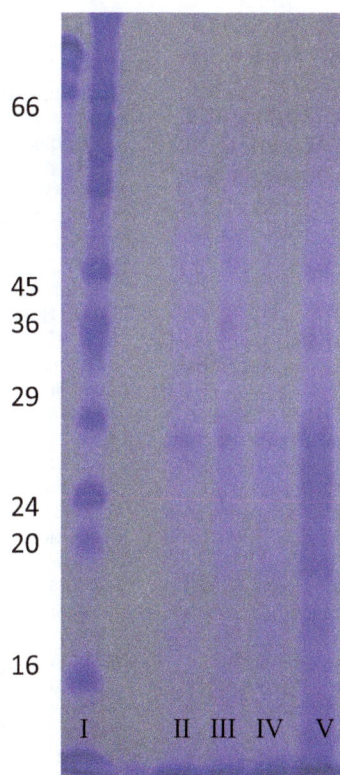

Figure 3. The electrophoretic pattern of antioxitative polypeptidesfrom winter wheat cultivars (*Triticum aestivum* L. Yildirim) after UV and BHA treatment (I: Standart; II: BHA; III:Control; IV: UV; V: UV+BHA).

Table 1. Effect of BHA, UV-B and UV-B+BHA treatments on the contents of chlorophyll, total phenol, total protein in winter wheat leaves.

Parameter	Total phenolic content (µg/ g leaf)		Chlorophyll content (µg g^{-1})		Total protein content (µg protein/ g leaf)	
	Mean ±SD	P	Mean ±SD	p	Mean ±SD	p
Control	2.39 ± 0.031	-	242.3 ± 2.16	-	94.04 ± 1.78	-
BHA	2.30 ± 0.02	p<0.05	235.8± 0.72	p<0.05	93.2 ± 1.05	p<0.05
UV	1.85 ± 0.02	p<0.05	226.4 ± 2.03	p<0.05	138.15 ± 1.20	p<0.05
UV + BHA	1.86 ± 0.016	p<0.05	206.5 ± 33.79	p<0.05	269.34 ± 0.91	p<0.05

protein, 138.15 ± 1.20 µg protein and 269.34 ± 0.91 µg protein, respectively. The results showed that the amount of protein increased in both treatments firstly UV-B and then BHA in wheat leaves. This causes an increase in the synthesis of the plant antioxidative enzymes as shown in previous studies (Agarwal, 2007). Amount of total phenolic substance, protein and chlorophyll contents were not changed significantly in winter wheat leaves only in BHA-treatment plants (Table 1).

While POX, PPO and PON enzyme activities were increasing, CAT activity inhibited after application of UV-B radiation on leaves (Figures 1 and 2). It is known that antioxidant enzymes in plants play an important role in response to the stress in plants. Previous studies have determined that UV-B enhanced oxidative stress and the generation of ROS (Mishra et al., 2011) and antioxidant defense system induced in *Arabidopsis thaliana* (Gao and Zhang, 2008), in *Oryza sativa* (Dai et al., 1997) and *Zea mays* seedlings (Carletti et al., 2003). Similar to our results, the increased levels of ROS trigger the activity of several antioxidant enzymes such as superoxide dismutase (SOD), catalase (CAT) and peroxidase (POD) (Mishra et al., 2011).

BHA, known as a food additive and antioxidant, enhanced antioxidative enzyme activities and accumulation of polypeptides in wheat after UV treatment. When polypeptides control, UV and UV+BHA were compared, accumulation of polypeptides was increased by BHA at wheat leaves. Amount of polypeptides (especially, 20, 24, and 29 kDa) slightly decreased with UV-B stress and later highly increased with BHA (Figure 3). According to the results, it could be thought that BHA influences the mechanism of the formation of antioxidant enzymes.

REFERENCES

Aebi H (1984). Catalase in vitro. Meth. Enzymol. 105:121-126.

Agarwal S (2007). Increased antioxidant activity in Cassia seedlings under UV-B radiation. Biol. Plantarum. 51(1):157-160.

Ambasht NK, Agrawal A (2003). Effects of Enhanced UV-B Radiation and Tropospheric Ozone on Physiological and Biochemical Characteristics of Field Grown Wheat. Biol. Plantarum. 47(4):625-628

Aviram M, Rosenblat M, Bisgaier CL, Newton RS, Primo-Parma SL, La Du BN (1998). Paraoxonase inhibits high-density lipoprotein oxidation and preserves its functions: A possible peroxidative role for paraoxonase. J. Clin. Invest.101:1581-1590.

Azzedine F, Gherroucha H, Bak M (2011). Improvement of salt tolerance in durum wheat by ascorbic acid application. J. Stress Physiol. Biochem. 7(1):27-37.

Bradford MM (1976). Rapid and sensitive method for the quantitation of microgram quantities of protein utilizing the principle of protein-dye binding. Anal. Biochem. 72:248-53.

Brian R (2011). Effects of UV-B radiation on plants: molecular mechanisms involved in UV-B responses. Handbook of Plant and Crop Stress (3rd Edition). pp. 565-576.

Carletti P, Masi A, Wonisch A, Grill A, Tausz M, Ferretti M (2003). Changes in antioxidant and pigment pool dimensions in UV-B irradiated maize seedlings. Environ. Exp. Bot. 50:149–157.

Castillo FI, Penel I, Greppin H (1984). Peroxidase release induced by ozone in Sedum album leaves. Plant Physiol. 74:846-851.

Dai Q, Yan B, Huang S, Liu X. Peng S, Miranda MLL, Chavez AQ, VergaraBS, Olszyk DM (1997). Response of oxidative stress defence systems in rice(Oryza sativa) leaves with supplemental UV-B radiation. Physiol. Plant. 101:301–308.

Demir N, Nadaroglu H, Demir Y (2011). Purification of Paraoxonase (PON1) from Olive (Olea Europaea L.) and effect of somec on paraoxonase activity in vitro. Asian J. Chem. 23(6):2584-2588.

Dėdelienė K, Juknys R (2010). Response of Several Spring Barley Cultivars to UV-B Radiation and Ozone Treatment. Environ. Res. Eng Manag. 4(54):13-19.

Escoubas JM, Lomas M, LaRoche J, Falkowski PG (1995). Light intensity regulation of cab gene expression is signaled by the redox state of the plastoquinone pool. P. Natl. Acad. Sci. USA 92:10237.10241.

Gao Q, Zhang L (2008). Ultraviolet-B-induced oxidative stress and antioxidant defense system responses in ascorbate-deficient vtc1 mutants of Arabidopsis thaliana. J. Plant Physiol. 165:138–148.

Fariduddin Q, Yusuf M, Hayat S, Ahmad A (2009). Effect of 28-homobrassinolide on antioxidant capacity and photosynthesis in Brassica juncea plants exposed to different levels of copper. Environ. Exp. Bot. 66(3):418-424.

Flurkey WH(1986). Polyphenol oxidase in higher plants. Plant Physiol. 81: 614-618.

Foyer CH, Graham N (2009). Redox Regulation in Photosynthetic Organisms: Signaling, Acclimation, and Practical Implications. Antioxid. Redox Sign. 11(4):861-905.

Foyer CH, Noctor G (2005). Redox homeostis and antioxidant signaling: a metabolic interface between stress perception and physiological responses. Plant Cell. 17:1866-1875.

Frohnmeyer H, Staiger S (2003). Ultraviolet-B Radiation-Mediated Responses in Plants. Balancing Damage and Protection. Plant Physiol.133:1420-1428.

Gan KN, Smolen A, Ecderson HW, La Du BN (1991). Purification of human serum paraoxonase/arylesterase. Evidence for one esterase catalyzing both activities. Drug Metab. Dispos. 19:100-106.

Hideg E, Barta C, K´alai T, Vass I, Hideg K, Asada K (2002). Detection of singlet oxygen and superoxide with fluorescent sensor in leaves under stres by photoinhibition or UV radiation. Plant Cell Physiol. 43:1154–1164.

Indrajith A, Ravindran KC (2009). Antioxidant Potential of Indian Medicinal Plant Phyllanthus Amarus L. Under Supplementary UV-B Radiation. Recent Res. Sci. Technol. 1(1):034–042.

Kakani VG, Reddy KR, Zhao D, Gao W(2004). Senescence and hyperspectral reflectance of cotton leaves exposed to ultraviolet-B radiation and carbon dioxide. Physiol. Plantarum 121:250-257.

Kuo CL, La Du BN (1995). Comparison of purified human and rabbit serum paraoxonases. Drug Metab. Dispos. 23:935-944.

Kuk YI, Shin JS, Burgos NR, Hwang TE, Han O, Cho B H, Jung S, Guh JO (2003). Antioxidative enzymes offer protection from chilling damage in rice plants. Crop Sci. 43:2109-2117.

Laemmli UK (1970). Cleavage of Structural Proteins during the Assembly of the Head of Bacteriophage T4. Nature 227:680–685.

La Du BN, Aviram N, Billecke S, Navab M, Primo-Parmo S, Sorenson RC, Standiford TJ (1999). On the physiological role(s) On the physiological role(s) of the paraoxonases. Chem-Biol Interact. 119-120:379-388.

Mackerness SAH, Surplus SL, Jordan BR, Thomas B (1998). Effects of supplementary ultraviolet-B radiation on photosynthetic transcripts at different stages of leaf development and light levels in pea (Pisum sativum L.): Role of active oxygen species and antioxidant enzymes. Photochem. Photobiol. 68:88-96.

Malanga G, Calmanovici G, Puntarulo S (1997). Oxidative damage to chloroplasts from Chlorella vulgaris exposed to ultraviolet-B radiation. Physiol. Plantarum 101:455-462.

Mishra V, Mishra P, Srivastava G, Prasad SM (2011). Effect of dimethoate and UV-B irradiation on the response of antioxidant defense systems in cowpea (Vigna unguiculata L.) seedlings. Pestic. Biochem. Phys. 100:118–123.

Mittler R, Vanderauwera S, Gollery M, Van Breusegem F (2004). Reactive oxygen gene network of plants. Trends Plant Sci. 9:490–498.

Nadaroglu H, Demir N (2009). In Vivo Effects of Chlorpyrifos and Parathion Methyl On Some Oxidative Enzyme Activities In Chickpea, Bean, Wheat, Nettle And Parsley Leaves. Fresen. Environ. Bull.18 (5): 647-652.

Olga B, Eija V, Kurt VF (2003). Antioxidants, oxidative damage and oxygen deprivation stress: a review. Ann. Bot..London .91: 179-194.

Parra-Lobato MC, Fernandez-Garcia N, Olmos E, Alvarez-Tinaut MC, Gomez-Jimenez MC (2009). Methyl jasmonate-induced antioxidant defence in root apoplast from sunflower seedlings. Environ. Exp. Bot. 66(1):9-17.

Reiner E, Radic Z (1985). Method for Measuring Human Plasma Paraoxonase Activity. Course on Analytical Procedures for Assesment of Exposure to Organophosphorus Pesticides, Manual of Analytical Methods Cremona (Italy). pp. 62-70.

Pal M, Sengupta UK, Srivastava AC, Jain V, Meena RC (1999). Changes in growth and photosynthesis of mung bean induced by UV-B radiation. Ind. J. Plant. Physiol. 4(5): 79-84.

Primo-Parmo L, Sorenson RS, Teiber J, La Du BN (1996). The human serum paraoxonase/arylesterase gen (PON1) is one member of a multigene family. Genomics 33:498-507.

Renault F, Chabrière E, Andrieu JP, Dublet B, Masson P, Rochu D (2006). Tandem purification of two HDL-associated partner proteins in human plasma, paraoxonase (PON1) and phosphate binding protein (HPBP) using hydroxyapatite chromatography. J. Chromatogr. B 836:15-21.

Santos I, Fernanda F, Jose MA, Roberta S (2004). Biochemical and ultrastructural changes in leaves of potato plants grown under supplementary UV-B radiation. Plant Sci. 167:925-929.

Stafforini DM, McIntyre TM, Prescott SM (1990). The platelet activating factor acetylhydrolase from human plasma. Method Enzymol. pp.187: 344.

Strid A (1993). Alteration in expression of defence genes in Pisumsativum after exposure to supplementary ultraviolet-B radiation. Plant Cell Physiol. 34:949-953.

Young E (1997). Prevalence of intolerance to food additives. Environ. Toxicol. Pharm. 4:111-114.

Verhagen H, Schilderman PAEL, Kleinjans JCS (1991). Butylated
  hydroxyanisole in perspective. Chem-Biol. Interact. 80:109-134.
Wu CR, Huang MY, Lin YT, Ju HY, Ching H (2007). Antioxidant
  properties of Cortex Fraxini and its simple coumarins. Food Chem.
  104:1464-1471.
Williams GM, Iatropoulos MJ, Whysner J (1999). Safety assessment of
  butylated hydroxyanisole and butylated hydroxytoluene as antioxidant
  food additives. Food Chem. Toxicol. 37:1027-1038.
Wintermans JF, De Mots A (1965). Spectrophotometric characteristics
  of chlorophylls *a* and *b* and their pheophytins in ethanol. Biochim.
  Biophys. Acta. 109:448-453.
Zancan S, Suglia I, Rocca NL, Ghisi R (2008). Effects of UV-B radiation
  on antioxidant parameters of iron-deficient barley plants. Environ.
  Exp. Bot. 63:71-79.

5

Effect of environmental factors on hybrid seed quality of Indian mustard *(Brassica juncea)*

Aniruddha Maity[1] and S. K. Chakrabarty[2]

[1]Seed Technology Division, Indian Grassland and Fodder Research Institute, Jhansi, UP- 284003, India.
[2]National Fund for Basic, Strategic and Frontier Application Research in Agriculture, 707, Krishi Anusadhan Bhavan-I, Pusa, New Delhi- 110012, India.

The trinity of parental line superiority, climatic conditions during crop growth and effective cross-pollination is the decisive factor to reap best quality hybrid seed. In order to study the effect of environmental factors during sowing time on hybrid mustard seed quality, freshly harvested hybrid seeds from three plots sown on 21st October (D1), 30th October (D2) and 18th November (D3), 2009-2010 were tested for quality traits before and after nine months of ambient storage. Percent germination had strong correlation with T_{max} at vegetative (r^2=0.881) and seed filling stage (r^2=0.88). Sunshine hours at vegetative and seed filling stage had significant correlation with percent germination (0.957 and -0.957 respectively), shoot length (0.898 and -0.870 respectively) and electrical conductivity of seed leachate after accelerated ageing (-0.880 and 0.856 respectively). No significant difference for germination and vigour indices was found between 1st (D1) and 2nd (D2) dates of sowing as the weather conditions during these two sowing periods was favourable and did not fluctuate much. Percent decrease in germination from D1 to D2 was 0.83/°C temperature decrease at vegetative stage and 1.33/°C temperature increase in seed filling stage. But percent decrease in germination from D2 to D3 was 2.31/°C temperature decrease at vegetative stage and 3.66/°C temperature increase in seed filling stage. A predictive model was developed as- Y= 234.545 - 1.688 × (±1.013) × RH_{pf} - 15.359 (±1.335) × SH_{sf}, where, Y= Percent germination, RH_{pf} = relative humidity at peak flowering stage and SH_{sf}= sunshine hours at seed filling stage. Seeds were subjected to accelerated ageing followed by electrical conductivity test after storage. D3 seeds had minimum germination percentage at 96 h ageing (5.1); significantly lower than that from D1 seeds (52.4). Grow out test result suggested standard isolation distance may be reconsidered to check out-crossing in optimum date of sowing. Seeds harvested from 21st October sowing (optimum time of sowing) gave the best result regarding seed quality parameters than that in hybrid seeds from 30th October and 18th November sowings.

Key words: Indian mustard, hybrid seed quality, environmental factors, percent germination, predictive model, correlation coefficient.

INTRODUCTION

Indian mustard *(Brassica juncea)* is an important oilseed crop in India, China and in south-western areas of the former Soviet Union. Indian mustard is cultivated worldwide as a vegetable, condiment and for oilseed. Due to its relatively greater tolerance to drought and heat that is, abiotic stress, pod shattering and blackleg

(*Leptosphaeria maculans*) than that of *B. napus/ B. rapa*, interest in growing Indian mustard (*B. juncea*) as an alternative to canola (*B. napus)* has recently increased worldwide (Anonymous, 2005; Banga et al., 2013). *B. juncea* is widely used for vegetable production particularly in Asia. Recently, it is being cultivated for biodiesel production. Since varieties with higher yield and enhanced oil content are becoming available through conventional breeding programs, it has found its growing way. In a significant effort, Directorate of Rapeseed-Mustard Research, Bharatpur, India successfully developed the first ever hybrid (released in 2008), NRC Sankar Sarson (NRCHB 506) of *Brassica juncea* (Indian mustard) through heterosis breeding using *Moricandia* CGMS system for cultivation primarily in Rajasthan and U.P., India to mitigate the ever-increasing demand of oilseed. This hybrid is of medium maturity duration, medium tall with high oil content and tolerant against diseases and pests.

How higher the seed yield is, its quality precisely depends on prevailing environmental factors. Environmental variation as a function of planting date is an important factor which significantly affects the timing and duration of the vegetative and reproductive periods as well as yield, its attributes and seed quality (Bhuiyan et al., 2008; Dornbos, 2002). Temperature, effective sunshine hours and light intensity differ with varying planting dates. Early date of planting results in significant increase in vegetative growth and produces more pods per plant consequently increases yield and quality of seeds. Delay in planting after October results in drastic reduction in seed quality (Thakur and Singh, 1998).

On the other hand, after harvest, seeds also deteriorate during prolonged storage and gradually lose viability (McDonald, 1999). Accelerated ageing test can give a simulated estimation of natural ageing. Its effect is manifested as reduction in percentage germination and if it germinates, produce weak seedling as in of case of natural ageing (Veselova and Veselovasky, 2003). Seed germination of the sunflower lines submitted to accelerated aging for three days was approximately equal to the seed germination rate measured after 12 months of natural aging ($R^2=0.93$). The soybean seed subjected to extreme conditions of accelerated aging for five days suffered stress that was comparable to that suffered by 12 months of natural aging (Bálešević-Tubić et al., 2010).

A few literatures are available regarding storage life of mustard varieties. Devi and Dadlani (2003) reported that mustard cv. Pusa Bold can last up to thirty one months in storage, in Delhi. Sharma and Singh (1997) suggested twelve months for cv. varuna in Palampur, Himachal Pradesh. Teari et al. (1998) reported eighteen months for cv. varuna and vardan in central plain zone of Uttar Pradesh. Kurdikeri et al. (2000) for thirteen months in Dharwad and Verma et al. (2003) reported three years for mustard variety RH-8113.

Some preliminary studies have been conducted to assess the effect of environmental factors during crop growth period on subsequent seed quality. But till now, limited information is available regarding hybrid seed quality and how the environmental factors influence the hybrid seed storability. The most marked and maximum values of seed storability expressed through percentage germination, shoot and root length, germination after ageing, electrical conductivity etc. differ in response to different planting dates (Veselova and Veselovasky, 2003). Hence, the objective of our experiment was to study the effect of different sowing dates and weather conditions on hybrid seed quality of first hybrid of Indian mustard (*Brassica juncea* (L.) Czern. and Coss.) and to compare the quality of hybrid seeds with that of an open pollinated variety, Pusa Bold.

MATERIALS AND METHODS

The study area is situated at 28.38° N, 77.20° E and with an altitude of 228.7 m above the mean sea level in the IARI Research Farm, New Delhi, India. The climate of the area is semiarid and sub-tropical with hot summer and cool winters. The mean monthly maximum and minimum temperature during the year ranges from 21.3 to 40.5°C and 7.3 to 28.7°C, respectively. The annual rainfall is 708.6 mm of which on an average 597 mm (84%) is received during the month from June to September. The female (MJA5) and male line (MJR1) seeds of the hybrid (NRCHB 506) were sown with a planting ratio of 16:2 (A: R) in plots with sufficient soil moisture (1 week after irrigation), in three different dates that is, 21st of October (D1), 30th of October (D2) and 18th of November (D3) in 2009 and was harvested in 3rd and 4th week of March and 1st week of April respectively. The experiment was laid out in a Randomized Block Design (RBD) with three replications each with three sets of plots. Weather data were recorded during the growing season from the agro-met observatory under Division of Agricultural Physics, Indian Agricultural Research Institute, New Delhi and have been presented in Figure 1.

Seed quality parameters were tested soon after harvest and after nine months of ambient storage. Seed moisture content (on fresh wet basis) was measured using an air- oven method at 103°C for 17 h, a thousand seed were weighed at 5-6% seed moisture content and germination test was conducted in a laboratory using paper methods in 8 replicates of 50 seeds following top of paper method at 20°C (ISTA, 2008). Seedling vigour was evaluated by taking seedling length (in cm for Vigour index I) and dry weight (in mg for Vigour index II after drying at 90°C for 18-24 h) of ten randomly selected 7-day old normal seedlings from each replication (Abdul–Baki and Anderson, 1972). According to available literature, effect of accelerated ageing for three consecutive days at around 41°C and 100% relative humidity (RH) gives same average effect of one year of ambient storage. Hence, accelerated ageing was achieved by treating the seeds at 41°C and ~ 100% RH for 24 h, 48 h, 72 h and 96 h followed by standard germination test (ISTA, 2008). The hybrid seeds collected from the female plants in each date of sowing were sown with at least 400 plants in 2010-2011 for grow out test. An authentic sample of NRCHB 506 was grown for comparison. Plants with male fertility were counted as pure and were compared with the authentic hybrid seed sample to check genetic purity. All the data were analysed using SAS (SAS Institute Inc., 1989). The model on germination and environmental factors was developed by SPSS software based on stepwise regression. Stepwise regression includes regression models in which the

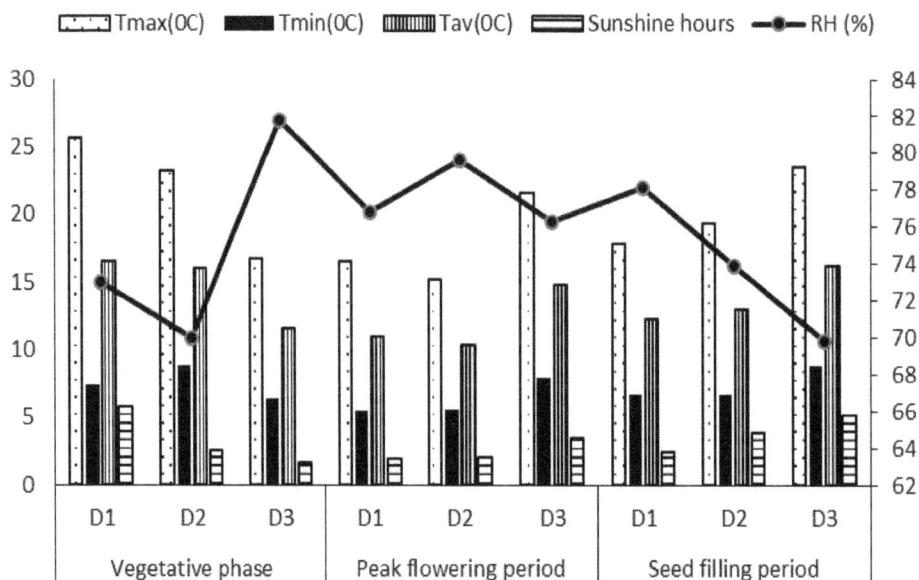

Figure 1. Weather data during crop growth period (winter season, 2009-10), RH (%) is presented by secondary axis.

Table 1. Effect of date of sowing on seed quality before ambient storage.

Quality parameters	21st Oct, 2009 (D1)	30th Oct, 2009 (D2)	18th Nov, 2009 (D3)
Off-type % in Grow out test	15.5	12.5	9.0
Moisture (%)	6.6^a	6.9^a	8.7^b
Germination (%)	96.7^a	94.7^a	79.7^b
Vigour Index I	1357^a	1343^a	1071.7^b
Vigour Index II	480^a	512^a	394^b
Test weight(g)	6.2^{ns}	5.7^{ns}	5.6^{ns}

Alphabets in each cell represent their correspondent group by CD value in all the tables; a, b different group at p<0.05; ns- non significant, D1= 21st October sowing; D2= 30th October sowing; D3= 18th November sowing.

choice of predictive variables is carried out by an automatic procedure.

RESULTS

Hybrid seed purity and quality

Results on grow out test showed non-significant difference for plant morphological characters among the seeds harvested from plants grown under different dates of sowing. Percent off-type plants from seeds produced under different dates of sowing had substantial difference, among these D1 maximum (15.5%) off-type plants in grow out test (Table 1).

Percent germination and vigour index I and II significantly reduced in hybrid seeds produced in D3 (79.7, 1071.7 and 394, respectively) as compared to those in 1st and 2nd dates of sowing (Table 1). Weather parameters had significant correlation with seed quality

traits. Weather conditions on D1, D2 and D3 have been presented in Figure 1. In case of D1 sown crop, T_{max} was 25.6°C, 16.5°C and 17.9°C at vegetative phase, peak flowering period and seed filling period respectively. In D2 T_{max} was 23.2°C, 15.2°C and 19.4°C and in D3 16.7°C, 21.6°C and 23.5°C at those stages. T_{max} at vegetative stage of D1, D2 and D3 were 16.5°C, 15.6°C and 17.9°C respectively. T_{max} and T_{min} at vegetative phase showed significant correlation with quality traits in almost all the cases (Table 6). Percent germination had strong correlation with T_{max} at vegetative ($r^2=0.881$) and seed filling stages ($r^2=0.88$). Sunshine hours at vegetative and seed filling stages had significant correlation with percent germination (0.957 and -0.957 respectively), shoot length (0.898 and -0.870 respectively) and solute lickage (-0.880 and 0.856 respectively). No significant difference for germination and vigour indices was found between 1st (D1) and 2nd (D2) dates of sowing as the weather conditions during

Table 2. Effect of accelerated aging on percent germination of harvested seeds from different dates of sowing

Accelerated ageing period (h)	Germination percentage			
	(D1)	**(D2)**	**(D3)**	**Pusa Bold**
Control	76.9(61.3)	66.0(54.5)	57.9(49.6)	78.7(62.5)
24	72.1(58.1)	53.7(47.2)	37.4(37.4)	70.5(57.2)
48	67.6(55.4)	38.9(38.5)	20.2(26.4)	59.6(50.6)
72	63.4(52.8)	28.7(32.1)	8.7(17.1)	42.6(40.6)
96	52.4(46.4)	13.9(21.9)	5.1(13.0)	20.7(27.0)
Mean	66.5	40.2	25.9	54.4
	AA period	**DOS**	**Interaction**	
CD at 5%	2.5	2.3	5.0	

Values in parentheses are arc shine conversion value; D1 = 21st October sowing; D2= 30th October sowing; D3 = 18th November sowing.

Table 3. Effect of dates of sowing on shoot length (mm) after accelerated ageing.

Accelerated ageing period (h)	Shoot length (mm)			
	(D1)	**(D2)**	**(D3)**	**Pusa Bold**
Control	51.0	43.9	38.6	74.8
24	42.0	29.4	25.0	55.6
48	32.8	20.0	16.5	27.2
72	23.9	15.6	11.4	17.4
96	20.2	10.5	10.3	17.1
Mean	34.0	23.9	20.4	38.4
	AA period	**DOS**	**Interaction**	
CD at 5%	3.2	2.9	6.4	

D1= 21st October sowing; D2= 30th October sowing; D3= 18th November sowing.

these two sowing periods were favourable and did not fluctuate much (Figure 1). Percent decrease in germination from D1 to D2 was 0.83°C temperature decrease at vegetative stage and 1.33°C temperature increase in seed filling stage. But percent decrease in germination from D2 to D3 was 2.31°C temperature decrease at vegetative stage and 3.66°C temperature increase in seed filling stage. A predictive model by stepwise regression was developed on the basis of predictive correlation coefficient. Stepwise regression includes regression models in which the choice of predictive variables is carried out by an automatic procedure.

$$Y = 234.545 - 1.688 \times (\pm 1.013) \times RH_{pf} - 15.359 (\pm 1.335) \times SH_{sf},$$
$$N = 24, r = 0.968, r^2 = 0.936, F = 66.254, SE = 24.$$

Where, Y = Percent germination, RH_{pf} = Relative humidity at peak flowering stage, SH_{sf} = Sunshine hours at seed filling stage, N = sample size, r = correlation coefficient, SE = standard error of estimate, F = ratio of correlation. Other environmental predictive variables except RH_{pf} and

SH_{sf} were automatically excluded from the model. Both RH_{pf} and SH_{sf} were negatively correlated with percent germination.

Accelerated ageing

Analysis of variance showed that effect of ageing period on all germination traits was significant. Interactions between ageing period and hybrid seeds were significant for all assessed traits. Mean comparison for percentage germination showed that hybrid D3 had minimum percentage germination at 96 h ageing (5.1), whereas maximum percentage germination was recorded from control seeds of Pusa Bold (78.7) followed by D1 (76.9) (Table 2). Evaluation of shoot length of D1, D2, D3 and Pusa Bold seeds under different ageing periods followed similar trend, that is, hybrid seeds which had high quality at the beginning of test (control) could maintain this superiority during the different ageing periods (Table 3). Analysing the effect of ageing period on root length (Table 4) showed that during all ageing periods D1 had the longest root (21.3 mm) after 96 h of ageing, while the

Table 4. Effect of dates of sowing on root length (mm) after Accelerated ageing

Accelerated ageing period (h)	Root length (mm)			
	(D1)	(D2)	(D3)	Pusa Bold
Control	64.9	61.0	52.9	60.6
24	48.9	48.6	49.2	41.7
48	43.2	33.0	31.3	30.6
72	28.0	15.4	19.5	19.6
96	21.3	10.4	8.2	15.0
Mean	41.3	33.7	32.2	33.5
	AA period	**DOS**	**Interaction**	
CD at 5%	2.9	2.6	5.8	

D1= 21st October sowing; D2= 30th October sowing; D3= 18th November sowing.

Table 5. Effect of dates of sowing on electrical conductivity of seed leachate after Accelerated ageing

Accelerated ageing period (h)	Electrical conductivity (dS/m)			
	(D1)	(D2)	(D3)	Pusa Bold
Control	21.6	29.0	33.5	25.1
24	22.7	32.1	42.6	27.9
48	25.5	37.3	46.8	30.7
72	27.3	44.5	51.3	35.9
96	28.7	60.1	56.8	34.4
Mean	25.2	40.6	46.2	
	AA period	**DOS**	**Interaction**	
CD at 5%	4.1	3.7	8.3	

D1= 21st October sowing; D2= 30th October sowing; D3= 18th November sowing.

shortest root length (8.2 mm) was observed in D3. *Pusa bold* had longest shoot followed by D1 during the prolonged ageing periods, but had shorter root than that from seeds of D1.

Electrical conductivity (EC)

Results of EC test showed that maximum solute leakage of the hybrid seeds was recorded for D3 and also revealed that ageing periods lower than 72 h were less efficient to separate hybrid seeds into different vigour levels (Table 5).

DISCUSSION

In later dates of planting, unavailability of compatible pollen source might have limited the extent of outcrossing resulting in better hybrid purity. Tunwar and Singh (1988) in Indian Minimum Seed Certification Standard, suggested the standard isolation distance as 25 m in self-compatible species and 50 m in self-incompatible species

for certified seed production. Some sporadic research results are available suggesting reconsideration of isolation distance for multiplication of hybrid parental lines, but no research is available in hybrid seed production aspect of Indian mustard. In optimum sowing date, minimum isolation distance for certification may certainly be revisited to check out-crossing.

The rate of reduction in seed germination and vigour varied with the different sowing seasons which is associated with changes in temperature and relative humidity conditions during the vegetative, reproductive and harvest stages and this is supported by Tekrony and Egly, 1980. The behaviour of seeds, in terms of percentage of normal seedlings in the germination and accelerated aging tests, as a function of sowing time for hybrid and cropping season, may be a result of physiological quality of seed (Ávila et al., 2003) at the time of harvest.

Mustard, like other angiospermous seeds, when approaching maturity, characteristically accumulates soluble sugars (Amuti and Pollard, 1977) through starch synthase activity and a wide range of Late Embryogenesis Abundance (LEA) proteins. These

Table 6. Correlation of seed quality traits and weather parameters.

Weather parameter	Seed quality traits			
Tmax (°C) vg	Germination %	Shoot length (mm)	Root length 9mm	Solute lickage (dS/m)
Tmin (°C) vg	0.881**	0.774**	0.562	-0.763**
Tav(°C) vg	0.250	0.138	0.040	-0.144
RH (%) vg	0.800**	0.685*	0.484	-0.677*
Sunshine hours vg	-0.573	-0.452	-0.290	0.450
Tmax(°C) pf	0.957**	0.898**	0.694*	-0.880**
Tmin(°C) pf	-0.614*	-0.493	-0.323	0.490
Tav(°C) pf	-0.768**	-0.651*	-0.455	0.645*

Weather parameter	Seed quality traits			
RH (%) pf	Germination %	Shoot length (mm)	Root length (9 mm)	Solute lickage (dS/m)
Sunshine hours pf	-0.025	-0.118	-0.158	0.108
Tmax(°C) sf	-0.782**	-0.665*	-0.467	0.658*
Tmin(°C) sf	-0.880**	-0.773**	-0.561	0.762**
Tav(°C) sf	-0.746**	-0.628*	-0.436	0.622*
RH (%) sf	-0.839**	-0.727**	-0.521	0.718**
Sunshine hours sf	0.955**	0.867**	0.650*	-0.852**

Vg = at vegetative stage; pf= at peak flowering stage; sf= at seed filling stage. *significant at (p=0.05); ** significant at (p=0.01).

solutes, enzymes and proteins are known to contribute to the development of grain followed by desiccation tolerance and longevity (Bernal-Lugo and Leopold, 1995). Seed quality is a genetically inherited trait whose intensity is modified by the environment during seed development and maturation. In this study, adverse weather condition during physiological maturity had strong correlation with seed quality parameters (Table 6), hence hampering this processes and conferring more susceptibility to ageing.

The result of accelerated ageing test could be attributed to differential sensitivity of the early growth traits against ageing treatments, the alternation in climatic conditions during entire growing period. As per knowledge, Indian mustard needs higher temperature during vegetative phase and lower temperature during reproductive phase. But in our case the condition was altered. Results of ageing periods on seedling vigour index and percentage germination showed same trend that is, seeds produced in D1 and D2 were better than that in D3. This test classified hybrid seeds produced in different time of sowing with the best and worst performance and evaluated the seeds vigour and their storability. Therefore, it was observed that D1 had high quality and was resistant against the deterioration, whereas D3 was identified as having a lower physiological potential. However, compared to open pollinated variety Pusa Bold, D1 performed worse.

Ripen mustard seed varies in moisture and oil content and this influences how good a seed lot can store. Decrease in oil content with increase in storage period has been reported in mustard and the decrease in oil content has been related to peroxidation of lipid by peroxidase, non-enzymatic lipid autoxidation and also to fungi invasion (Verma et al., 2003). The protective effect of these poly-ols is thought to occur through maintaining the structural integrity of membranes, and providing stability for macromolecules such as proteins (Crowe and Crowe, 1986). Through much of the literature there is an assumption that oxidative reactions are responsible for the deteriorative changes observed in aged seeds.

Four types of oxidations are known which might reasonably contribute to the progress of seed ageing. These include free radical oxidations, enzymic dehydrogenation, aldehyde oxidation of proteins, and Maillard reaction (Bernal-Lugo and Leopold, 1998). Free radicals attack membrane lipids, and cause major disruption of their viscosity and their permeability. Increase of solute leakage at this situation would be due to damaged membrane (Gupta and Aneja, 2004). In addition, as water is withdrawn from a solution of sugars form a glassy state, this can serve as a physical stabilizer. Viscosity obtained in the glassy state, suppresses deteriorative reactions. Different responses of hybrid seeds to accelerated ageing treatments are explained by their differential sensitivity to environmental factors during seed filling period. These chains of interlinked physiological phenomenon are subject to environmental variation and more pronounced under unfavourable condition than that in case of open pollinated varieties.

In this study, the 72 and 96 h periods had a higher sensitivity in identifying the optimum sowing time to produce hybrid seeds that possessed different levels of physiological potential. Seeds harvested from 21st October sowing (optimum time of sowing) gave the best result regarding seed quality than that in hybrid seeds from 30th October and 18th November sowings in Delhi and adjoining areas.

The effect of climatic variables on the parental lines during hybrid seed production and subsequently on seed quality with respect to its regeneration has not been well studied in Indian mustard. Alarming investigations are being reported now-a-days in this regard. In the changing climatic era, this type of studies is highly needed to adopt some mitigation strategies to supply good quality seeds to the farmers.

ACKNOWLEDGEMENTS

Facilities provided by Division of Seed Science and Technology, Indian Agricultural Research Institute, New Delhi during standardisation of this method are duly acknowledged. The authors are greatly thankful to Dr. K. H. Singh, senior scientist, Directorate of Rapeseed-mustard Research, Bharatpur, and Rajasthan, India who provided the seeds for this research.

REFERENCES

Abdul-Baki AA, Anderson JD (1972). Physiological and biochemical deterioration of seeds. In: Kozlowski, T.T. (ed.). Seed Biol. Academic Press, New York. 2:283-315.

Amuti KS, Pollard CJ (1977). Soluble carbohydrates of dry and developing seeds. Phytochemistry 16:529-32.

Anonymous (2005). Questions and Answers DIR 057/2004 under Office of the gene technology regulator, Ministry of Health, Australian Government. http://www.ogtr.gov.au/internet/ogtr/publishing.nsf/Content/dir057-3/$FILE/dir057qanda.pdf as on 11.11.2013.

Ávila M, Braccini RADL, Motta IDS, Scapim CA, Braccini MDCL (2003). Sowing seasons and quality of soybean seeds. Sci. Agric. 60:245-252.

Balešević-Tubić S, Tatić M, Đorđević V, Nikolić Z, Đukić V (2010). Seed viability of oil crops depending on storage conditions. Helia 33:153-160.

Banga SK, Kaur G, Vishal (2013). Breeding canola quality Indian mustard (Brassica juncea L.). http://www.australianoilseeds.com/__data/assets/pdf_file/0006/6864/30_Breeding_canola_quality_Indian_mustard_Brassica_juncea.pdf as on 11.11.2013.

Bernal-Lugo I, Leopold AC.(1998). The dynamics of seed mortality. J. Exp. Bot. 49:1455-1461.

Bernal-Lugo I, Leopold AC (1995). Seed stability during storage: raffinose and seed glassy state. Seed Sci. Res. 5:75-80.

Bhuiyan MS, Mondol MRI, Rahaman MA, Alam MS, Faisal AHMA (2008). Yield and Yield Attributes of Rapeseed as Influenced by Date of Planting. Int. J. Sustain. Crop Prod. 3:25-29.

Crowe JH, Crowe LM (1986). In: Leopold AC, ed. Stabilization of membranes in anhydrobiotic organisms. pp. 188-209. Membranes, metabolism, and dry organisms. Ithaca, NY: Cornell University Press.

Devi C, Kant K, Dadlani M (2003). Effect of size grading and ageing on sinapine leakage, electrical conductivity and germination percentage in the seed of mustard (Brassica juncea L.). Seed Sci. Tech. 31:505-509.

Dornbos Jr. D (2002). Production environment and seed quality. In: Seed quality: Basic mechanism and agricultural implications, pp.119-152. CBS Publishers, New Delhi.

Gupta A, Aneja KR (2004). Seed deterioration in soybean varieties during storage: Physiological Attributes. Seed Res. 32:26-32.

ISTA (2008). International Rules for Seed Testing. International Seed Testing Association, Bassersdorf, Switzerland.

Kurdikeri MB, Merwade MN, Channaveeraswamy (2000). Maintenance of viability in different crop species under ambient storage. Seed Res. 28:109-110.

McDonald MB (1999). Seed deterioration: physiology, repair and assessment. Seed Sci. Tech. 27: 177-237.

SAS Institute Inc. (1989). SAS/STAT User's Guide: Version 6.12th edn. SAS Institute Inc., Cary, North Carolina.

Sharma JK, Singh HB (1997). Relative storability of the seeds of some crop species under ambient conditions. Seed Res. 25:37-40.

Teari N, Singh P, Vaish CP, Katiyar RP, Kanaujia VP (1998). Viability of mustard seeds (Brassica juncea L.) stored under ambient conditions of central plain zone of Uttar Pradesh. Seed Res. 9:75.

Tekrony DM, Egly DB, Phillips AD (1980). Effects of field weathering on the viability and on vigor of soybean seed. Agron. J. 72:749-753.

Thakur KS, Singh CM (1998). Performance of Brassica species under different dates of sowing in mid hills of Himachal Pradesh. Ind. J. Agron. 43:464-468.

Tunwar NS, Singh SV (1988). Indian Minimum seed Certification Standards. The Central Seed certification Board. Dept. of Agriculture and Cooperation, Ministry of Agriculture, Govt. of India, New Delhi, India.

Verma SS, Tomer RPS, Verma U (2003). Loss of viability and vigour in Indian mustard seeds stored under ambient conditions. Seed Res. 31:98-101.

Veselova TV, Veselovasky VA (2003). Investigation of a typical germination changes during accelerated ageing of pea seeds. Seed Sci. Tech. 31:517-530.

6

Seed dry dressing with botanicals to improve physiological performance of fresh and aged seeds of blackgram (*Vigna mungo* L.)

S. Sathish and M. Bhaskaran

Department of Seed Science and Technology, Seed Centre, Tamil Nadu Agricultural University, Coimbatore, Tamil Nadu, India.

An experiment was conducted in Department of Seed Science and Technology during 2012 to standardize suitable seed dry dressing treatments, using botanicals that can alleviate the deleterious effects of accelerated ageing and improve the physiological performance of both fresh and aged seeds in blackgram variety TNAU blackgram CO 6. Both fresh and four days accelerated aged seeds were dry dressed with fenugreek seed powder, custard apple and moringa leaf powder at 2, 3 and 4 g/kg of seeds with 1, 2 and 3 h of shaking and evaluated for its physiological performance of seedlings under lab condition. The result revealed that seeds treated with 3 g/kg of fenugreek seed powder with 1 h shaking registered an increased physiological performance in terms of germination percentage, shoot length, root length, dry matter production and vigour index for both fresh and aged seeds. This, which was on par with 4 g/kg of custard apple leaf powder treated seeds. Interestingly, the increase in physiological performance of identified seed treatments was more pronounced in aged seeds, indicating the efficacy of botanicals in alleviating the deleterious effect of accelerated ageing. All the botanicals found to contain higher antioxidant activity and rich in minerals such as titanium, molybdenum and iron, apart from other trace elements.

Key words: Blackgram, custard apple leaf powder, fenugreek seed powder, moringa leaf powder, seed ageing.

INTRODUCTION

Blackgram (*Vigna mungo* L.) is a protein rich food, containing about 26% protein, which is almost three times that of cereals and ranks third among the major pulses cultivated in India. Blackgram supplies a major share of protein requirement of vegetarian population of the country. It is consumed in the form of split pulse as well as whole pulse, which is an essential supplement of cereal based diet. In India, blackgram occupies 12.7% of total area under pulses and contribute 8.4% of total pulses production. However, area and production of blackgram has declined from 3.01 million ha and 1.30

million tons in 2000 to 2001 to 2.97 million ha and 1.23 million tons, respectively in 2009 to 2010 (ASSOCHAM, 2012). This emphasized the need to increase the performance of pulse crops, particularly in developing countries, where most grain legume production is for human consumption and demand is increasing due to population increase (Jeuffroy and Ney, 1997). The poor performance of blackgram may be attributed to several factors, of which level of seed deterioration is of great importance. Invariably, the seeds have to be stored through the monsoon for the next sowing during which period the rise in the ambient relative humidity coupled with the prevailing high temperature can accelerate the ageing process of the seed, resulting in loss of vigour and viability.

Using the seed lot with reduced vigour and viability for sowing may affect the field performance and productivity of the resultant crops (Layek et al., 2007; Pati and Bhattacharjee, 2011). Therefore, there is a need to improve the performance of deteriorated seeds which is possible through seed treatments. Seed treatments with synthetic chemicals had effectively managed to alleviate the deteriorative effect of seed ageing (Sathish and Sundareswaran, 2010; Sathish et al., 2011). Though, many of the synthetic chemicals look effective but they are not readily degradable and yield more toxic residues (Ames et al., 1990; Moore and Waring, 2001). However, the use of chemical is still in vogue. Hence, the safest and feasible approach is the treatment of seeds with botanicals which are safe, economical, ecofriendly and non-harmful to seed, animal and human beings. It was proven that the deteriorating effect of seed ageing was mainly due to the production of free radicals (Bailly, 2004; Bailly et al., 2008) and use of antioxidants can quench the free radicals and prolong the storability of seeds (Maeda et al., 2005; Sattler et al., 2006).

Among the several botanicals available, presence of antioxidant property along with high nutrient content was pharmacologically proved in fenugreek seed powder (Bukhari et al., 2008; Toppo et al., 2009), custard apple leaf powder (Baskar et al., 2007; Pandey and Brave, 2011) and moringa leaf powder (Fahey, 2005; Ferreira et al., 2008). Therefore, these botanicals could serve as a potential source of natural antioxidants which can be used for seed treatments to improve vigour and viability. However, dry dressing of seeds will be more effective rather than wet treatment; since wet treatment leads to soaking injury due to the hygroscopic nature of pulses seeds (Kalavathi, 1985).

Accelerated ageing has been widely used to study the pattern of seed deterioration in various crops instead of waiting to obtain naturally aged seeds (Jatoi et al., 2001; Scialabba et al., 2002). Hence, with this backdrop, the present investigation was formulated to study the effect of botanicals in improving the performance of aged seeds in comparison with fresh seeds of blackgram variety TNAU blackgram CO 6.

MATERIALS AND METHODS

Seed materials

Genetically, pure, freshly harvested seeds of blackgram variety TANU blackgram CO 6 obtained from the Department of Pulses, Centre for Plant Breeding and Genetics, Tamil Nadu Agricultural University, Coimbatore were graded using a BSS 7 × 7 wire mesh sieve and used for the study. Standardization of seed dry dressing with botanicals was carried out in Department of Seed Science and Technology, Tamil Nadu Agricultural University, Coimbatore during 2012.

Preparation of two seed lots with different vigour levels

In order to obtain two seed lots with different germination potential and vigour level, half the quantity of fresh seeds having 98% germination was subjected to four days accelerated ageing. It was carried out by packing seeds in paper bag with uniform pin head size perforation all over and placed in ageing jar containing 100 ml double distilled water to maintain 98 ± 2% relative humidity and the whole unit was kept in incubator maintained at 40 ± 1°C for four days to reduce germination approximately around 80% (Delouche and Baskin, 1973).

Seed dry dressing with botanicals

Fenugreek seed obtained from local market, custard apple and moringa leaves collected from the orchards of Tamil Nadu Agricultural University were freeze dried at -80°C and grounded finely to obtain fine powder. Both fresh seeds (FS) and four days accelerated aged seeds (AS) were dry dressed with finely ground botanicals namely, fenugreek seed powder (FSP), custard apple leaf powder (CALP) and moringa leaf powder (MLP) at the rate of 2, 3 and 4 g/kg of seeds and shaken for 1, 2 and 3 h to find effective duration for uniform seed dressing. Treated seeds along with untreated, fresh and aged seeds which served as control were assessed for the following physiological seed quality parameters.

Seed quality evaluation

The laboratory germination test was carried out in quadruplicate using 100 seeds each with 4 sub replicates of 25 seeds in the paper medium (ISTA, 2009). The test conditions of 25 ± 2°C temperature and 95 ± 3% relative humidity were maintained in a germination room. Germination was counted in 24 h intervals and continued until no further germination occurred. At the end of seven days, the number of normal seedlings was counted and the mean was expressed as a percentage. Speed of germination was calculated based on the following formula of Maguire (1962):

$$\text{Speed of germination} = \frac{X_1}{Y_1} + \frac{X_2 - X_1}{Y_2} + \ldots\ldots\ldots + \frac{X_2 - X_{n-1}}{Y_n}$$

Where X_1, X_2 and X_n are number of seeds germinated on first, second and n^{th} day, respectively and Y_1, Y_2 and Y_n are number of days from sowing to first, second and n^{th} count, respectively.

Root and shoot lengths were measured at the time of germination count from ten normal seedlings selected at random from each replication and the mean was expressed in centimetre and those seedlings used for growth measurement were placed in a paper cover and dried in shade for 24 h and then they were kept in an oven maintained at 85 ± 2°C for 48 h. The dried seedlings were weighed to estimate the dry matter production and the mean values were expressed in gram per 10 seedlings. The vigour index was computed as described by Abdul-Baki and Anderson (1973) with slight modifications as follows and expressed in whole numbers.

VI 1 = GP × (RL + SL)
VI 2 = GP × DMP

Where VI 1 is vigour index one, VI 2 is vigour index two, GP is germination percentage, RL is root length (cm), SL is shoot length (cm) and DMP is dry matter production (gram per 10 seedlings).

Properties of botanicals

The properties such as antioxidant activity and mineral content were analyzed as follows. The total potential antioxidant activity

Table 1. Effect of botanical treatment on speed of germination of fresh seeds of blackgram variety TNAU blackgram CO 6.

Botanicals	1 h				2 h				3 h				B × C interaction mean			Grand mean
	2 g	3 g	4 g	Mean	2 g	3 g	4 g	Mean	2 g	3 g	4 g	Mean	2 g	3 g	4 g	
FSP	24.0	24.5	24.5	24.3	23.9	23.9	24.1	24.0	23.8	24.1	24.3	24.1	23.9	24.2	24.3	24.1
CALP	24.0	24.5	24.0	24.2	24.1	24.1	24.0	24.1	23.6	23.4	24.2	23.7	23.9	24.0	24.1	24.0
MLP	24.3	25.0	23.8	24.4	23.6	24.1	24.1	23.9	22.9	23.8	24.1	23.6	23.6	24.3	24.0	24.0
Mean	24.1	24.7	24.1	24.3	23.9	24.0	24.1	24.0	23.4	23.8	24.2	23.8	23.8	24.2	24.1	24.0
Control	23.3															

	Control versus rest	D	B	C	DB	DC	BC	DBC
SEd	0.432	0.141	0.141	0.141	0.245	0.245	0.245	0.424
CD (P = 0.05)	NS	NS	NS	NS	NS	NS	NS	NS
CD (P = 0.01)	NS	NS	NS	NS	NS	NS	NS	NS

D – Shaking duration, C – Botanicals concentration, B – botanicals, FSP, fenugreek seed powder; CALP, custard apple leaf powder; MLP, moringa leaf powder.

of the aqueous acetone extract of powdered botanicals were assessed based on their scavenging of 1,1-diphenyl-2-picrylhydrazyl (DPPH) free radicals, using a modified DPPH assay of Koleckar et al. (2007). The extract was prepared by dissolving 0.3 g of flour in 10 ml of 70% (v/v) acetone. After continuous shaking for 30 min at room temperature, the solution was centrifuged for 20 min at 13,000 rpm. An aliquot of 100 µl extract was mixed with the ethanol DPPH solution (0.5 mM, 0.25 ml) and the acetate buffer (100 mM, pH 5.5, 0.5 ml). After standing for 30 min in the dark, the absorbance was measured at 517 nm against a blank containing absolute ethanol instead of a sample aliquot. DPPH-radical scavenging activity is expressed as % of blank:

$$\text{Per cent inhibition activity} = \frac{\text{Blank OD} - \text{Sample OD}}{\text{Blank OD}} \times 100$$

About 18 elements namely, Al, B, Ba, Be, Bi, Cd, Co, Cr, Cu, Fe, Mn, Mo, Ni, Se, Sr, Te, Ti and Zn were estimated by following the method described by McQuaker et al. (1979) and the results of minerals having more than 0.1 ppm were presented here. Finely ground seed samples weighing 0.5 g was digested in 15 ml of concentrated nitric acid using kelplus infra digestion system (Model: KES 12IL) at 100°C for half an hour and later increased to 300°C till the solution turns colourless. To the digested solution, 100 ml of double distilled water was added and filtered through whatman's filter paper number 40 to obtain a clear colourless solution and mineral content were quantified using inductively coupled plasma spectrometer (ICP) and iTEVA software and expressed in ppm. The ICP multi-element standard solution VIII (24 elements in dilute nitric acid) obtained from Merck Chemicals, Germany was used as standard.

Statistical analysis

The experiment was carried out with four replications in factorial completely randomized block design. The data obtained were analysed by the 'F' test of significance following the methods described by Rangaswamy (2002). The percent values were transformed to arc-sine values and used for analysis. The critical differences (CD) were calculated at 1 and 5% probability level. The data were tested for statistical significance by three ways ANOVA. If the F test is non-significant, it was indicated by the letters NS.

RESULTS

Speed of germination and germination percentage

The results of the present investigation revealed that there was no significant difference in speed of germination between treatments of fresh seeds. Whereas, aged seeds registered significant difference in speed of germination between the control and other treatments but the difference was not significant within treatments and their interactions (Tables 1 and 2). Similarly, the difference was not significant for germination percentage among treatments and their interactions in fresh seeds. Whereas, the difference was highly pronounced and significant for germination percentage of aged seeds to all the factors and their interactions over control (81%). FSP (87%) registered maximum germination followed by CALP (86%). Among the shaking duration, 1 and 2 h of shaking are on par with each other and registered higher germination (86%) than 3 h of shaking (84%). Among the concentration, 4 g/kg of seeds (88%) registered maximum germination followed by 3 g/kg of seeds (86%). However, interaction between botanicals and concentration registered maximum germination in FSP at 3 and 4 g/kg of seeds and CALP at 4 g/kg of seeds (89%) which were on par with

Table 2. Effect of botanical treatment on speed of germination of aged seeds of blackgram variety TNAU blackgram CO$_6$.

Botanicals	1 h				2 h				3 h				B × C interaction mean			Grand mean
	2 g	3 g	4 g	Mean	2 g	3 g	4 g	Mean	2 g	3 g	4 g	Mean	2 g	3 g	4 g	
FSP	19.8	20.6	20.4	20.3	19.8	20.4	20.5	20.2	19.1	20.0	20.2	19.8	19.6	20.3	20.4	20.1
CALP	19.3	19.6	20.9	19.9	19.4	20.0	20.6	20.0	19.4	19.9	20.1	19.8	19.4	19.8	20.5	19.9
MLP	18.9	19.1	19.6	19.2	18.9	19.0	19.6	19.2	18.9	19.4	19.9	19.4	18.9	19.2	19.7	19.3
Mean	19.3	19.8	20.3	19.8	19.4	19.8	20.2	19.8	19.1	19.8	20.1	19.7	19.3	19.8	20.2	19.8
Control	17.8															

	Control versus rest	D	B	C	DB	DC	BC	DBC
SEd	0.948	0.310	0.310	0.310	0.537	0.537	0.537	0.931
CD (P = 0.05)	1.884	NS	NS	NS	NS	NS	NS	NS
CD (P = 0.01)	NS	NS	NS	NS	NS	NS	NS	NS

D – Shaking duration, C – Botanicals concentration, B – Botanicals, FSP, fenugreek seed powder; CALP, custard apple leaf powder; MLP, moringa leaf powder.

each other.

Interaction between duration of shaking (D) and botanicals (B); D and concentration of botanicals (C) and interaction between D, B and C were not significant (Figure 1).

Shoot and root length

In fresh seeds, difference was significant over control (21.07 cm) for shoot length in all the three factors, except their interactions. Higher shoot length was observed in seeds dry dressed with FSP (23.03 cm) and CALP (22.66 cm) which was on par with each other. Among concentration, 3 and 4 g/kg of seeds registered maximum shoot length and on par with each other; among the duration of shaking, 1 h (23.02 cm) and 2 h (23.17 cm) were on par and produced significantly higher shoot length than 3 h (21.71 cm). Similar trend was also observed in shoot length of aged seeds. In addition, significant difference in the interaction between botanicals and its concentration of aged seeds were also observed by producing significantly longer shoot length in seeds dry dressed with FSP at 3 g/kg of seeds (22.26 cm), 4 g/kg of seeds (22.29 cm) and CALP at 4 g/kg of seeds (22.26 cm) which are on par with each other (Figure 2). Root length of fresh seeds exhibited different trend from shoot length where no significant differences were observed due to shaking duration and interactions between D and C, D and B, and D, B and C. But significant difference was observed in seeds dry dressed with botanicals, its concentration and their interaction and also all the treatments as a whole registered significantly higher root length than control.

Interactional effect of botanicals and its concentration revealed that seeds dry dressed with FSP at 3 g/kg of seeds (18.03 cm), 4 g/kg of seeds (18.03 cm) and CALP at 4 g/kg of seeds (18.02 cm) registered higher root length which were on par with each other. In case of aged seeds, root length depicted similar trend as like that of its shoot length (Figure 3).

Dry matter production

Dry matter production of fresh seeds was significantly influenced by botanicals, concentration and duration of shaking and performed better than control (0.218 g / 10 seedlings). But there was no significant difference for interactional effect. Among the botanicals, FSP registered higher dry matter production (0.273 g / 10 seedlings) which was on par with CALP (0.268 g / 10 seedlings) while dry matter production was maximum in concentration of 3 and 4 g/kg of seeds (0.265 and 0.271 g / 10 seedlings, respectively) which was on par with each other. Among the duration of shaking, 1 h (0.268 g / 10 seedlings) and 2 h (0.270 g / 10 seedlings) was on par and produced significantly higher dry matter than 3 h (0.247 g / 10 seedlings). Similar trend was also observed in dry matter of aged seeds. In addition, a significant difference in the interaction between botanicals and its concentration were also observed by producing significantly higher dry matter in seeds dry

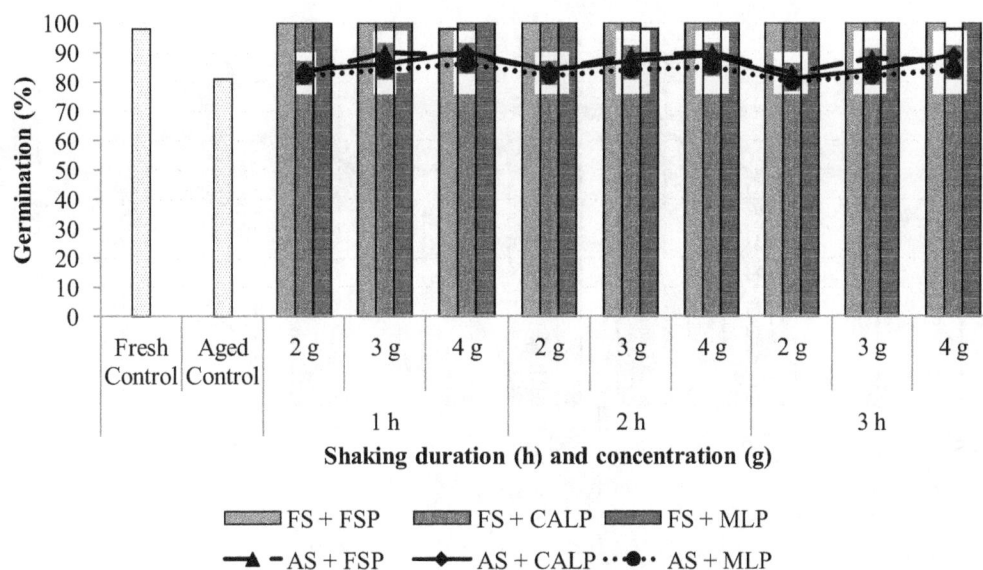

Figure 1. Effect of botanicals dry dressing on germination (%) of fresh and aged seeds of blackgram.

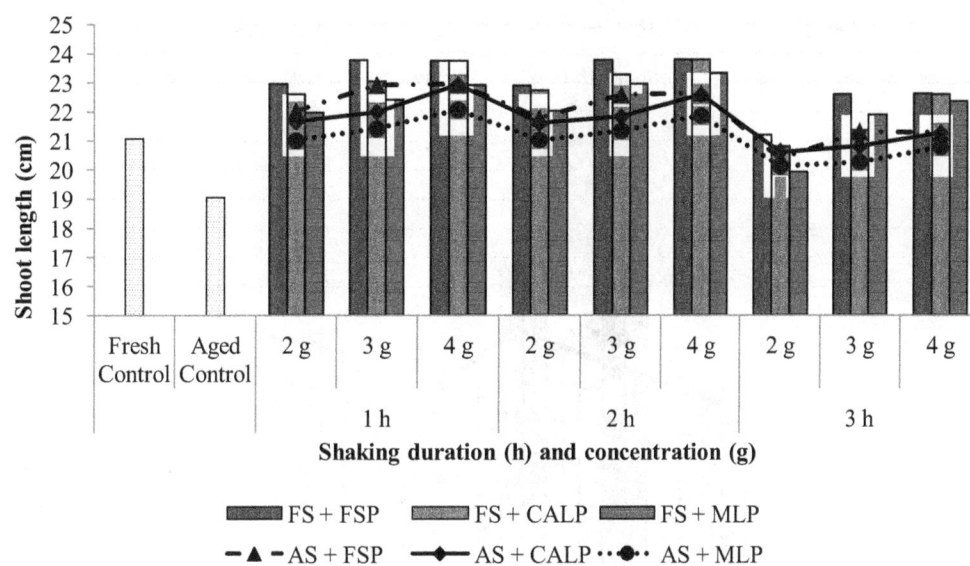

Figure 2. Effect of botanicals dry dressing on shoot length (cm) of fresh and aged seeds of blackgram.

dressed with FSP at 3 g/kg of seeds (0.262g / 10 seedlings) and 4 g / kg of seeds (0.262 g / 10 seedlings) and CALP at 4 g / kg of seeds (0.263 g / 10 seedlings) which are on par with each other (Figure 4).

Vigour index

Interestingly, vigour index 1 and 2 of both fresh and aged seeds were significantly influenced by all the three factors and their interactions. Seeds dry dressed with FSP at 3 g/kg of seeds and CALP at 4 g/kg of seeds and shaked for 1 h registered maximum vigour index 1 and 2 for both fresh and aged seeds (Figures 5 and 6).

Properties of botanicals

Antioxidant activity was higher in all the three botanicals among which FSP (95.9%) and CALP (96.0%) registered higher antioxidant activity than MLP (91.7%) (Table 3). In case of mineral content, FSP contain higher amount of Mo, Ti, Fe, Al, Zn and traces of Mn, B, Cu, Sr, Ba; while CALP was rich in Ti, Mo, Fe, Al, Sr, B and traces of Ba,

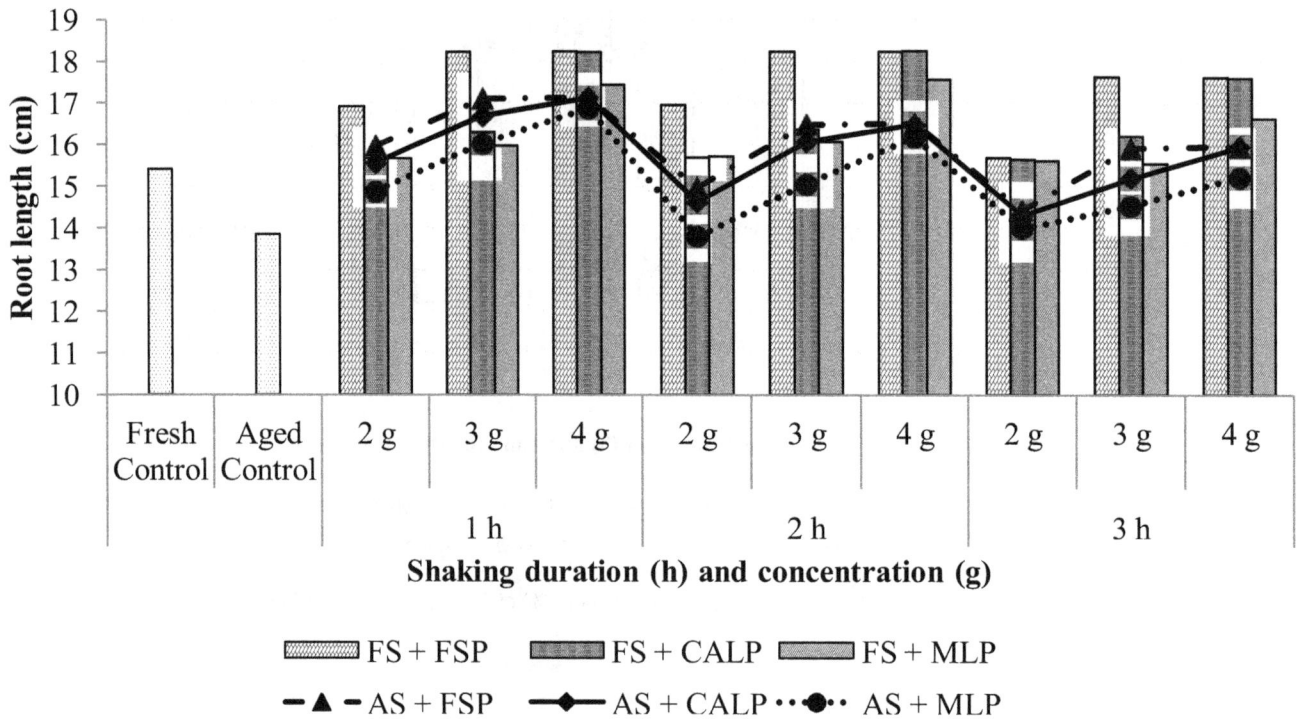

Figure 3. Effect of botanicals dry dressing on root length (cm) of fresh and aged seeds of blackgram.

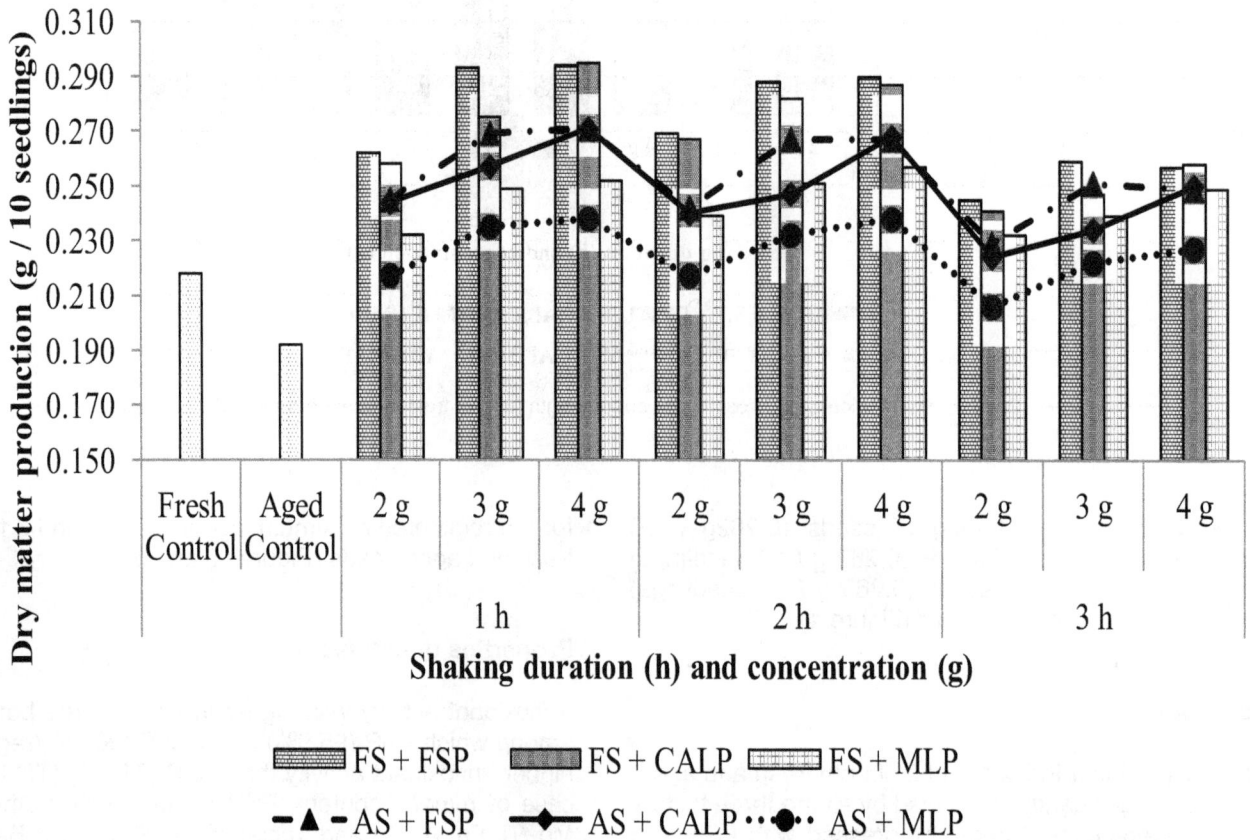

Figure 4. Effect of botanicals dry dressing on dry matter production (g / 10 seedlings) of fresh and aged seeds of blackgram.

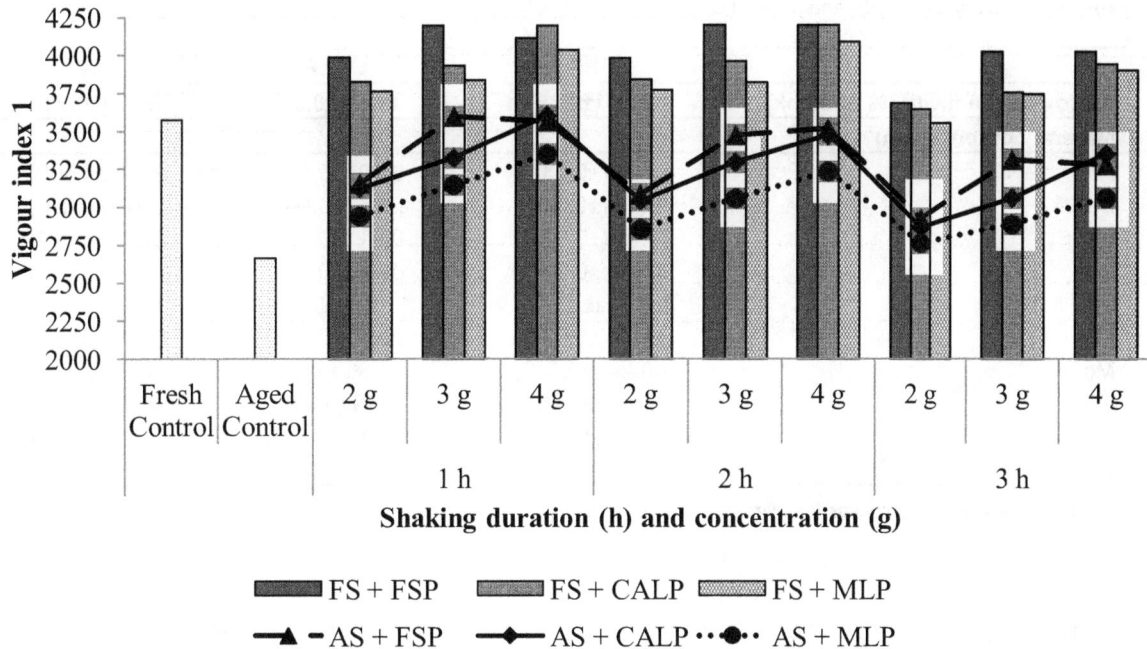

Figure 5. Effect of botanicals dry dressing on vigour index 1 of fresh and aged seeds of blackgram.

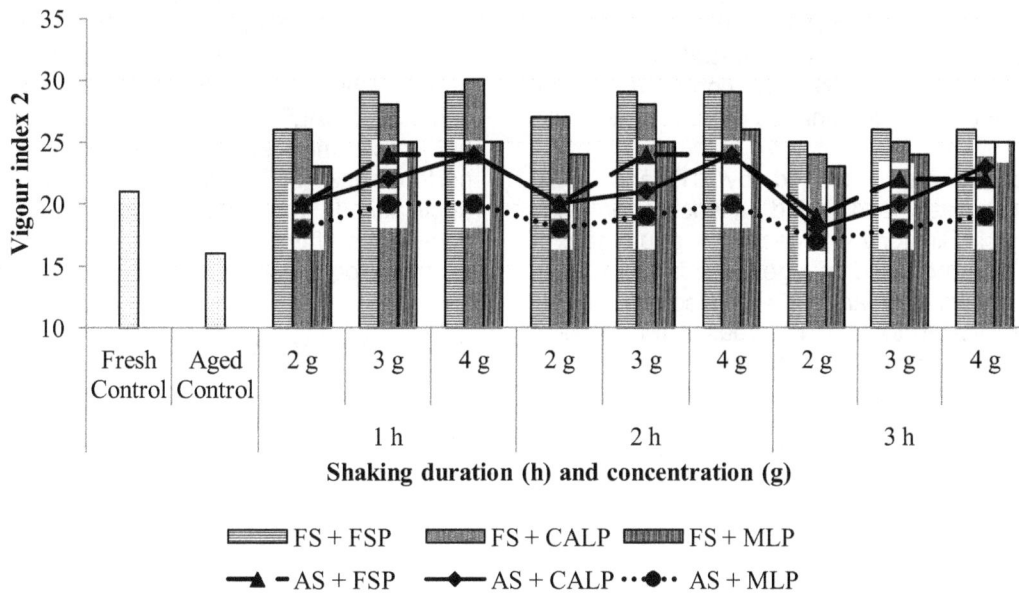

Figure 6. Effect of botanicals dry dressing on vigour index 2 of fresh and aged seeds of blackgram.

Zn, Cu, Mn. MLP was rich in Ti, Mo, Fe, Al, Sr and traces of Zn, B, Mn, Ba, Cu. However, MLP contains lesser mineral content than FLP and CALP (Table 3).

DISCUSSION

Availability of good quality seed is the key for successful agriculture and their use is an important factor for increased productivity. The seeds with good physiological potential act as catalyst for all agricultural inputs. All the botanicals, its concentration and shaking duration performed better than control of both fresh and aged seeds. In fresh seeds though, speed of germination and germination percentage was not significantly influenced by treatments by all the other parameters namely, shoot

Table 3. Antioxidant property and mineral content (ppm) of botanicals.

	FSP	CALP	MLP
Antioxidant activity (% of blank)	95.9 (±0.36)	96.0 (±0.27)	91.7 (±0.28)
Mineral content (ppm)			
Al	3.20 (±0.037)	6.10 (±0.512)	2.79 (±0.100)
B	0.49 (±0.006)	2.36 (±0.008)	0.55 (±0.010)
Ba	0.15 (±0.001)	0.91 (±0.003)	0.28 (±0.010)
Cu	0.40 (±0.002)	0.44 (±0.001)	0.27 (±0.004)
Fe	20.51 (±0.218)	11.68 (±0.045)	7.59 (±0.130)
Mn	0.62 (±0.004)	0.44 (±0.002)	0.49 (±0.010)
Mo	40.24 (±0.543)	24.86 (0.196)	13.73 (±0.550)
Sr	0.26 (±0.002)	3.19 (0.009)	2.78 (±0.050)
Ti	38.34 (±0.443)	107.10 (±1.852)	47.71 (±3.050)
Zn	1.35 (0.008)	0.86 (±0.004)	0.87 (±0.010)

Note: Each value in the table was the average of three replications± standard deviation.

length, root length, dry matter production, vigour index 1 and 2, were significantly influenced by the botanicals. In case of aged seeds, except speed of germination, all the other parameters were highly influenced by the seed dry dressing treatments; both in fresh and aged seeds, dry dressing with FSP at 3 g/kg of seeds and CALP at 4 g/kg of seeds with 1 h shaking performed better than other treatments; both the treatments increased germination percentage of aged seeds to 9% than control (81%) but no significant influence in germination of fresh seeds was observed.

Vigour index 1 was increased by 17 and 32%, respectively for fresh and aged seeds over their respective control. Similarly, increase in vigour index 2 was also much pronounced in aged seeds than fresh seeds. Similar results of botanicals were also reported by Vijayalakshmi (2012) in tomato. The reason for increase in germination, vigour index and other parameters can be substantiated from the result obtained by the analysis of properties of botanicals. All the botanicals have higher antioxidant property among which FSP and CALP topped (Table 3). The presence of higher antioxidant activity in FSP and CALP was also reported by several researchers (Baskar et al., 2007; Bukhari et al., 2008; Toppo et al., 2009; Pandey and Brave, 2011; Bose et al., 2011). Both these powders proposed to contain poly phenolics and flavonoids, namely vitexin, tricin, naringenin and quercetin, which act as a hydrogen donor and the OH⁻ scavenger (Kaviarasan et al., 2007). In addition, both FSP and CALP were rich in titanium, molybdenum and iron apart from other trace elements (Table 3). Ti plays a major role in biomass production and participate in cell metabolism as redox catalyst (Tlustos et al., 2005). Molybdenum is utilized by selected enzymes to carry out redox reactions and helps in vigorous seedling growth (Kaiser et al., 2005). Iron is also utilized by several enzymes and participates in the energy-yielding electron

transfer reactions of respiration during germination (Guerinot and Yi, 1994).

The pronounced effect of FSP and CALP could be attributed to excellent proton radical scavenging property as described earlier and subsequent alleviation of deteriorative effect (Bhatia et al., 2002; Chandrashekar and Kulkarani, 2011) with substantial mineral supplement required for seed germination and vigourous seedling growth. Hence, seed dry dressing with FSP at 3 g/kg of seeds and CALP at 4 g/kg of seeds with 1 h shaking can be recommended to improve the physiological performance of blackgram seeds. Especially, it is applicable in the case of aged seeds to alleviate the deleterious effect of ageing and subsequently to improve its physiological performance during germination.

ACKNOWLEDGEMENTS

We express our deep gratitude to Professor R. Umarani, TNAU for careful perusal of the manuscript. We thank Professor N. Senthil and Assistant Professor S. Vellaikumar, TNAU for assisting in mineral analysis using 'inductively coupled plasma spectrophotometer'.

REFERENCES

Abdul-Baki AA, Anderson JD (1973). Vigour determination of soybean seed by multiple criteria. Crop Sci. 13:630-633.
Ames BN, Profet M, Gold LS (1990). Nature's chemicals and synthetic chemicals: Comparative toxicology. Proc. Natl. Acad. Sci. USA. 87:7782-7786.
ASSOCHAM (2012). Study paper on "Emerging pulses scenerio in 2015 - productivity, awareness and affordability to drive pulses economy in India". The Associated Chambers of Commerce and Industry of India, New Delhi, India. pp. 1-18.
Bailly C (2004). Active oxygen species and antioxidants in seed biology. Seed Sci. Res. 14(2):93-107.
Bailly C, Bouteau HEM, Corbineau F (2008). From intracellular signaling networks to cell death: the dual role of reactive oxygen species in

Seed dry dressing with botanicals to improve physiological performance of fresh and aged seeds...

57

seed physiology. C. R. Biol. 331:806-814.

Baskar R, Rajeswari V, Sathish Kumar T (2007). *In vitro* antioxidant studies in leaves of *Annona* species. Indian J. Exp. Biol. 45:480-485.

Bhatia VS, Tiwari SP, Pandey S (2002). Soybean seed quality scenario in India - A Review. Seed Res. 30(2):171-185.

Bose A, Bhattacharya S, Pandey J, Biswas M (2011). Thin layer chromatographic studies and in vitro free radical scavenging activity of *Annoasquamosa* leaf extracts. Pharmacol. Online 1:1223-1229.

Bukhari SB, Bhanger MI, Memon S (2008). Antioxidative activity of extracts from fenugreek seeds (*Trigonellafoenum graecum*). Pak. J. Anal. Environ. Chem. 9(2):78-83.

Chandrashekar C, Kulkarni VR (2011). Isolation, characterizations and Free radical scavenging activity of *Annona squamosa* leaf. J. Pharm. Res. 4(3):610-611.

Coolbear P (1995). Mechanisms of seed deterioration, In: Seed Quality Basra A. S. (Eds.). Basic Mechanisms and Agricultural Implications, Food Products Press, New York. pp. 223-277.

Delouche JC, Baskin CC (1973). Accelerated ageing techniques for predicting the relative storability of seed lots. Seed Sci. Technol. 12:427-452.

Fahey JW (2005). *Moringaoleifera:* A Review of the Medical Evidence for Its Nutritional, Therapeutic, and Prophylactic Properties. Part 1. Trees for Life Journal: http://www.tfljournal.org/images/articles/20051201124931586_3.pdf.

Ferreira PMP, Farias DF, Oliveira JTA, Carvalho AFU (2008). *Moringaoleifera*: bioactive compounds and nutritional potential. Revista de Nutrição 21(4):431-437.

Guerinot ML, Yi Y (1994). Iron: Nutritious, Noxious, and Not Readily Available. Plant Physiol. 104:815-820.

ISTA (2009). International Rules for Seed Testing. International Seed Testing Association, Switzerland.

Jatoi SA, Afzal M, Nasim S, Anwar R (2001). Seed deterioration study in pea, using accelerated ageing techniques. Pak. J. Biol. Sci. 4(12):1490-1494.

Jeuffroy MH, Ney B (1997). Crop physiology and productivity. Field Crop. Res. 53:3-16.

Kaiser BN, Gridley KL, Brady JN (2005). The Role of Molybdenum in Agricultural Plant Production. Ann. Bot. 96:745-754.

Kalavathi D (1985). Studies on seed viability and vigour in soybean. M.Sc. Thesis, Tamil Nadu Agricultural University, Coimbatore.

Kaviarasan S, Naik GH, Gangabhagirathi R, Anuradha CV, Priyadarsini KI (2007). *In vitro* studies on antiradical and antioxidant activities of fenugreek (*Trigonellafoenum graecum*) seeds: Food Chem. 103:31-37.

Koleckar V, Jun D, Opletal L, Jahodár L, Kuča L (2007). Assay of radical scavenging activity of antidotes against chemical warfare against by DPPH test using sequential injection technique. J. Appl. Biomed. 5:81-84.

Layek N, De BK, Mishra SK, Mandal AK (2007). Seed invigoration treatments for improved germinability and field performance of gram (*Cicer arietinum* L.). Legume Res. 29(4):257-261.

Maeda H, Sakuragi Y, Bryant DA, DellaPenna D (2005). Tocopherols Protect *Synechocystis* sp. Strain PCC 6803 from Lipid Peroxidation. Plant Physiol. 138:1422-1435.

Maguire ID (1962). Speed of germination aid in selection and evaluation for seedling emergence and vigour. Crop Sci. 2:176-177.

McQuaker NR, Brown DF, Fluckner PD (1979). Digestion of environmental materials for analysis by inductively coupled plasma atomic emission spectrometry. Anal. Chem. 51:1082-1084.

Moore A, Waring CP (2001). The effects of a synthetic pyrethroid pesticide on some aspects of reproduction in Atlantic salmon (*Salmosalar* L.). Aquat. Toxicol. 52:112.

Pandey N, Brave D (2011). Phytochemical and Pharmacological Review on *Annona squamosa* Linn. Int. J. Res. Pharm. Biom. Sci. 2(4):1404-1412.

Pati CK, Bhattacharjee A (2011). Sunflower seed invigoration by chemical manipulation. Afr. J. Plant Sci. 5(15):867-872.

Rangaswamy R (2002). A text book of agricultural statistics. New Age International Ltd., India. pp. 244-433.

Sathish S, Sundareswaran S (2010). Biochemical evaluation of seed priming in fresh and aged seeds of maize hybrid [COH(M) 5] and its parental lines. Curr. Biotica. 4(2):162-170.

Sathish S, Sundareswaran S, Ganesan N (2011). Influence of seed priming on physiological performance of fresh and aged seeds of maize hybrid [COH(M) 5] and it's parental lines. ARPN J. Agric. Biol. Sci. 6(3):12-17.

Sattler SE, Mene-Saffrane L, Farmer EE, Krischke M, Mueller MJ, DellaPenna D (2006). Nonenzymatic lipid peroxidation reprograms gene expression and activates defense markers in *Arabidopsis*tocopherol-deficient mutants. Plant Cell 18:3706-3720.

Scialabba A, Bellani LM, Dell'Aquila A (2002). Effects of ageing on peroxidase activity and localization in radish (*Raphanus sativus* L.) seeds. Eur. J. Histochem. 46:351-358.

Tlustos P, Cigler P, Hruby M, Kuzel S, Szakova J, Balík J (2005). The role of titanium in biomass production and its influence on essential elements' contents in field growing crops. Plant Soil Environ. 51(1):19-25.

Toppo, FA, Akhand R, Pathak AK (2009). Pharmacological actions and potential uses of *Trigonellafoenum graecum*: A review. Asian J. Pharm. Clin. Res. 2(4):29-38.

Vijayalakshmi V (2012). Seed vigour and viability studies in TNAU tomato hybrid CO_3 (*Lycopersicon esculentum* Mill.). Ph.D. Thesis, Tamil Nadu Agricultural University, Coimbatore.

7

Biomass allocation and nitrogen distribution of *Pinus wallichiana* Jackson seedlings under different nitrogen fertigation levels

Hillal Ahmad[1], T. H. Masoodi[1], Altamash Bashir[1], G. I. Hassan[2], S. A. Mir[3], P. A. Sofi[1], G. M. Bhat[1] and M. Maqsood[4]

[1]Faculty of Forestry, SK University of Agricultural Sciences and Technology of Kashmir, Shalimar, Srinagar, J and K, 19112, India.
[2]Division of Pomology, SK University of Agricultural Sciences and Technology of Kashmir, Shalimar, Srinagar, J and K, 19112, India.
[3]Division of Statistics, SK University of Agricultural Sciences and Technology of Kashmir, Shalimar, Srinagar, J and K, 19112, India.
[4]Department of Botany, University of Kashmir, Srinagar, India.

Pinus wallichiana seeds were sown in polybags filled with sterilized riverbed sand in 2010. Seedlings were thinned out to one seedling per polybag, which were treated weekly with Ingestad pretreatment nutrient solution for 4 weeks and then were fertigated with nitrogen levels of 0.99, 1.98, 2.97, 3.96, 4.95, 5.94, 6.93, 7.92, 8.91 and 9.90 mg and fixed levels of phosphorus, potassium, calcium and magnesium dissolved in 50 ml of water seedling^{-1} week^{-1} upto or until 28 weeks. Biomass allocation to needles, (separated by hand from shoot axis), stem (Includes hypocotyls and epicotyls) and roots increased significantly with the increase in nitrogen level and seedling age. At 7.92 mg level of nitrogen fertigation biomass allocation and nitrogen concentration partitioning at 28 weeks got increased by 190, 32 and 44%, and 179, 101 and 147% in needles, stem and roots, respectively as compared to control. Nitrogen concentration in shoot and root decreased at lower nitrogen level as seedlings age increased, but increased at higher nitrogen fertigation levels in needles, stem and roots. To the end of the experiment, nitrogen uptake per seedling increased, whereas, nitrogen use efficiency (NUE) decreased with the increasing nitrogen levels.

Key words: *Pinus wallichiana*, Biomass allocation, nitrogen partitioning, nitrogen uptake, nitrogen use efficiency (NUE).

INTRODUCTION

Pinus wallichiana Jackson (Kail pine) is a one of the important timber species of wealthy coniferous forests of Kashmir. It is found in moist and dry temperate forest types of Western and Central Himalayas from Kashmir to Bhutan with an altitudinal zonation of 1500 to 3700 m a.m.s.l. *P. wallichiana* Jackson is a tall evergreen tree species with average height of 30 to 36 m and girth 2.5 to 3 m (Troup, 1921). The poor natural regeneration of this species has been a major concern and has thus, caused constant attention of researchers in Kashmir. A number of factors are considered responsible for the failure or patchy regeneration of this species, that is, early slow growth, poor regeneration and competition etc. may be the main causes for replacing this valuable timber species of Jammu and Kashmir natural forests by the other species like *Cedrus deodara* at an alarming rate.

Seedling quality, when defined as out planting performance is logically related to its mineral nutrient status. Nutrient loaded seedling performed better than conventionally produced seedlings when out planted especially in nutrient poor sites (Malik and Timmer, 1995; Xu and Timmer, 1999). Planting nutrient loaded seedlings is a more effective method of stimulating growth than post transplanting fertilizer application (Xu and Timmer, 1999). The improved seedling quality as a result of applied nutrients allows better interaction of planted seedlings with the planting site and therefore fuller expression of site potential (Fry and Poole, 1980). Fertilization is thus an essential component of producing high quality nursery stock for reforestation and enhances plant growth, nutrient storage reserves, root growth potential, and resistance to drought stress, freezing temperatures and diseases (Landis, 1985; Rook, 1991). Limited nitrogen supply usually inhibits shoot growth proportionally more than root growth (Ledig, 1983). Under high nitrogen condition; genotypes that allocate relatively more dry matter to shoots and may grow faster because they reinvest the high proportion of their assimilates in above ground biomass and vice-versa (Ledig, 1975). Leaf photosynthetic capacity has often been highly correlated with leaf nitrogen content. The reason for this includes the paramount role of nitrogen in light harvesting and CO_2 fixation (Evans, 1989).

Fertilization is one of the most important cultural practices for producing container seedlings when the limited volume seriously hinders seedling growth (Landis, 1989). and it also accelerate shoot and root growth, modify tissue nutrient contents and hence the amount of available reserves, improve post-transplant rooting and growth capacity, and increase resistance to water stress, low temperature and disease (Landis, 1985; Haase and Rose, 1997; Shaw et al., 1998; Grossnickle, 2000), However, such properties are of vital importance for successful seedling establishment under unfavorable conditions (Puttonen, 1997; Birchler et al., 1998). The morphological studies have shown that nutrient status of nursery seedlings played a significant role in post planting response. Keeping this in view, the present investigation has been undertaken to know the optimal range of nitrogen for *P. wallichiana* with the following objectives: the first is to study the pattern of biomass allocation in *P. wallichiana* seedlings and second is to study nitrogen partitioning, nitrogen uptake efficiency and nitrogen use efficiency (NUE) in *P. wallichiana* seedlings.

MATERIALS AND METHODS

The experiment was conducted at the Wadura, Sopore, Faculty of

Forestry, SK University of Agricultural Sciences and Technology of Kashmir, Srinagar, Jammu and Kashmir. The site is located at 34°17′ N latitude and 74°33′ E longitude with altitude of 1524 m a.m.s.l. Ingested and Lund (1979, 1986) and Ingested (1987) developed a method to regulate relative growth rates of seedlings by controlling nutrient supply. The technique is based on the concept that, it is the nutrient amount supplied per unit time (relative addition rate) and not the nutrient concentration in the growing medium that is important in controlling nutrient uptake and the subsequent growth. The Kail pine seeds were collected from Bandipora Forest Division, Kashmir, soaked in cold water for 24 h and treated with Captan (50% WP) at 2.5 g kg^{-1} seed. The seeds were sown in polythene bags (10 × 15 cm) filled with sterilized river bed sand and kept in a mist chamber for germination in the last week of February, 2010. Three seeds were then sown in each polybag. Weeding was carried out manually as and when required. Polybags were irrigated once a week when fertilizer was not applied. Germinates were thinned out to one per polybag, 3 weeks after germination and fertigated at weekly interval with Ingestad pretreatment nutrient solution (Ingestad and Lund, 1979) that included all major nutrients by weight proportions of nitrogen, phosphorus, potassium, calcium and magnesium (in 100:13:65:7:8.5 ratio), 25 ml $seedling^{-1}$ for 1 month as pretreatment and each application supplied 1.982 mg nitrogen, 0.258 mg phosphorus, 1.288 mg potassium, 0.139 mg calcium and 0.168 mg magnesium $seedling^{-1}$. The pretreated seedlings were shifted from mist chamber to nursery and treated weekly with nitrogen fertigation which was continued till the termination (upto 28 weeks of age) of experiment.

The nitrogen levels were 10 and total fertigation treatments were 12 viz., 0 (with and without fertigation), 0.99, 1.98, 2.97, 3.96, 4.95, 5.94, 6.93, 7.92 8.91 and 9.90 mg N $seedling^{-1}$. All the treatments, except without fertigation treatment, were given fixed levels of other nutrients viz., P, K, Ca and Mg $seedling^{-1}$ $week^{-1}$. Nutrient fertigation was applied weekly at 50 ml $seedling^{-1}$ $week^{-1}$. The control treatment received only plain water. The experiment was conducted with a randomized complete block design with three replications and per treatment level was represented by 21 seedlings at each of the replications. Nine seedlings from each treatment were selected randomly and harvested at 4 weeks interval to measure physical parameters on monthly basis, that is, dry weight of needles, stem and roots. Total nitrogen (Jackson, 1973) was also measured on those same samples. Nitrogen uptake was determined as per procedure of Burgess (1991). NUE was estimated by dividing seedling biomass at a sample date by the amount of nitrogen taken up at that date (Prescott et al., 1989).

Statistical analysis

The experimental data were statistically analyzed using the two-way analysis of variance (ANOVA) approach following Gomez and Gomez (1984). The differences between treatments and the interaction between experimental factors (for example, nitrogen and seedling age) were tested at the 0.05 significance level. Data analysis was carried out using the software IBM SPSS Statistics.

RESULTS AND DISCUSSION

Pattern of biomass allocation in *P. wallichiana* seedlings

P. wallichiana seedlings allocated more carbohydrates to shoot at higher levels of nitrogen fertigation. Increase in nitrogen fertigation upto 7.92 mg nitrogen level

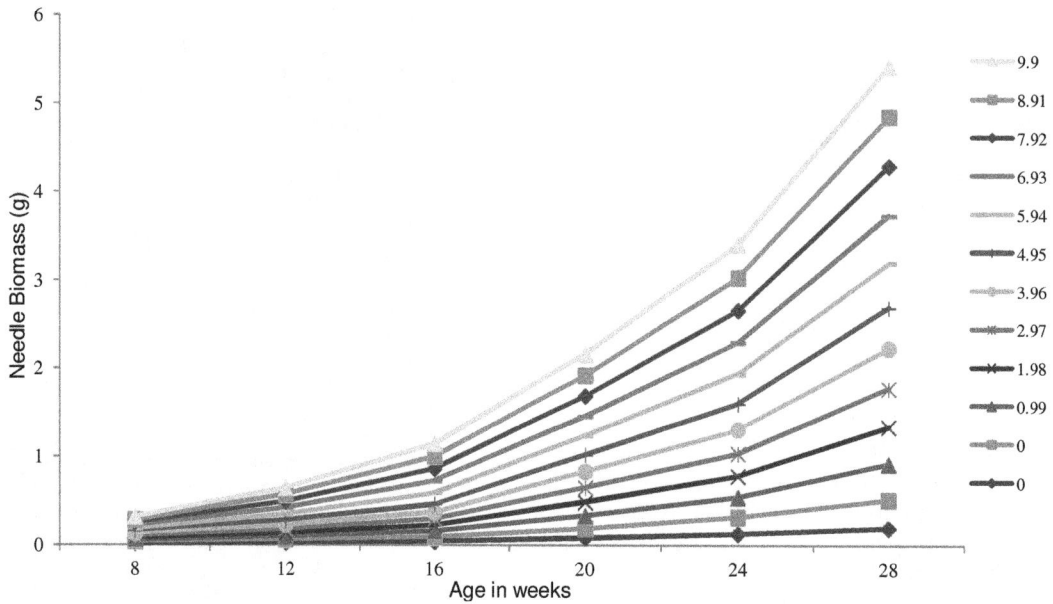

Figure 1. Effect of nitrogen fertigation and seedling age on allocation of biomass (g) to needles.

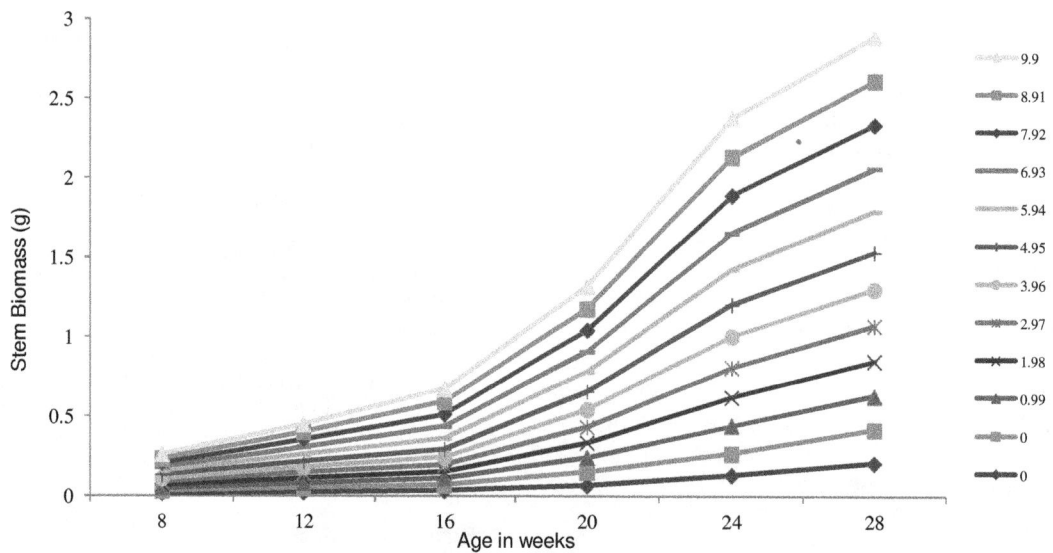

Figure 2. Effect of nitrogen fertigation and seedling age on allocation of biomass to stem (g).

enhances mean needles (0.232 g), stem (0.132 g) and root (0.179 g) biomass (Figures 1 to 3) in Kail pine seedlings. There was no significant difference among the higher doses. There was significant difference in seedling mean needle, stem and root biomass in case of nitrogen-free and control treatment. The effect of age on different components of seedling (needle, stem and root) biomass was conspicuous with highest (0.452, 0.241 and 0.265 g) observed at 28 weeks age, which were 17, 11 and 6 times more than the seedling needle, stem and root biomass at 8 weeks age, respectively. With increase in age from the 8 to 28 weeks, the biomass showed significant increase. Interaction between fertigation and seedling age of Kail pine was maximum at the 9.90 mg nitrogen treatment and seedling age of 28 weeks shown in seedling biomass of needle (0.566 g), stem (0.280 g) and roots (0.320 g). Ledig (1983) suggested that under high nitrogen condition, plants allocate relatively more biomass to shoots and may grow faster by inverting a high proportion of their photosynthetic capital to shoots. The shift in relative total plant dry weight from the roots to the shoots might be due to redirection of the relative

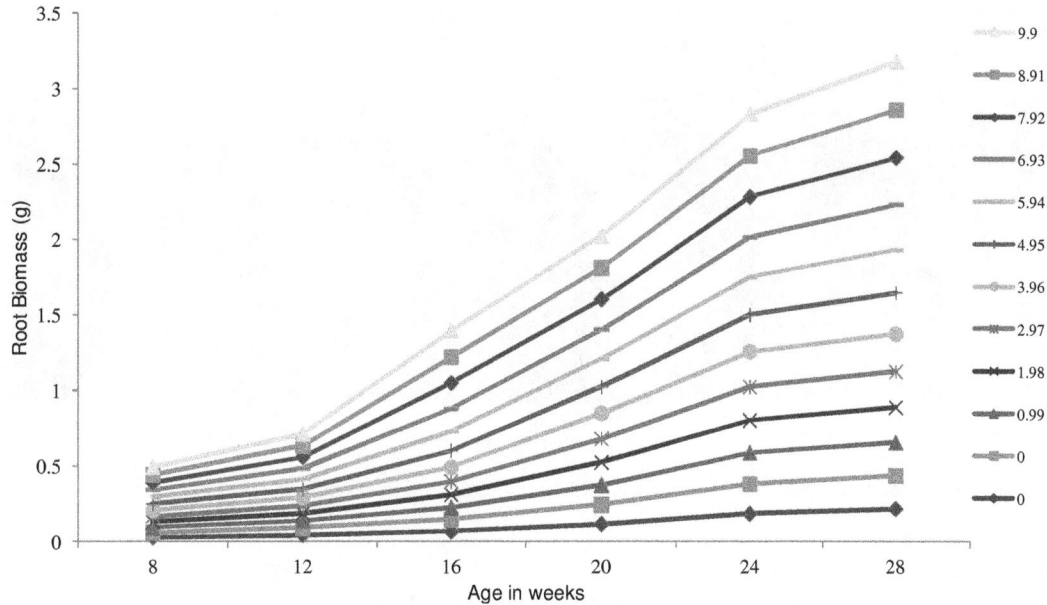

Figure 3. Effect of nitrogen fertigation and seedling age on allocation of biomass to roots.

proportion of total plant nitrogen from the root system to leaves (Grime, 1979; Rose and Biernacka, 1999). Genetic variability is higher within individual of long leaf pine seedlings than among stands or sources, both in years and nitrogen rate (1 to 3 mg) root biomass increased about 30%, compared with the 230% increase in shoot biomass which may be due to this factor that root biomass appears less plastic in response than shoot biomass (Boyer, 1990). Our results were confirmed with Jackson et al. (2007) who revealed that shoot (2.06 g) and root biomass (1.52 g) increased at 4 mg nitrogen seedling^{-1} week^{-1} in long leaf pine (*Pinus palustris*) for a period 30 week. Similar findings have been reported for sugar (*Pinus lambertiana*) and Jeffrey pine (*Pinus jeffreyi*) by Walker (2001). However, our results were also supported by Li et al. (1991) who revealed that nitrogen had significant effects on seedling growth and biomass allocation to needles, stem and roots.

Tissue nitrogen concentration and relative nitrogen distribution in *P. wallichiana* seedlings

Increase in nitrogen fertigation upto 7.92 mg nitrogen level significantly enhances the mean needle (Figures 4 to 7) nitrogen percent (1.97), stem (1.02) and roots (1.42) in Kail pine. There was significant difference in mean seedling needle, stem and root nitrogen concentration for nitrogen-free and control treatments. As age advances mean nitrogen concentration significantly increased in different components [needle (1.53%), stem (0.79%) and roots (1.19%)] of seedlings, which was observed at 28 weeks age, nitrogen concentration was highest in needles, then in roots and lowest in stem. Interaction

between fertigation and seedling age of Kail pine again indicate that the maximum interaction occurred between 9.90 mg nitrogen level and the age of 28 weeks with nitrogen concentration in needle (2.08%), stem (1.14%) and roots (1.50%). As expected, with increase in the rate of nitrogen applied, the seedling nitrogen level increased significantly. The increase in total nitrogen content of *P. wallichiana* seedling with increased nitrogen concentration may be due to the combination of increase in dry weight and nitrogen concentration in tissues. Because nitrogen is a major constituent of enzymes, changes in tissue nitrogen concentration generally reflects a change in enzymes concentration. Both gross primary products and plant respiration represents biochemical processes that are catalyzed by nitrogen rich enzymes as the rate of these processes depend on the nitrogen content in tissues. Higher level of nitrogen availability generally increases the nitrogen concentration in leaves, which increases the growth of the plants (Mc Guire and Joyce, 1995). Fertilization can modify tissue nutrient contents and the amount of available reserves, improves post transplant rooting and growth capacity of seedling (Landis, 1985; Van den Driessche, 1992). Our results were similar to the optimal range of conifers since 2.5% of needle nitrogen concentration is the upper limit of optimum range for conifers as defined by Landis (1989) and Van den Driessche (1988) reported maximum survival of Douglass fir seedling with a 2.1% nitrogen concentration in needles at the time of planting. Burgess (1991) recorded 2.41% nitrogen in Douglas fir 14 weeks of age under 6% nitrogen fertigation rate, which subsequently decreased to 1.61% at 18 weeks of age, while in Western hemlock at same fertigation rate it increased from 1.94% at 14 week age to 2.23% at 18

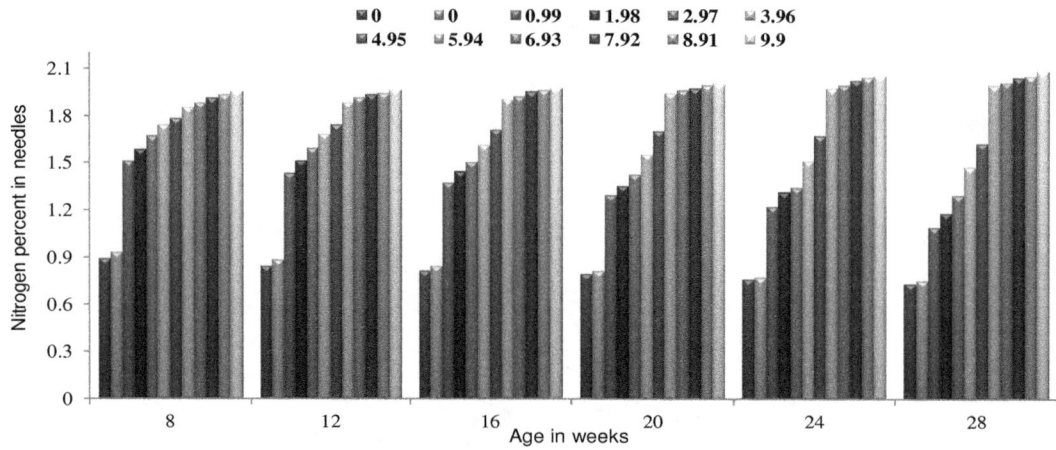

Figure 4. Effect of nitrogen fertigation and seedling age on nitrogen concentration of needles.

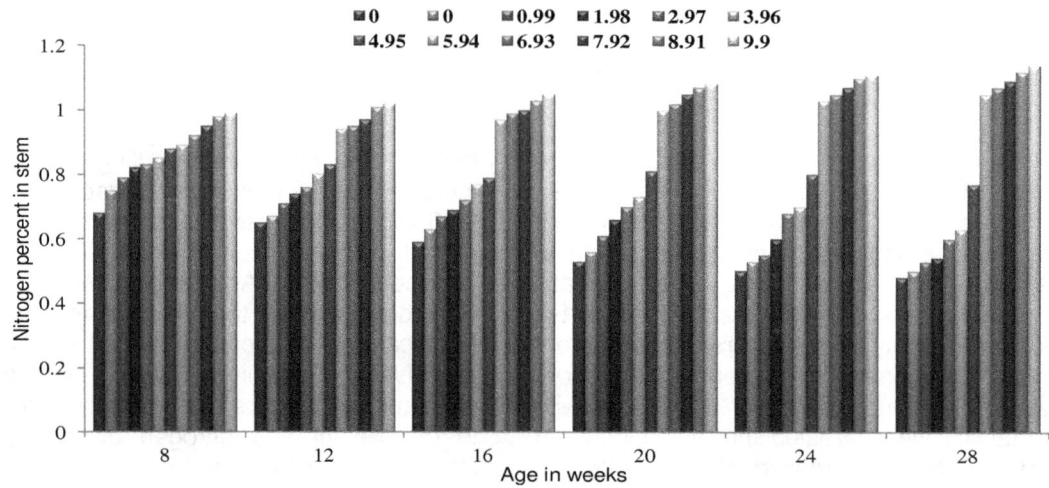

Figure 5. Effect of nitrogen fertigation and seedling age on nitrogen concentration of stem.

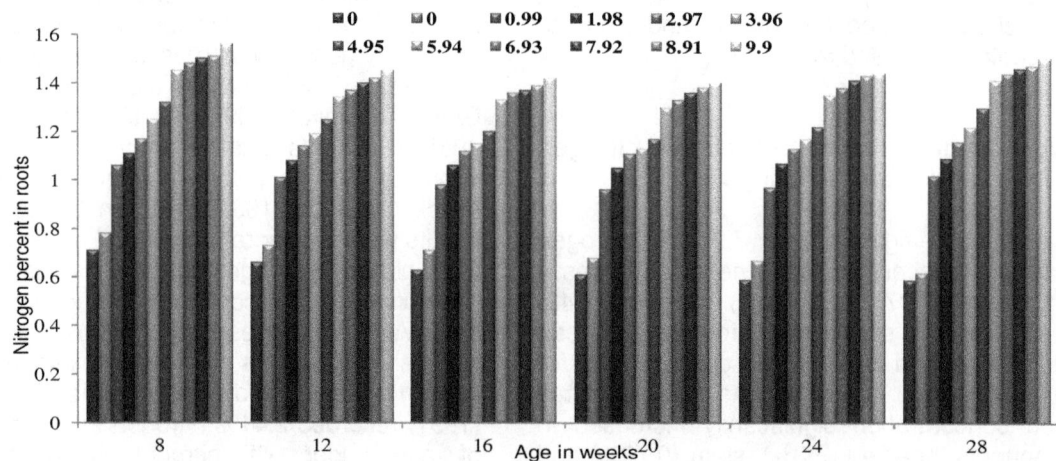

Figure 6. Effect of nitrogen fertigation and seedling age on nitrogen concentration of roots.

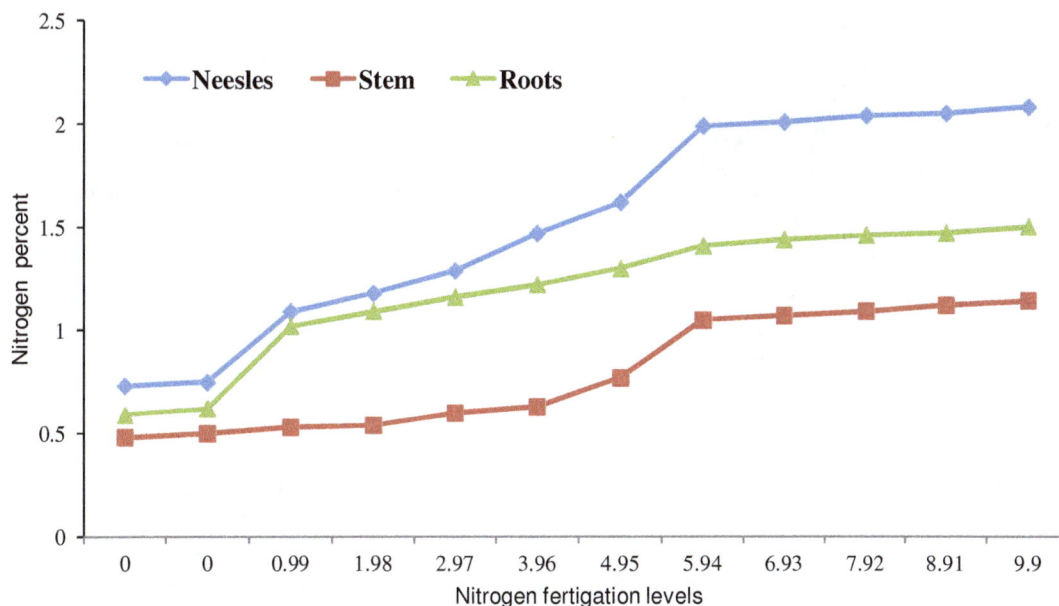

Figure 7. Nitrogen distribution in *P. wallichiana* seedlings at 28 weeks of age.

weeks age. So, the percentage of nitrogen in plants depends upon the nitrogen addition rate and species response towards the nitrogen fertilizers. Increasing trend of nitrogen content has also been reported by Dumroese et al. (2005) in long leaf pine under the application of 40 mg and 66 mg nitrogen seedling[-1].

Nitrogen uptake of *P. wallichiana* seedlings under different fertigation levels

The cumulative amount of nitrogen added by treatments varied from 19.82 to 198.2 mg seedling[-1] by the time the seedlings were 28 weeks old (Figure 8). Nitrogen uptake by *P. wallichiana* seedlings at the 9.90 mg nitrogen fertigation rate accumulated only 19.76 mg of nitrogen (9.97%). At basal dose (7.92 mg seedling[-1]), *P. wallichiana* seedlings absorbed 60.9% of the nitrogen applied to medium. As concentration of nitrogen in treatments increased, the nitrogen uptake efficiency decreased from 60.92% in the treatment with 0.00 mg nitrogen to 9.97% at the 9.90 mg nitrogen which means it decreased as the nitrogen addition levels increased. The decrease in the nitrogen -uptake efficiency at initial nursery stage definition may be due to its intrinsically low relative growth rate. Matching plant growth with nutrient uptake and maintain stable internal nutrient concentration has been a topic of interest. In this regard, Ingestad (1974), and Ingestead and Lund (1986) improved fertilizer efficiency and prevent nutrient deficiency and toxicity in different studies and concluded that it occurs due to low or high nutrient addition levels. These results are consistent with that of Burgess (1991) who observed nitrogen uptake efficiency of 30.60 and 22.42% at highest

(6% addition) rate in Douglas fir and Western hemlock, respectively. Burgess (1991) further reported that although both Douglas fir and Western hemlock grew fastest under 6% nitrogen fertigation rate, Western hemlock was more efficient in nitrogen uptake. Contrary to these findings, Jackson et al. (2007) observed increase in total nitrogen content seedling[-1] of long leaf pine with increase in nitrogen fertigation rate. Burgess (1990) also reported similar results in Black and White Spruce seedlings.

NUE of *P. wallichiana* seedlings under different fertigation levels

Figure 8 reveals that total biomass of seedlings (oven dry weight) at 28 weeks of age varied from 618 mg at lowest nitrogen application rate to 1166 mg at highest addition rate. The cumulative amount of nitrogen added by treatments varied from 19.82 to 198.20 mg seedling[-1] up to the time seedlings were harvested at the 28 weeks of age. NUE in *P. wallichiana* seedlings decreased from lowest addition rate 95.19 mg of biomass per mg of nitrogen in the treatment 0.00 per mg nitrogen to 5.88 mg per mg nitrogen at the highest nitrogen level at 9.90 mg. NUE CO_2 gained per unit of nitrogen absorbed from medium by *P. wallichiana* seedlings was lowest (5.88%) at highest nitrogen addition rate (Figure 8) emphasizing the ability of slow growing species to retain nutrients with in plant body at nursery stage. Leaves with maximum photosynthesis rate may invest a large proportion of the leaf nitrogen in rubisco (Field and Mooney, 1986). Low NUE may be as a result of inefficient allocation of nitrogen among photosynthetic compounds, such that

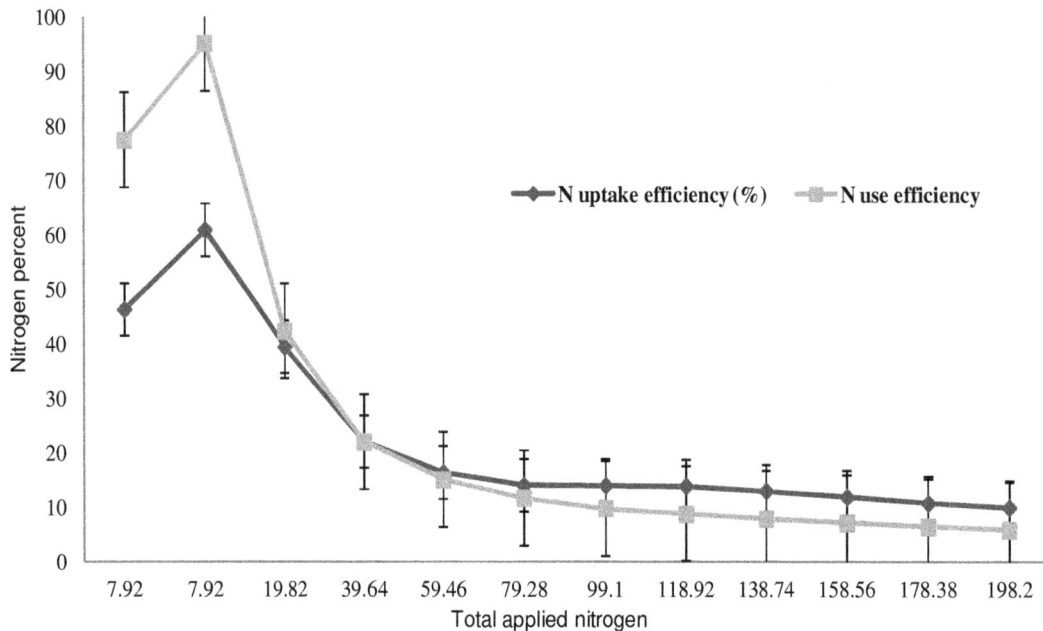

Figure 8. NUE and nitrogen uptake efficiency of *P. wallichiana* seedlings at the end of first growing season.

some compounds are present in excess, while the rate limiting compounds are underrepresented. For example, shade plants invert larger quantity of nitrogen in light harvesting pigments and proteins, but make only small investment in rubisco and others CO_2 processing enzymes (Bjorkman, 1981; Evans, 1989). Our results are consistent with Riech and Schoettle (1988) who reported that in white pine (*Pinus strobus*) seedlings NUE decreased with increasing concentration of nitrogen in tissues. Difference in NUE may also result from difference in allocation levels. Contrary to this, Masoodi et al. (2007) reported NUE of 57.47% in *C. deodara* and 100.14% in *Cupressus torrulosa* at 9 mg nitrogen addition rate. However, both these species grow relatively faster than *P. wallichiana*.

Conclusion

All the species have an optimum range of nitrogen requirement. However, the optimum range for biomass allocation and nitrogen distribution in *P. wallichiana* seedling is 7.92 mg level of nitrogen addition rate for first growing season at nursery, at this rate optimum biomass and nitrogen content is recoded. However, the below nitrogen addition rates have non-significance (low concentration have non significance effect on response) and the above nitrogen levels adds minute biomass (more fertilizer was used and less gains was obtained) and nitrogen percent to Kail pine seedlings, which has statistically and economically no importance. NUE in *P. wallichiana* seedlings was found inverse of nitrogen

uptake efficiency.

REFERENCES

Birchler T, Rose R, Royo A, Pardos M (1998). La planta ideal: revisión del concepto parámetros definitoriose implementación práctica. Investigación Agraria: Sistemas y Recursos Forestales. 7:109-121.

Bjorkman O (1981). Responses to different quantum flux densities. PP. 57-107 In: Physiological plant ecology. Encycl. Plant Physiol. P. 12A. new series, Lange OL, Nobel PS, Osmond CB, Ziegler H (eds.). Springer-Verlag, Berlin.

Boyer WD (1990). Longleaf pine. Pages 405-412 in: Silvics of North America, 1. Conifers. U.S.D.A. Forest Service, Agriculture Handbook. P. 654.

Burgess D (1990). Controlling white and black spruce seedling development using the concept of relative addition rate. Scand. J. For. Res. 5:471-480.

Burgess D (1991). Western hemlock and Douglas fir seedling development with exponential rates of nutrient addition. For. Sci. 37:54-67.

Dumroese RK, Parkurst J, Barnett JP (2005). Controlled release fertilizer improves quality of container longleaf pine seedling. USDA Forest Service Proceeding. RMRS. P. 50.

Evans JR (1989). Photosynthesis and nitrogen relationship in leaves of C3 plants. Oecologia 78:9-19.

Field C, Mooney HA (1986). The photosynthesis-nitrogen relationship in wild plants. In: On The Economy of Plant Form and Function, T.J. Givnish (ed.). Cambridge University Press, New York, USA. pp. 25-55.

Fry G, Poole BR (1980). Evaluation of planting stock quality several years after planting (a discussion). New Zea. J. For. Sci. 10: 299-300.

Gomez KA, Gomez AA (1984). Statistical Procedure for Agricultural Research. (2nd edn.) John Wiley and Sons, Inc., New York, USA.

Grime JP (1979). Plant Strategies and Vegetation Processes. John Wiley and Sons, Chichester. In: (Growth, Dry Weight and Nitrogen Distribution of Red Oak and 'Autumn Flame' Red Maple Under Different Fertility Levels. Jill Larimer J. Environ. Hort. 20(1):28–35.

Grossnickle SC (2000). Ecophysiology of Northern Spruce Species: The

Performance of Planted Seedlings. NRC Research Press, Ottawa, Ontario, Canada. P. 409.

Haase DL, Rose R (Eds.), (1997). Proceedings of the Symposium on Forest Seedling Nutrition from the Nursery to the Field. Nursery Technology Cooperative, Oregon State University.

Ingestad T (1987). New concepts on soil fertility and plant nutrition as illustrated by research on forest trees and stands. Geoderma. 40:237-252.

Ingestad T (1974). Towards optimum fertilization, Ambio. 3(2):49-54.

Ingestad T, Lund AB (1979). Nitrogen stress in Birch seedlings. I. Growth technol. Growth Physiol. Plan. 45:137-148.

Ingestad T, Lund AB (1986). Theory and techniques for steady state mineral nutrition and growth of plants. Scand. J. For. Res. 1: 439-453.

Jackson DP, Dumroese R, K, Barnett JP, Patterson WB (2007). Container longleaf pine seedling morphology in response to varying rates of nitrogen fertilization in the nursery sand subsequent growth after outplanting. USDA Forest Service Proceeding RMRS. p. 50.

Jackson ML (1973). Soil chemical analysis (2nd edition). Prentice Hall of India, Private Limited, New Delhi. pp. 1-498.

Landis TD (1989). Mineral nutrients and fertilization. In: Landis TD, Tinus, RW, McDonald SE, Barnett JP. (Eds.), The Container Tree Nursery Manual, Agriculture Handbook No. 674. USDA For. Serv. 4:1-70.

Landis TD (1985). Mineral nutrition an index of seedling quality. P. 29-48 In: Evaluating seedling quality; principles, procedures and predictive abilities of major tests. M.L. Duryea (ed.) For. Res. Lab. Oregon State University, Corvallis.

Ledig FT (1983). The influence of family and environment on dry matter distribution in plants. pp. 427-454 In: Plant research and agroforestry. Proc. Consultative meeting held in Nairobi. Huxley, P.A (ed.). International Council. Research for Agroforestry. Nairobi, Kenya.

Ledig FT (1975). Increasing the productivity of forestry trees. In: Forest tree improvement-the third decade. Thielges, B.A. (ed.). Louisiana State Univ., Baton Roag. pp. 189-206.

Li B, Allen HL, Mckeand SE (1991). Nitrogen and family effects on Biomass allocation of loblolly pine seedlings. For. Sci. 37(1):271-283.

Malik V, Timmer VR (1995). Interaction of nutrient loaded black spruce seedlings with neighboring vegetation in greenhouse environments. Can. J. For. Res. 25:1017-1023.

Masoodi NA, Masoodi TH, Gangoo SA, Islam A (2007). Effect of nitrogen on whole plant carbon gain and nitrogen use efficiency in Cedrus deodara and Cupressus torrulosa. Appl. Biol. Res. 9:1-8.

Mc Guire AD, Joyce LA (1995). Responses of net primary production in temperate forests to potential changes in carbon dioxide and climate. In: Gen. Tech. Report for the (1993). RPA Assessment Update, ed. Joyce. L.A. Fort. Collins: USDA Forest Service.

Prescott CE, Corbind JP, Parkinson D (1989). Biomass, productivity, and nutrient-use efficiency of above ground vegetation in four Rocky Mountain coniferous forests. Can. J. For. Res. 19:309-317.

Puttonen P (1997). Looking for the "silver-bullet"—can one test do it? New For. 13(1-3):9-27.

Rook DA (1991). Seedling development and physiology in relation to mineral nutrition. In: van den Driessche, R. (Ed.), Mineral Nutrition of Conifer Seedlings. CRC Press. pp. 86-112.

Rose MA, Biernacka B (1999). Seasonal patterns of nutrient and dry weight accumulation in Freeman maple. Hort. Sci. 34:91-95.

Shaw TM, Moore JA, Marshall JD (1998). Root chemistry of Douglas-fir seedlings grown under different nitrogen and potassium regimes. Can. J. For. Res. 28:1566-1573.

Troup RS (1921). The Silviculture of Indian Trees. Oxford University Press, Oxford. 3: 81-83 [cf : The Wealth of India, vol. 1 revised-1995, Publication and Information Directorate, CSIR, Hillside Road, New Delhi.

Van den Driessche R (1992). Changes in drought resistance and root growth capacity of container seedlings in response to nursery drought, nitrogen and potassium treatments. Can. J. For. Res. 22:740-749.

Van den Driessche R (1988). Nursery growth of conifer seedlings using fertilizers of different solubilities and application time, and their forest growth. Can. J. For. Res. 18:172-180.

Walker RF (2001). Growth and nutritional responses of containerized sugar and Jeffrey pine seedlings to controlled release fertilization and induced mycorrhization. For. Ecol. Manage. 149:163-179.

Xu X, Timmer VR (1999). Growth and nitrogen nutrition of Chinese fir seedlings exposed to nutrient loading and fertilization. Plant Soil. 216:83-91.

8

Effects of decapitation on dry fruit and seed yield of early (*Abelmoschus esculentus*) and late okra (*Abelmoschus caillei*) cultivars in South Eastern Nigeria

O. S. Udengwu

Department of Plant Science and Biotechnology, Faculty of Biological Sciences, University of Nigeria Nsukka, Enugu State, Nigeria. E-mail: obi.udengwu@unn.edu.ng, obiudengwu@gmail.com

The effects of decapitation and non-decapitation on dry fruit and seed production by 12 cultivars of early and late okra cultivars were studied for two cropping seasons. The analysis of variance showed that year was very highly significant while decapitation-non decapitation, as well the interaction between year and decapitation- non decapitation, for the four treatments for both early and late okra types, were not significant. The highly significant difference in years was attributed to the differences in the rainfall distribution patterns as well relative humidity, air and soil temperatures for the two cropping seasons. Though the differences in seed and dry fruit yield between early and late okra was not statistically significant, late okra showed greater promise for okra seed production in the region. The non-significant differences in dry fruit and seed yield between the early and late okra types as well as between the decapitated and non-decapitated plants were attributed to the physiological and reproductive attributes of both okra types. Comparatively, the seed yield of the evaluated cultivars showed that yield was below the optimum reported for some other okra growing regions of the world, hence the urgent need for more intensified effort to improve the seed yield of the promising cultivars. It was concluded that decapitation as a horticultural practice may not be an effective way of increasing dry fruit and seed yield in okra in South Eastern Nigeria.

Key words: Okra, early, late, decapitation, non-decapitation, dry fruit, seed production.

INTRODUCTION

Okra is a multipurpose crop due to its various uses (Martin, 1982; Adeniji et al., 2007). It is grown in many countries and cultivars from different countries have certain adapted distinguishing characteristics specific to the country to which they belong (Dhankhara and Mishrab, 2005).

In the West African region, two distinct species of okra exist; *Abelmoschus esculentus* (L.) Moench, which is the conventional or early okra and *Abelmoschus caillei* (A. Chev) Stevels, late okra (Siemonsma and Hamon, 2004). Late okra is also called the West African okra, since its distribution is restricted to the West African region (Njoku, 1958; Martin et al., 1981, Siemonsma, 1982; Hamon and Hamon, 1991). The major difference between the two okra types lies in their physiological response to changes in natural photoperiods (Njoku 1958; Oyolu, 1977; Nwoke, 1980; Siemonsma, 1982; Udengwu, 1998).

Okra is a kind of special vegetable, which is rich in amino acids, vitamins, minerals, polysaccharides, flavonoids and unsaturated fatty acids, with multiple nutritional and health-care functions (Dong and Liang, 2007). Notwithstanding the great value of okra, information on characterization is either not accessible or simply unavailable. Characterization and evaluation of crops is done to provide information on diversity within or among crops. This permits the identification of unique entries (accessions) necessary for curators of gene banks and plant breeders (Torkpo et al., 2006).

The world today is concerned with exploitation of alternative sources of clean energy in a bid to stem global pollution occasioned by the burning of fossil fuel. As the demand for vegetable oils is rapidly increasing due to the growing human population and the expanding oleo-chemicals industry, the exploration of some

Non-conventional and newer resources of vegetable oils is of much concern (Anwar et al., 2011). Studies with okra had been concentrated more in the direction of its production, improvement and utilization as a vegetable crop (Martin and Ruberte 1978; Mangual-Crespo and Martin, 1980), with little attention paid to its huge potentials as a seed and oil crop. Karakoltsidis and Constantinides (1975) stimulated interest in okra as a seed crop, yielding protein and oil, which had previously been unrecognized. Crossley and Hilditch (1951), Martin and Ruberte (1979), Mangual-Crespo and Martin (1980) and Anwar et al. (2010) have written on the importance of okra as both an oil and protein crop.

The economic and nutritive importance of the oil, proteins and amino acids of dehulled okra seeds have been highlighted by Camciuc et al. (1998), Savello et al. (1980) and Oyelade et al. (2002). Okra seeds contain gossypol or gossypol-like phenolic compound, which has been used effectively as an anti-male sterilant in China and the USA (Karakoltsides and Constantinides, 1975; Martin et al., 1979; Martin and Ruberte, 1981; Martin, 1982; Martin and Rhodes, 1983; Fogg, 1984). Consequent, upon the enormous economic potentials of okra seed, especially for the poverty stricken sub-Saharan African region, serious attention to germplasm documentation, evaluation, characterization and development of effective production protocols for large scale production is considered as an urgent research imperative.

Techniques for production of okra as a vegetable are so different from those for the production of seeds (Mangual-Crespo and Martin, 1980). "Topping" or decapitation is an ethno-agricultural practice, often carried out by local farmers to increase the yield potentials of their crops. Incidentally, hardly any reports exist about scientific investigations into the potential contributions or otherwise of such a practice to dry fruit and seed production in okra in the region. Udeogalanya and Muoneke (1985) as well as Udengwu (2009) reported on the effects of this ethno-agricultural practice on the production of marketable fresh fruits in okra.

This present study reports on the comparatively evaluation of the dry fruit and seed yield potentials of 12 cultivars of the two okra species, adopting the decapitation or topping ethno-agricultural practice, over a two-year period, in South Eastern Nigeria; with a view to determining its suitability for large scale okra seed production, which is necessary for the exploitation of the neglected economic potentials of the crop.

MATERIALS AND METHODS

Field studies on dry fruit and seed yield of twelve cultivars of early and late okra

Germplasm used for the studies

Seeds of the 12 early and late okra cultivars (Figures 12 and 13), which form part of the germplasm retrieved mostly from local farmers were used for the studies. The early okra cultivars were-

Awgu early, "Ogba mkpi", "Ogbu oge"., Iloka, Nnobi fat and Lady finger(exotic okra) while the late okra cultivars included, "Ogolo", "Oru ufie", "Alanwaghoho", "Ojo ogwu", "Ebi ogwu" and "Tongolo".

Land preparation

Tractor ploughing as well as the first harrowing of the soil took place at the onset of the early rains, in March/April of the first year, 2002 after earlier preliminary studies in the late nineties. The second harrowing was carried out at the onset of the steady rains. Well-cured poultry dropping was worked into the soil at the rate of 13.5 t ha^{-1}. Due to the acidity of the soil, in the experimental area, lime was broadcast at the rate of 20 t ha^{-1}. After two days of application, the lime was worked into the soil with the third harrowing.

Design and general layout of experiment

A split plot, in randomized complete block (RCBD) was the experimental design used for the two years studies. The total experimental units were 12 with 4 treatments each and three replications. The following treatments were applied- early non-harvest (ENH), early nipped (decapitated), non harvest (ENiNH), late non harvest (LNH) and late nipped (decapitated) non harvest (LNiNH). Fisher's table of random numbers was used for the assignment of each of the experimental units. The cultivars were likewise randomly assigned within each sub plot.

Field planting

Pre-germination of seeds

Only pre-germinated seeds were used for the plantings for the two years, as a measure to overcome the problem of uneven seed germination, caused by possession of hard and impenetrable seed coats by some seeds, especially the late okra cultivars. The seeds with thin seed coats were soaked in tap water for two days with the water being changed every day. At the end of the second day, the radicle of all viable seeds emerged. The seeds of the cultivars with hard seed coats were scarified either by treatment with concentrated H_2SO_4 for 15 min and then washed several times in running tap water to wash off all the acids or cutting off a small portion of the seed coat (on the broader end of the seeds only) using the conventional nail cutter. In each case, the seeds were later soaked in water for the same two days for the radicles of the viable seeds to emerge.

Pre-planting application of Furadan

Consequent upon the high incidence of insect pests and nematode in the area, l.5 g of Furadan, a combined nematicide and insecticide, was introduced into each of the planting holes at the depth of about 2.5 cm. The pre-germinated seeds were introduced into the planting holes after covering the applied Furadan with 1 cm thick layer of soil.

Planting procedure

Three pre-germinated seeds were introduced into each prepared planting hole and covered with soil. Thinning down to one plant per stand took place two weeks after planting. There were six plants per row. The within and between row spacing were each 30 cm. In each experimental plot or unit, there were 36 plants belonging to 6 different cultivars. For the entire experimental area, there were 432 plants excluding discard border plants.

Table 1. ANOVA table for dry weight (ENH, ENiNH, LNH and LNiNH treatments) for 2 years.

Parameter	df	MS	VR
Total	23	840.020	
Block	2	230.430	2.077[NS]
Year (Y)	1	16986.889	153.170***
Treatment (T)	3	16.095	-
Y x T	3	90.410	-
Error	14	110.940	

Significant levels used. ***$P<.001$; **$P<.01$; *$P<.05$; NS, Not significant. LNiH, Late nipped (decapitated) harvest; ENH, early non harvest; ENiNH, early nipped (decapitated) non harvest; LNH, late non harvest; LNiNH, late nipped (decapitated) non harvest.

Decapitation of terminal buds to induce branching in cultivars of early and late okra

Removal of apical bud or nipping was carried out, using a pair of secateurs, to allow for at least two weeks of vegetative growth for all the cultivars prior to the initiation of flower buds. Preliminary studies had shown that this was a good stage to decapitate leading to the formation of not too tall branches that can bear fruits without staking or support, which will increase labour in-put.

Nipping for late okra cultivars was carried out in mid August, because of their critical day length (CDL) of 12¼ h. Flower initiation in late okra usually took place early in September when natural day length usually shortened considerably to 12¼ h or less.

Harvesting of dry fruits

Harvesting of dry fruits for the four treatments was done using a pair of secateurs once the fruits turned yellowish brown in colour and exhibited the characteristic longitudinal splits along the ridges of the fruits.

For standardization of dry weight, fruits harvested from the field were dried in a Gallenkamp oven at a temperature of 35°C for 24 h before the dry weights were recorded on per plant basis. The seeds were later shelled and seed yield estimated as 50% of the dry fruit weight, following the method of Akoroda (1986). Dry seeds were stored in the deep freezer or dessicator using anhydrous Calcium chloride pellets for other studies.

Meteorological data collection

The following meteorological data were collected from the University of Nigeria Meteorological Centre for the two years of studies: Mean monthly relative humidity, mean monthly air temperature, mean monthly soil temperature and monthly rainfall (Figures 8 to 11).

Field studies repeated over time only

Field studies, as were earlier described, were repeated on another location within the same experimental farm during the planting season of the following year. All the protocols followed in the previous year were precisely repeated for this second planting.

The data for the two years planting were combined and analyzed employing the analysis of variance (ANOVA) statistical method and bar charts with standard errors to compare the fruit dry weight and seed yield among the cultivars of each okra species and between cultivars of the two species for the two years of study.

RESULTS

Analysis of yield of dry weight of fruits produced by 12 cultivars of both early and late okra cultivars for 2 years for 4 treatments; ENH, ENiNH, LNH, and LNiNH

From the ANOVA in Table 1, the difference between years was highly significant (P < 0.001). The treatment effect was not significant while the interaction between year (Y) and treatment (T) was also not significant. Additionally, Figure 1 shows that the yield for the first year was very much higher than the second year for all the treatments, with the cultivars differing in their responses. For some cultivars, decapitation resulted to production of more fruits than the non- decapitated ones. In some other cases, reverses were observed, these differences were however not significant.

Analysis of dry weight produced by 6 early okra cultivars for 2 treatments, ENH and ENiNH, for the 2 years of study

The ANOVA (Table 2) showed that year was very highly significant (P < 0.001) while treatment was not significant. Cultivar was highly significant (P < 0.01), while the interaction between year (Y) and cultivar (C) was very highly significant, the interaction between year and treatment was not significant. The interaction between cultivar and treatment (C × T) was very highly significant (P< 0.001), while the interaction (Y × C × T) was significant. Figure 2 indicates that yield for the first year was higher than the second year, for both treatments for all the cultivars except Lady finger, the only exotic okra. It showed little variation in performance for the non-harvest and nipped non-harvest treatments for the two years. Non significant differences were observed in the performance of the non-harvest and nipped non-harvest treatments for the two years for other cultivars. For yield measured as total number of fruits (Figure 3), the number of fruits produced for the first year, for both ENH and ENiNH treatments were significantly much higher than the second year, for Nnobi fat, "Ogba mkpi", Awgu early and "Ogbu oge". Decapitation resulted to higher number of fruit production by Nnobi fat, "Iloka" and "Ogbu Oge". Lady finger, the exotic okra, showed slight variation.

Analysis of dry weight produced by 6 late okra cultivars for 2 treatments, LNH and LNiNH, for the 2 years of study

The ANOVA shown in Table 3 indicate that year was very highly significant (P < 0.001), while treatment and cultivar were not significant. The interaction between year and cultivar (Y × C) was very highly significant (P<0.001), while the interaction between year and treatment (Y × T); cultivar and treatment (C × T) as well as between year, cultivar and treatment (Y × C × T) were all none significant.

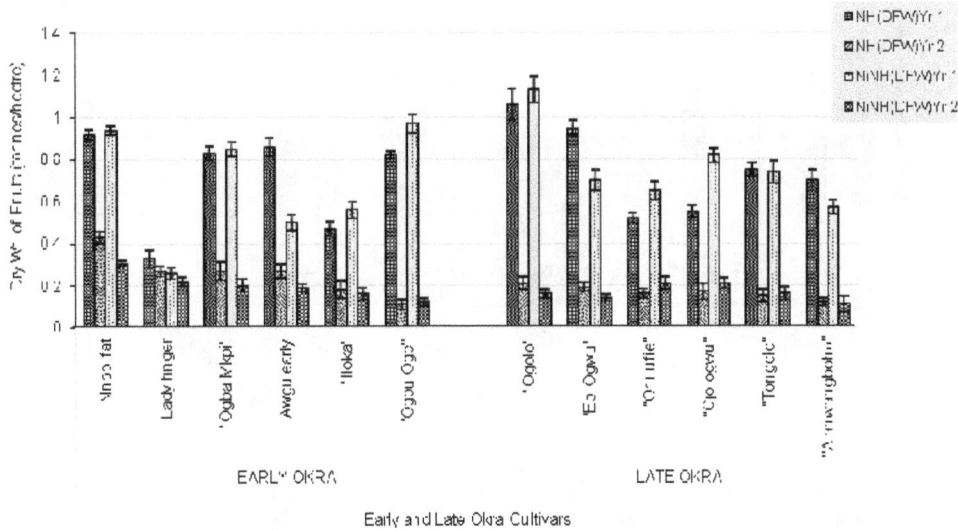

Figure 1. Yield of dry fruits (tonnes per hectare) by 12 cultivars of early and late okra cultivars over 2 planting seasons for 2 treatments (Non harvest and non nipped harvest).

Table 2. ANOVA based on fruit dry weight of 6 cultivars of early okra for 2 treatments (ENH and ENiNH) for 2 Years (Log$_{10}$ data transformation).

Source	df	SS	MS	F cal.	5%	1%	0.5%
Total	71	6.92	0.097				
Block	2	0.018	0.009	-	3.23	5.18	6.07
Year (Y)	1	3.915	3.915	391.5***	4.08	7.31	8.83
Treatment (T)	1	0.064	0.064	3.368 NS	4.08	7.31	8.83
Cultivar (C)	5	0.361	0.072	3.789**	2.45	3.51	3.99
Y x C	5	0.632	0.126	6.631***	2.45	3.51	3.99
Y x T	1	0.012	0.012	-	4.08	7.31	8.83
C x T	5	0.732	0.146	7.684***	2.45	3.51	3.99
Y x C x T	5	0.300	0.060	3.157*	2.45	3.51	3.99
ERROR	46	0.886	0.019				

Significant levels used. ***P<.001; **P<.01; *P<.05; NS, not significant.

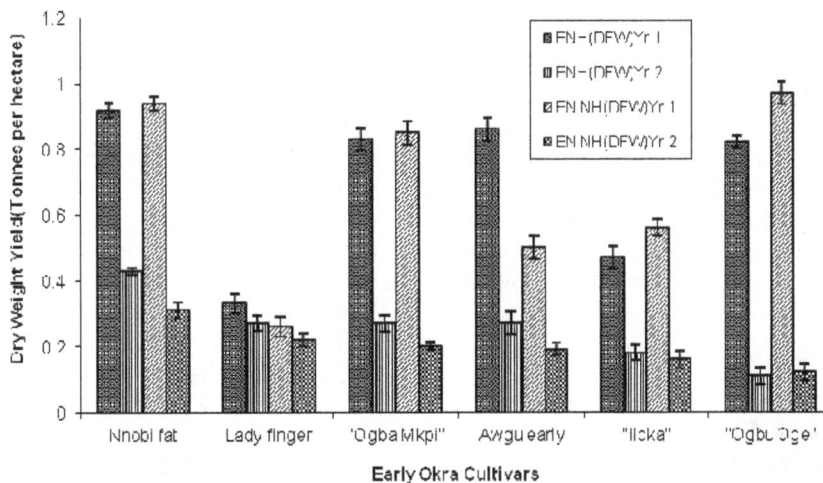

Figure 2. Yield of dry fruits (tonnes per hectare) by 6 cultivars of early okra for 2 treatments (ENH and ENINH) for 2 cropping seasons.

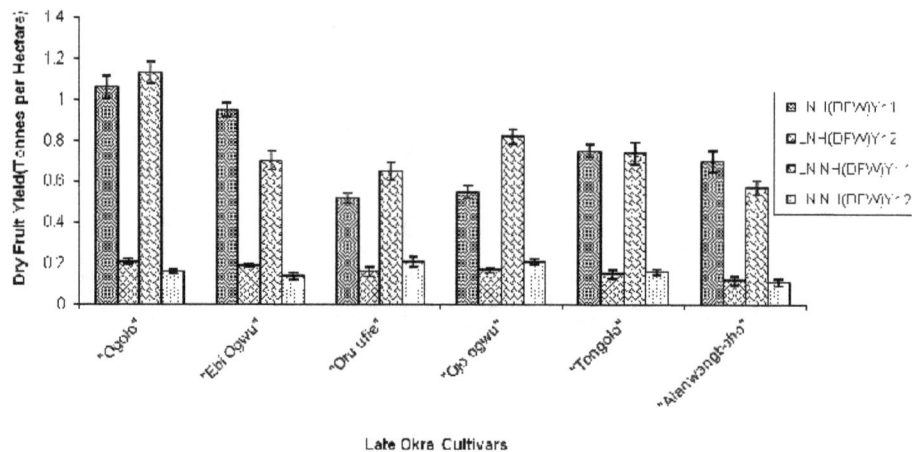

Figure 3. Yield of dry fruits (tonnes per hectare) of 6 late okra cultivars for 2 treatments (LNH and LNINH) for 2 cropping seasons.

Table 3. ANOVA based on fruit dry weight of 6 cultivars of late okra for 2 treatments (LNH and LNiNH) for 2 Years (Log_{10} data transformation).

Source	df	SS	MS	F cal	5%	1%	0.5%
Total	71	8.917	0.125				
Block	2	0 .103	0.051	4.25*	3.23	5.18	6.07
Year (Y)	1	7.696	7.696	641.33***	4.08	7.31	8.83
Treatment(T)	1	0.000094	0.000094	-	2.45	3.51	3.99
Cultivar (C)	5	0.090	0.018	1.50 N.S	4.08	7.31	8.83
Y × C	5	0.310	0.062	5.166***	2.45	3.51	3.99
Y × T	1	0.00168	0.016	-	4.08	7.31	8.83
C × T	5	0.082	0.012	1.33 N.S	2.45	3.51	3.99
Y × C × T	5	0.064	0.012	1.00 N.S	2.45	3.51	3.99

Significant levels used. ***$P<.001$; **$P<.01$; *$P<.05$; NS, not significant.

From Figure 4, the yield of dry fruit for the first year was significantly much higher than the second year. In the first year, while decapitation resulted to more fruit yield in Ogolo, Oru Ufie, Ojo ogwu, the reverse was the case for Ebi ogwu and Alanwangboho. Tongolo showed no difference in yield for the two treatments for the two years. For the decapitated and non-decapitated plants for all the cultivars for the second year planting, there were no significant differences in yield. Figure 5 shows that fruit production (based on total number of fruits) for the first year was much higher than the second year. Decapitation resulted to production of more fruits for both treatments for the two years. This was significant, very much higher for the first year when compared with the second year.

Analysis of yield of dry seed produced by 12 cultivars of both early and late okra cultivars for 2 years for 4 treatments; ENH, ENiNH, LNH, and LNiNH

The ANOVA on Table 4 showed that year was very highly significant, while treatment and the interaction (Y × T)

were all non-significant. Seed yield for ENH treatment for the first year ranged from 170 to 430 kg ha^{-1}, while for ENiNH, it ranged from 180 to 490 kg ha^{-1}. For the second year planting, seed yield for ENH ranged from 60 to 230 kg ha^{-1}, while for ENiNH, it ranged from 60 to 160 kg ha^{-1}. For LNH, seed yield for the first year ranged from 260 to 530 kg ha^{-1} while for LNiNH, it ranged from 290 to 570 kg ha^{-1}. For the second year planting, it ranged from 60 to 110 kg ha^{-1} for both LNH and LNiNH treatments (Figures 6 and 7). For the ENH and ENiNH treatments, seed yield was much higher for the first year than the second year (Figure 6). This was however not the case with the exotic okra, Lady finger, which showed slight higher increase in seed production for the first year than the second year for both treatments. The lowest seed yield for ENH was 170 kg ha^{-1} for first year while it was 60 kg ha^{-1} for the second year. The highest seed yield for ENH for the first year stood at 430 kg ha^{-1} while it was 160 kg ha^{-1} for the second year (Figure 6).Figure 7 shows that seed yield for the two treatments, for the first year, for all the cultivars, was significantly higher than that of the second year. The lowest seed yield for LNH and LNiNH for the first year stood at 260 and 290 kg ha^{-1}, respectively, while the

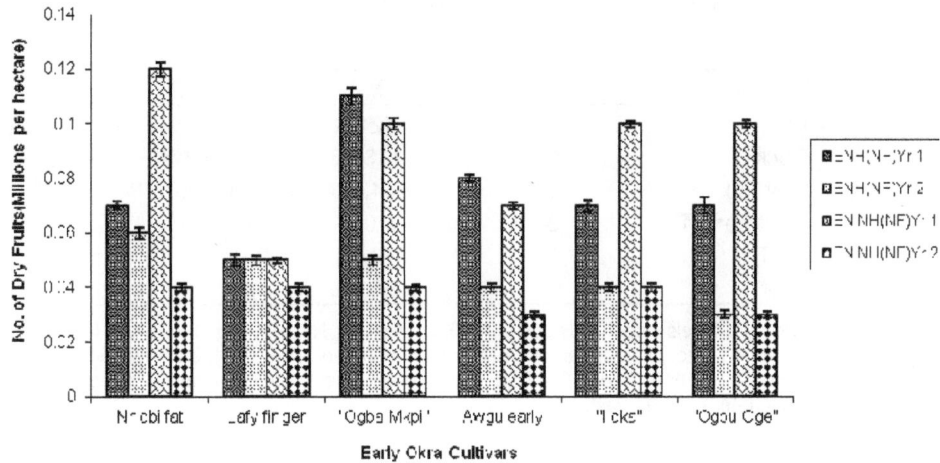

Figure 4. Yield of dry fruits as total no. of fruits (millions per hectare) of 6 early okra cultivars for 2 m treatments (ENH and ENINH) for 2 cropping seasons.

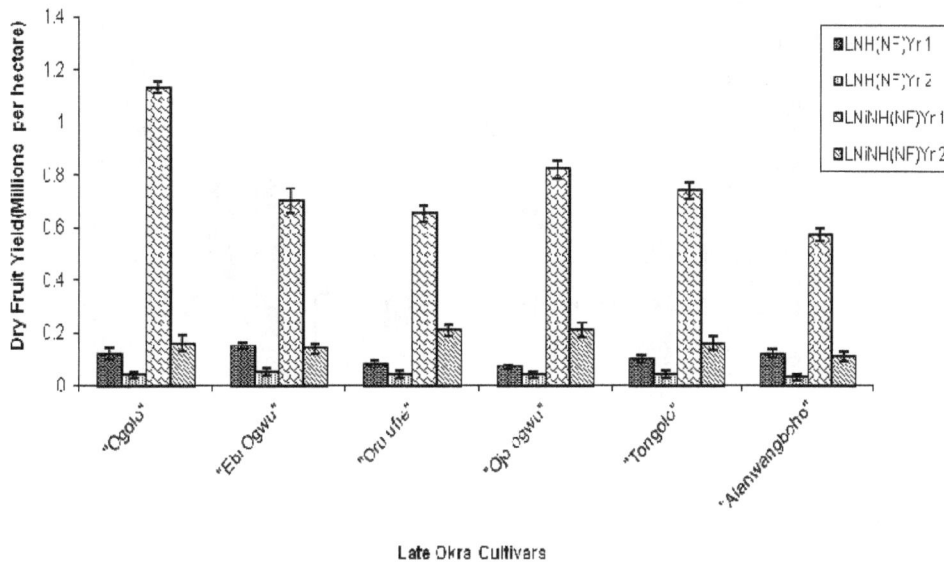

Figure 5. Dry fruit yield (millions of fruit per hectare) of 6 late okra ciltivars for 2 treatments (LNH and LNINH) for 2 planting seasons.

highest yield for both treatments was 530 and 570 kg ha[-1], respectively   (Figure 7). Differences in seed yield between the decapitated and the non-decapitated plants were not significant, for each of the years.

DISCUSSION

Dry fruit yield

The non significance of the four treatments ENH, ENiNH, LNH and LNiNH for the two years of planting can be ascribed to the limited number of flowers buds that set fruit and attain maturity when okra fruits are grown for seed production as against the numerous fruits produced when okra is grown for fresh fruits, with regular fruit very highly significant variance ratio for year which is an

indication of the very sensitive nature of okra to changes in mean monthly soil temperature, mean monthly air temperature, mean monthly relative humidity and mean monthly rainfall, abiotic environmental conditions that were monitored for the two years. The most dramatic fluctuation was in the annual rainfall. While the distribution for the first year started in March (75 cm) and showed an almost normal distribution that of the second year started in April (5 cm) and showed wide fluctuations. The first year rainfall rose steadily and eventually peaked in July (352 cm), which is normally the peak of rainy season in the South Eastern part of Nigeria, before ending in December.The second year, on the other hand, showed a decline in July (146 cm) and a much lower and unusual peak (250 cm) in September, with a sharp drop (80 cm) during the August break, much lower than that of

Table 4. ANOVA table for dry seed weight for (ENH, ENiNH, LNH and LNiNH) treatments for 2 Years.

Parameter		MS	VR
Total	23	421.5	
Block	2	116.25	2.077NS
Year (Y)	1	8443.44	151.587***
Treatment (T)	3	8.067	-
Y x T	3	46.35	-
Error	14	55.47	

Significant levels used. ***P<.001; **P<.01; *P<.05; NS, not Significant LNiH, late nipped (decapitated) harvest; ENH, early non harvest; ENiNH, early nipped (decapitated) non harvest; LNH, late non harvest; LNiNH, late nipped (decapitated) non harvest.

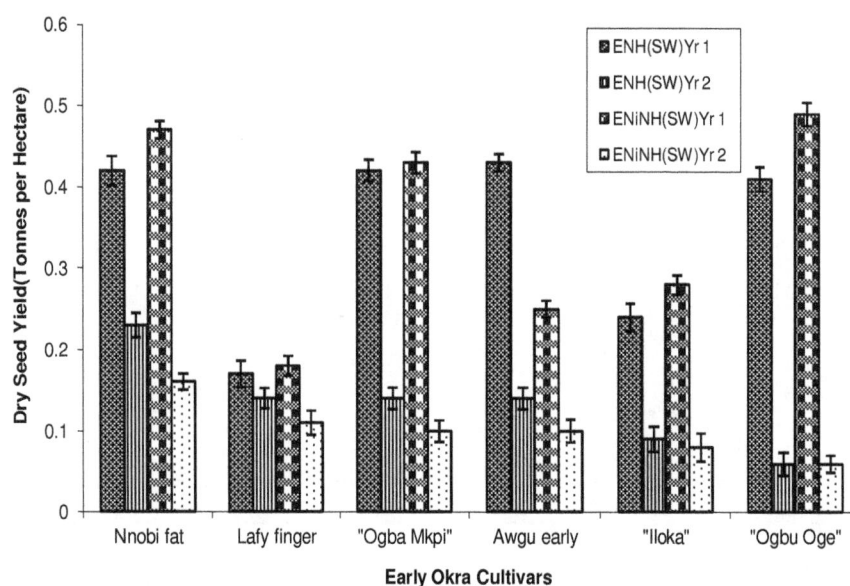

Figure 6. Dry seed yield (tonnes per hectare) by 6 early okra cultivars for 2 treatments for 2 cropping seasons.

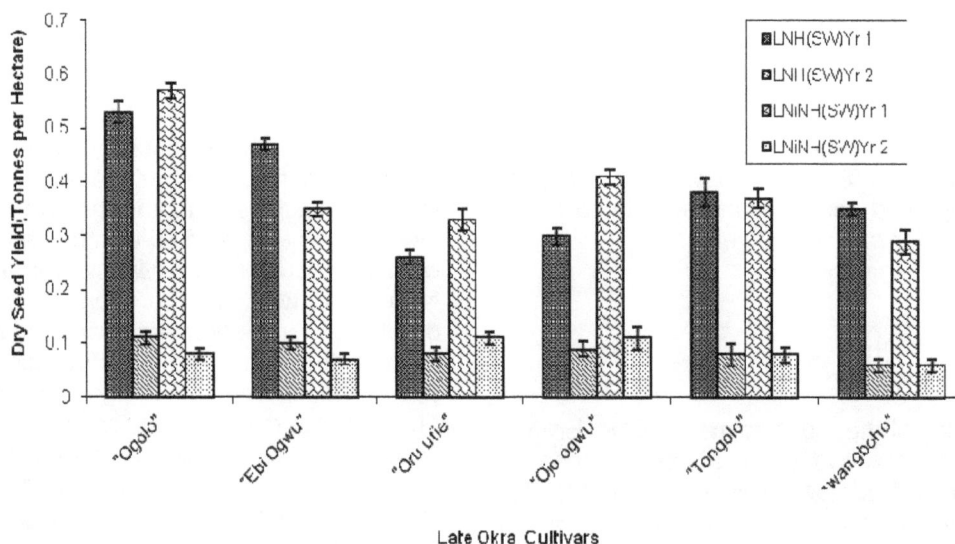

Figure 7. Yield of dry seed (tonnes per hectare) for 2 treatments (LNH and LNINH) for 2 cropping seasons.

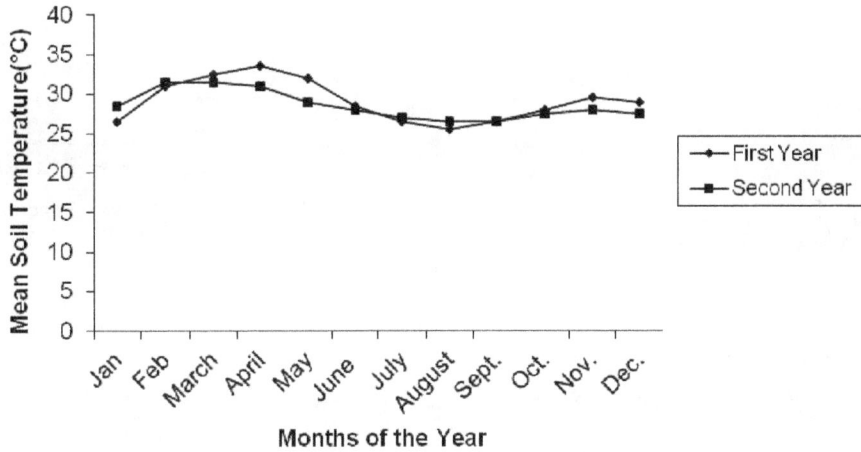

Figure 8. Mean monthly soil temperature at Nsukka for the two different years.

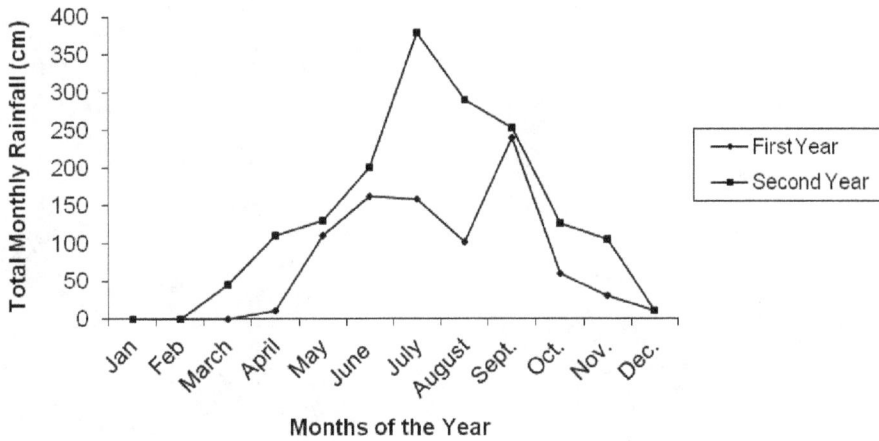

Figure 9. Total monthly rainfall at Nsukka for the 2 different years.

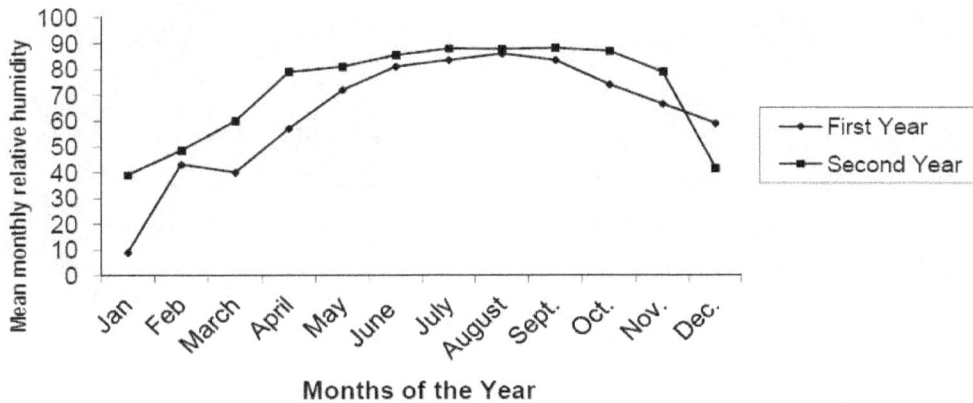

Figure 10. Mean monthly relative humidity at Nsukka for the 2 different years.

harvesting (Akoroda, 1986; Udengwu, 2009). There are the previous year (255 cm). These observations are similar to the findings of Asif and Greig (1972), Adelana (1985) and Zimmerman (2006). Of special interest is the yield pattern of Lady finger, an introduced exotic *Abelmoschus esculentus*, cultivar popularly grown in the region. For the harvest and non harvest treatments, as well as the decapitated and nondecapitated treatments, there were no observable differences in the yield of seed fruits by this cultivar for the two years of study. Most

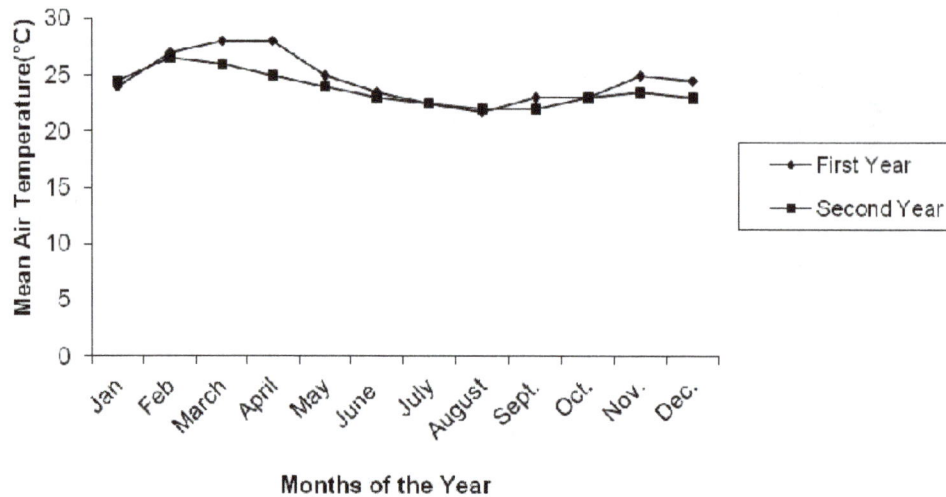

Figure 11. Mean monthly air temperature at Nsukka for the 2 different years.

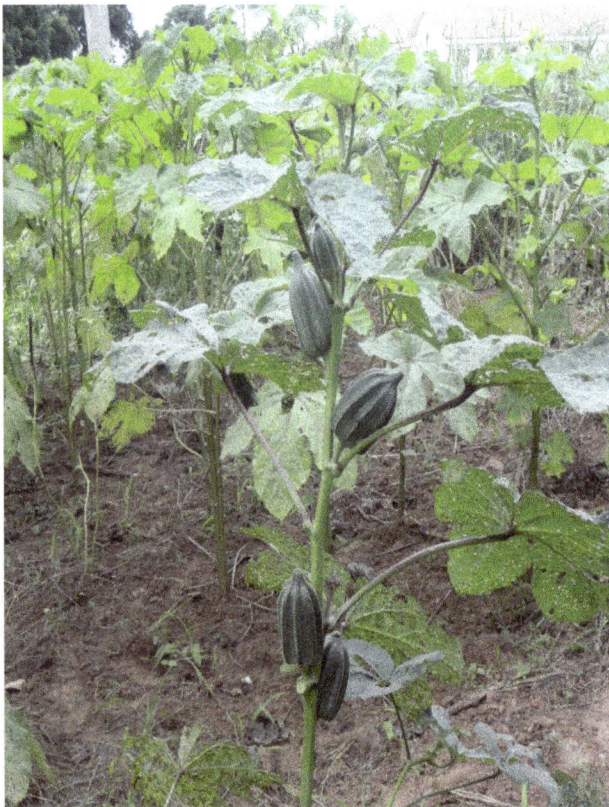

Figure 12. A typical early okra, *Abelmoschus esculentus* (L.) Moench cultivar.

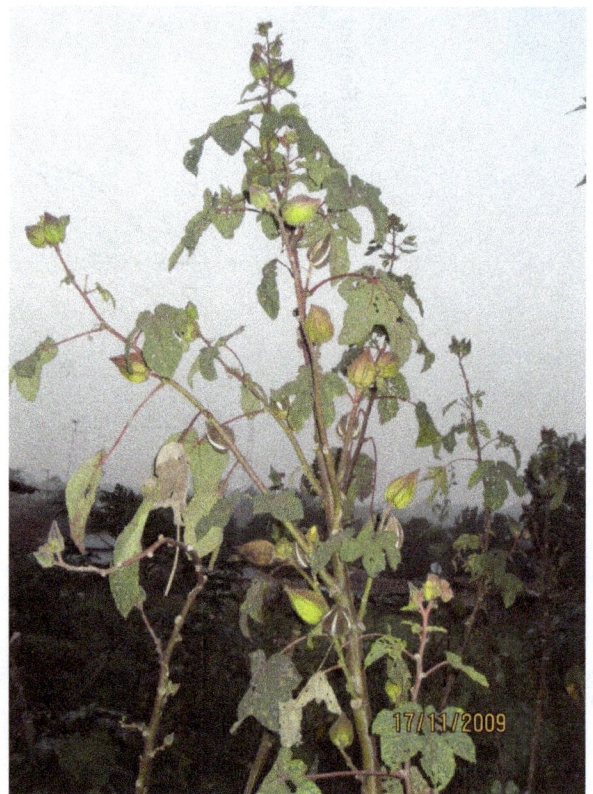

Figure 13. A typical late okra, *Abelmoschus callei*, cultivar.

probably, this cultivar believed to be an introduction from America might have been bred to withstand diverse environmental conditions. Further simulated experiments over different environmental conditions will be carried out to determine whether it can serve as a possible source of the much needed genes for the improvement of local okra cultivars for resistance to changing environmental conditions.

Martin (1982) reported that okra is adapted to climates with relatively short rainy season, hence its special acceptance in north-east Brazil where it is considered a crop that never fails. This however differs from the situation in South Eastern Nigeria, where under natural conditions normal duration of rainfall which ranges from March/April to October/November, is ideal for okra growth under the tropical rain forest condition. It is suspected that late okra evolved under this environmental condition.

Okra ecotypes therefore does exist and it is necessary for different areas to carry out *in situ* research to develop okra cultivars that will suite their conditions. Uncontrolled introduction of okra germplasm into any area may result to pollution rather than improvement of the local germplasm.

The highly significant difference between the *A. esculentus* cultivars as well as the very highly significant difference between the interactions (Y × C) and (C × T) as well as the significant difference between (Y × C × T) are indicators of a high degree of variation among the cultivars used which resulted to different responses to the changes in the environmental conditions over this study years as well as the decapitation and non decapitation treatments. Fatokun et al. (1979) and Akoroda (1986) had hinted on the existence of a wide range genetic variability in okra germplasm in Nigeria. Such variability was found to be higher and significantly different from those found in exotic okra which provides further evidence to support Harlan (1971) submission that West Africa is a centre of diversity for okra.

The response of late okra cultivars differed from that of the early okra cultivars. The reasons for the very highly significant variance ratio for year for late okra are similar to what has been discussed under early okra. The non significant variance ratio of cultivar is indicative of the existence of less variation among the late okra cultivars used in the studies. This is a curious development because it appears that the higher chromosome number of 197 for late okra as against the 130 for early okra (Singh and Bhatnagar, 1976), does not necessarily result to higher degree of variation and diversification of late okra. Further studies are needed to understand why such a phenomenon could exist among other peculiarities of late okra which is restricted in distribution to the West African region. However, it has been introduced to many okra growing regions of the world as a source of genes for resistance against okra viral diseases (Siemonsma and Hamon, 2004). Udengwu (2009) reported that the duration for the production of fresh marketable fruit by early okra cultivars ranged from 7 to 12 weeks while that of late okra cultivars ranged from 3 to 8 weeks under natural planting conditions in South Eastern Nigeria. This marked difference in duration of fresh fruit production was linked to the greater sensitivity of the late okra cultivars to changes in natural photoperiod due to their 12¼ h Critical Day Length. Invariably, late okra spends most of its time on vegetative growth and has very short time for reproductive growth. It is suspected that the genes responsible for photoperiodic control in late okra may be epistatic in nature, compelling the cultivars to get into the reproductive phase once the CDL of 12¼ h is reached.

Dry seed yield

The non significant response of both early and late okra cultivars to the decapitation of non harvest treatments over the two years, as well as the interaction between year and treatment (Y × T) can be attributed to the limited number of seed fruits the plants could produce because of serious competition for assimilates by the retained old fruits on the plant (Akoroda, 1986; Udengwu, 2009).

Dry seed yield by early okra cultivars equally indicate slight variation in the performance of Lady finger for the two treatments over the two growing seasons, thus confirming the stability of the cultivar over varying abiotic environmental conditions. The exotic okra cultivar is therefore a potential source of genes for the improvement of the other early okra cultivars which are sensitive to changing environmental conditions. Other cultivars showed little difference between seed yield for ENH and ENiNH treatments suggesting again that the treatments may not be quite suitable for early okra cultivars. The cultural practice of nipping or decapitation should therefore be limited to the production of fresh marketable fruits (Udengwu, 2009). Similar views were expressed by Kolhe and Chavan (1965).

The response of late okra cultivars is equally similar to that of early okra with narrow differences between the LNH and LNiNH treatments for the two years. The determinate nature of okra growth coupled with the photoperiodic sensitivity of late okra cultivars meant that once the CDL is reached, irrespective of treatment, fruit production will commence and once the optimum number of fruits is produced; whether on the main axis or on the induced branches, no more fruits will be produced because all subsequent fruits will abort.

From the results of seed yield, for the early okra cultivars, three groups of seed yielders stood out based on the first year planting which was more promising and the focus of this discussion. Awgu early, Nnobi Long, Ogba mkpi and Ogbu Oge belong to the first group of fairly good yielders (FGY), with production up to 400 kg ha^{-1} while Iloka belongs to the group of the fairly medium yielders (FMY) with up to 200 kg ha^{-1}. Lady finger belongs to the poor yielder (PYD) group with up to 170 kg ha^{-1}. The same trend was repeated for the ENiNH treatment with the exception of Awgu early which yielded 250 kg ha^{-1}.

For the LNH treatment, three categories of seed yielders equally emerged. Ogolo and Ebi Ogwu with 530 and 470 kg ha^{-1}, respectively, are viewed as the FGH group, while Tongolo (380 kg ha^{-1}), Alanwangboho (350 kg ha^{-1}) and Ojogwu (300 kg ha^{-1}) belong to the FMY group. Oru ufie (260 kg ha^{-1}) is classified as a PYD. For the LNiNH treatment, Ogolo (570 kg ha^{-1}) and Ojo ogwu(410 kg ha^{-1}) are classified as FGY, while Ebi ogwu (350 kg ha^{-1}), Tongolo (370 kg ha^{-1}) and Oru ufie (330 kg ha^{-1}) are considered the FMY group. The cultivar, Alanwagboho (290 kg ha^{-1}) is considered a PYD. In West Africa, information on okra seed yield is limited because of the long negligence of the crop as a seed crop at the expensive of its production as a fruit crop. Mangual-Crespo and Martin (1980) reported that seed yield in eight *A. esculentus* cultivars ranged between 422 to 1032 kg ha^{-1}., depending on plant density.

Their average seed yield at 30.5 cm spacing was 558

kg ha^{-1}, compared to 713 kg ha^{-1} for the 23 cm spacing. In this present report at a planting distance of 30 cm, the highest seed production by the FGY for early okra was 400 kg ha^{-1} and for late okra, 530 kg ha^{-1} for the ENH and LNH treatments. These figures are comparable to the lowest seed yielders (422 kg ha^{-1}) reported by Mangual Crespo and Martin (1980).

Though late okra cultivars showed greater promise for seed production in the region, seed yield as reported in this present study is low when compared with yield from other reports in advanced okra production areas, selections could still be made to identify cultivars that will be involved in further breeding studies for enhanced okra seed production in the region. In their own report, Siemonsma and Hamon (2004) reported that seed yield in okra in West Africa was in the range of 500 to 1000 kg ha^{-1}.

This equally indicates that the yield of the cultivars used in this present study were low and calls for an urgent need to search for better seed producers in the region for broader selection of elite high yielding cultivars. The better performance of the late okra cultivars (which are restricted to the West African region) over the early ones is indicative of the fact that greater attention should be given to this okra group for commercial seed production in the region. This view is supported by the Siemonsma and Hamon (2004), who noted that selection and breeding of West African okra have not been carried out by the commercial sector, but African farmers have selected an enormous diversity of forms which fit into a great variety of cropping systems.

The stable performance of Lady finger (an introduced exotic early okra cultivar) over the recorded changing environmental conditions during the two years of study suggests that it could be a potential source of environmental stable gene(s) that could be used to stabilize the environmentally unstable cultivars. Further simulated changing environmental studies will however be needed to confirm this. The following four late okra cultivars have been ear marked for further seed yield improvement studies, Ogolo, Ebi Ogwu, Tongolo, and Ojogwu. The elite late okra cultivar, "Ogolo" outperformed all others cultivars by the production of the highest number of fruits due to decapitation. Subsequent studies will show the best age and height of decapitation for higher fruit production by the cultivar. For early okra also four cultivars, Awgu early, Nnobi Long, Ogba mkpi and Ogbu Oge, have been selected for further seed yield improvement programme while further search for better performers continues.

ACKNOWLEDGEMENTS

Special thanks go to Prof. (Dr.) O.C. Nwankiti for his invaluable help during the initial course of this study. The author also sincerely thanks Prof. E.E. Ene- Obong for his help with the experimental design and statistical analysis. To the staff of the Botanic Garden, Field staff of Crop Science Department, as well as staff of the Meteorological Unit, all of the University of Nigeria, Nsukka, the author say thanks for your numerous assistance during the course of this study.

REFERENCES

Adelana BO (1985). Effects of NPK fertilizer on the yield of okra in South-West Nigeria. Samaru J. Agric. Res. 3:67-72.

Adeniji OT, Kehinde OB, Ajala MO, Adebisi MA (2007). Genetic studies on seed yield of West African okra [Abelmoschus caillei (A. Chev.) Stevels]. J. Trop. Agric. 45(1-2):36–41.

Akoroda MO (1986). Relationships of plantable okra seed and edible fruit production. J. Hort. Sci. 61:233-238.

Anwar F, Rashid U, Ashraf M, Nadeem M (2010). Okra (Hibiscus esculentus) seed oil for biodiesel production. Appl. Energy 87(3):779-785

Anwar F, Rashid U, Mahmood Z, Iqbal T, Sherazi HT (2011). Inter-varietal variation in the composition of Okra (Hibiscus esculentus L.) Seed oil. Pak. J. Bot. 43(1):271-280, 2011.

Asif MI, Greig JK (1972). Effects of N,P and K fertilization on fruit yield, macro and micronutrient levels and nitrate accumulation in okra. (Abelmoschus esculentus (L) Moench) J. Am. Soc. Hort. Sci. 97:440-442.

Camciuc M, Deplagne M, Vilarem G, Gaset A (1998). Okra—Abelmoschus esculentus L. (Moench.) a crop with economic potential for set aside acreage in France. Ind. Crops Products 7(2-3):257-26.

Crossley A, Hilditch TP (1951). The fatty acids and glycerides of okra seed oil. J. Sci. Food Agric.2(6):251-255.

Dhankhara BS, Mishrab JP (2005).Objectives of Okra Breeding. J. New Seeds 6 (2-3):195-209.

Dong C, Liang S (2007). Function characteristic and comprehensive developing and utilization of okra. Available at www.cnki.com.cn. [DOI]: CNKI: ISSN: 1005-6521.0.2007-05-058.Accessed on June,13,2011.

Fatokun CA, Chheda HR, Oken'ova ME (1979). Potentials for the Genetic improvement of okra (Abelmoschus esculentus (L) Moench) In Nigeria. Nig. J. Genet. 3:43-52.

Fogg E (1984). Revival of the male pill. New Scientist 20/27 December 1984:1435

Hamon S, Hamon P (1991). Future prospects of the genetic integrity of two species of okra (Abelmoschus esculentus and A. caillei) cultivated in West Africa. Euphytica 58(2):101-111.

Harlan JB (1971) Agricultural origins: Centres and Non centers. Science 174:463-474.

Karakoltsidis PA, Constantinides SM (1975). Okra seeds: A new protein source. J. Agric. Food Chem. 23:1204-1207.

Kolhe AK, Chavan NM (1965). Development of fruit, yielding capacity and influence of fruit maturity on the reproductive and vegetative behaviour in okra Abelmoschus esculentus (L) Moench. Indian J. Agric. Sci. 37(3):155-166.

Mangual-Crespo G, Martin FW (1980). Effect of spacing on seed, protein and oil production of four okra varieties. J. Agric. Univ. Puerto Rico 64:450-459.

Martin FW (1982). Okra, potential multiple purpose crop for the temperate zones and tropics. Econ. Bot. 36:340-345.

Martin FW, Rhodes AM (1983). Seed characteristics of okra and related Abelmoschus species. Qual. Plant Foods Hum. Nutr. 33:41-49.

Martin FW, Rhodes AM, Ortiz M, Diaz F (1981). Variation in okra Euphytica 30:697-705.

Martin FW, Ruberte R (1978). Vegetables for the hot humid tropics. Part 2 Okra, Abelmoschus esculentus, Science and Education Administration, United States Department of Agric. (USDA) New Orleans. p. 22.

Martin FW, Ruberte R (1979). Milling and use of okra seed meal at the household level. J. Agric. Univ. Puerto Rico 63:1-7

Martin FW, Ruberte R (1981). Variability in okra seed quality. J. Agric. Univ. Puerto Rico. 65:205-212.

Martin FW, Telek L, Ruberte R, Santiago AG (1979). Protein, oil and Gossypol contents of a vegetable curd made from okra seeds. J. Food Sci. 44:1517-1519,1529.

Njoku E (1958). The photoperiodic response of some Nigerian plants. J.

W. Afr. Sci. Assoc. 4:99-111.

Nwoke FIO (1980). Effect of number of photoperiodic cycles on flowering and fruiting in early and late varieties of okra (*Abelmoschus esculentus* (L.) Moench). J. Exp. Bot. 31 (125):1657-1664.

Oyelade OJ, Ade-Omowaye BIO, Adeomi VF (2002). Influence of variety on protein, fat contents and some physical characteristics of okra seeds. J. Food Eng. 57(2):111-114.

Oyolu C (1977). Variability in photoperiodic response in okra (*Hibiscus esculentus*) Acta Hort. 53:207-214.

Savello AP, Martin FW, Hill JM (1980). Nutritional composition of okra seed meal. J. Agric. Food Chem. 28:1153-1166.

Siemonsma JE (1982). West African okra- morphological and cytogenetical indications for the existence of a natural amphidiploid of (*Abelmoschus esculentus* (L.) Moench) and *A. manihot* (L.) Medikus. Euphytica 31:241-252.

Siemonsma JS, Hamon S (2004). *Abelmoschus caillei* (A.Chev.) Stevels [Internet] Record from Protabase. Grubben, G.J.H. & Denton, O.A. (Editors). PROTA (Plant Resources of Tropical Africa / Ressources végétales de l'Afriquetropicale), Wageningen, Netherlands. < http://database.prota.org/search.htm>.

Singh HB, Bhatnagar A (1976). Chromosome number in an okra from Ghana. Indian J. Genet. Plant Breed. 36(1):26-27

Torkpo SK, Danquah EY, Offei SK, Blay ET (2006). Esterase, Total Protein and Seed Storage Protein Diversity in Okra (*Abelmoschus esculentus* L. Moench). West Afr. J. Appl. Ecol. 9 (1):1-7

Udengwu OS (1998). Photoperiodic response of Early and Late Okra types *Abelmoschus esculentus* and application to accelerated Gene Transfer. Nig. J. Bot.11:151-160.

Udengwu OS (2009). Studies on the effects of continuous harvesting and non harvesting; decapitation and non decapitation; on fruit yield of early and late okra cultivars in South Eastern Nigeria. African Crop Science Conference Proceedings Cape Town South Africa, 28th Sept-1st October. 9:107-116.

Udeogalanya ACC, Muoneke CO (1985). Effect of irrigation levels and pruning heights on the yield of okra (*Hibiscus esculentus* L.) in Southeastern Nigeria. Beitrage trop. Landwirtsch. Veterinarmed. 21:437-443.

Zimmerman M (2006). Environment. Microsoft ® Encarta ® 2006. © 1993-2005 Microsoft Corporation.

9

Emergence and initial development of Cape gooseberry (*Physalis peruviana*) seedlings with different substrate compositions

Andre Luiz Piva[1], Eder Junior Mezzalira[1], Anderson Santin[1], Daniel Sschwantes[1], Jeferson Klein[1], Leandro Rampim[1], Fabíola Villa[1], Claudio Yuji Tsutsumi[1] and Gilmar Antônio Nava[2]

[1]Universidade Estadual do Oeste do Paraná/ Centro de Ciências Agrárias/ Programa de Pós Graduação em Agronomia. Rua Pernambuco, 1777, CEP: 85960-000. Marechal Cândido Rondon, Paraná, Brazil.
[2]Universidade Tecnológica Federal do Paraná, Campus Dois Vizinhos, Brazil.

The seedlings quality is one of the most important factors in the implantation of new orchards. In this sense, this work aimed to evaluate the emergence and initial development of Cape gooseberry seedlings with different substrate compositions. The experiment was performed in the experimental area of Federal Technological University of Parana, Campus of Dois Vizinhos, Brazil. The experimental design used was a randomize block design, with four replications, being evaluated the following substrates: Soil + chicken manure (2:1 v/v); soil + chicken manure + sand (2:1:1); soil + humus + vermiculite (2:1:1); soil + humus (2:1) and Macplant® (commercial substrate). The emergence percentage, number of leaves, height of plants, dry mass of root system and dry mass of the shoot system was evaluated. The substrate compound by soil + humus + vermiculite and the substrate compound by soil + humus allowed higher emergence of plants, however not contributing for the development of the Cape gooseberry seedlings. Nevertheless, the substrate with soil + chicken manure and the substrate with soil + chicken manure + sand only provided the initial development of the seedlings. On the other hand, the use of the substrate Macplant® must be avoided in the propagation of this culture.

Key words: Exotic fruits, propagation, germination, Macplant®, vermiculite.

INTRODUCTION

The progressive increase of the demand for food is one of the greatest incentives for farmers to increase and diversify their production, in this context, the orcharding has great contribution in the Brazilian scenario, with the addiction of many exotic species (Muniz et al., 2011). The group of small fruits is causing great interest because of the high market value and the low cost of production (Mota, 2006; Rodrigues et al., 2009). In this case, the Cape gooseberry (*Physalis peruviana* L.) is one of the

more recent plants, occupying a considerable area in the production (Lima et al., 2010).

The Cape gooseberry is a plant of the solanaceae family, with indeterminate grow habit and it has caused great interest for production in the south, southwest and northwest states of Brazil (Kuhn et al., 2012; Velasquez et al., 2007; Andrade, 2008). This fruit offer great nutraceutical value, with great quantity of vitamin A and C, among others minerals. In different regions of

Colombia, the fruit is used in the popular medicine, for purification of blood and relief of throat problems (Chaves et al., 2005).

In relation to the cultivation, for the obtaining more homogeneous orchards the use of seedlings of quality is necessary, however, studies show that, the method of propagation more used for annual fruits like Cape gooseberry is by using seeds, even with high genetic variability (Lima et al., 2010). Díaz et al. (2010) evaluating the emergence of Cape gooseberry seedlings in different substrates, observed that, the coconut fiber is able to replace the peat in the emergence and development of Cape gooseberry.

In order to present seeds with a homogeneous germination and emergence, some factors must be optimized, like the relation between the substrate quantity, hydric availability, thermal proprieties and absence of physical obstacles for the emergence of some species (Ferraz et al., 2005; Castro et al., 2005; Wagner-Júnior et al., 2006).

When these factors are improved, the seeds have better conditions for the germination and emergence, and the seedlings better conditions for initial development, creating conditions for new and quality orchards. However, the choice of the substrate composition must be according to the species characteristics (Dias et al., 2008; Brasil, 2009).

Frequently farmers are limited by the use of one or few commercial substrates, normally of high cost and hard access, something very similar occurs in the Southwest region of Parana State, Brazil, where there is a weak technical and commercial support for the horticulture activities. However, it should be noted as favorable aspect to the production of fruit seedlings that in the Southwest region of Parana, it is possible to find many byproducts from the traditional agricultural and livestock activities, as production of grains or animals, which could be used in the formulation of alternatives substrates, meeting the existent requirements for an easy access and low cost product.

There is little technical information about the production of quality seedlings for Cape gooseberry. In this way, this work aimed to evaluate the emergence and initial development of *P. peruviana* L. seedlings with different substrate compositions.

MATERIALS AND METHODS

The experiment was conducted in the experimental station of orcharding at the Federal Technological University of Parana, Campus of Dois Vizinhos, Dois Vizinhos – Parana, from September to December 2009. The geographic coordinates were 42° 25' S of latitude and 53° 06' W of longitude, with an altitude of 520 m, in the eco-climatic region of Southwest Parana, Brazil, with an average temperature for the period of 21°C (Inmet, 2012).

The study was conducted in a screen house, and the randomized blocks design was used, with four replications. The plots were formed by 10 poly tubes with capacity of 150 cm^3, with five seeds in

each tube. The used seeds were collected from mature fruits (yellow/orange color) from *P. peruviana* L. plants.

The seeds were sowed in poly tubes with substrate, being evaluated five substrate compositions. The tested substrates were: Soil + chicken manure (2:1 v/v); soil + chicken manure + sand (2:1:1 v/v/v); soil + humus + vermiculite (2:1:1 v/v/v); soil + humus (2:1 v/v) and Macplant® (commercial substrate). The used soil was a Red Alfisol (USDA classification), which was sieved for the standardization of the particles size. Besides, the manure used was originated from a chicken farm, being also submitted to a sieve for standardization of the particles sizes. Moreover, the humus used was originated from vegetal residues (lettuce, cabbage, kale, etc) and vermiculite (Plantmax®) used in a medium granulometry.

These substrates were chosen because these are widely used in the production of seedlings of many different fruit and forestry species in Brazil (Severino et al., 2006). The commercial substrate Macplant® has the following characteristic: porosity of 0.20 cm^3, humidity of 0.77 cm^3 (saturated), density of 0.17 g cm^{-3}, 440 mg dm^{-3} of K, 74 mg dm^{-3} of P, percentage of organic matter (% OM) of 20.7 m v^{-1} (Pacheco and Franco, 2008).

After the sowing, the tray with the poly tubes was maintained in the screen house during the experiment time. The irrigation was performed on a daily basis according to the observed needs, being held with a Sprinkler System.

The seedlings emergence was monitored trough the counting of the emerged seedlings daily in each experimental unit. Besides, it was considered the beginning of emergence at 14 days after sowing (DAS) when the first seedlings emerged, and the final percentage was obtained in the 25 DAS, according to the norms of analysis of Brasil (2009).

After the end of the seedlings emergence (that is, 25 DAS), the thinning of the seedlings was performed, maintaining only one seedling per poly tube. This moment was considered the start point for the initial develop of the seedlings of Cape gooseberry.

The height of plants was performed using a graduated scale, being also performed the counting of completely expanded leaves. At 90 DAS, the last evaluation of height and number of leaves was performed.

Later, the seedlings were carefully removed from the poly tubes and cut in shoot system and root system. The roots were washed to remove the substrate particles, and the two parts of the plant were dried in an air circulation oven at 65°C for 96 h. At the end of this process, the dry mass from the shoot and root systems were obtained. The data were submitted to the variance analysis and the means compared by Tukey test at 5% of probability, with the statistical software SAS (SAS Institute, 1999).

RESULTS AND DISCUSSION

In the evaluation of the emergence data, it was verified a higher percentage of emergence of Cape gooseberry seedlings with the substrate vermiculite + soil + humus and with soil + humus, that did not differ statistically among each.

Inferior results were obtained with the commercial substrate Macplant®; soil + chicken manure and with soil + sand + chicken manure (Table 1). Possibly the high seedlings emergence obtained with the substrate soil + vermiculite + humus and with soil + humus can be explained by the excellent capacity of water storage performed by the vermiculite and the humus (Alvino and Rayol, 2007). Also the characteristics of the used soil can influence, because it is a clay soil with great water

Table 1. Emergence percentage of Cape gooseberry in response to different substrate compositions at Federal Technological University of Parana, Campus of Dois Vizinhos, 2009.

Substrate	Emergence (%)
Soil + chicken manure (2:1)	64.5[c]
Soil + chicken manure + sand (2:1:1)	70.5[bc]
Soil + humus + vermiculite (2:1:1)	94.5[a]
Soil + humus (2:1)	88.0[ab]
Macplant® (commercial substrate)	52.5[c]
Pr>Fc	0.0001
CV (%)	13.08

Means followed by the same letter do not differ statistically by the Tukey test at 5% of probability.

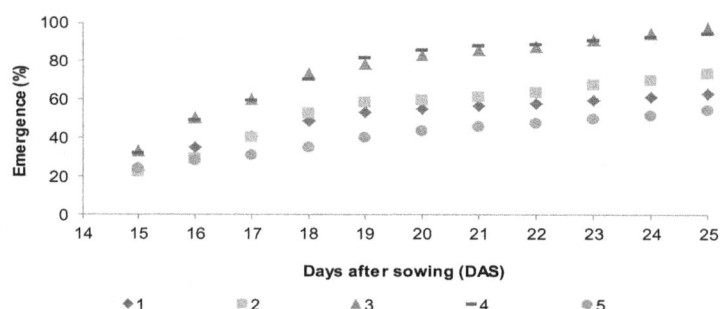

Figure 1. Emergence of Cape gooseberry seedlings in response to different substrate compositions, Federal Technological University of Parana, Campus of Dois Vizinhos, 2009. Substrate 1: Soil + chicken manure (2:1 v/v) 2: Soil + chicken manure + sand (2:1:1 v/v/v); 3: Soil + humus + vermiculite (2:1:1 v/v/v); 4: Soil + humus (2:1 v/v); 5: Macplant® (commercial substrate).

absorption. Besides, the vermiculite and the humus also have a good drainage of the excess of water. These factors possibly allowed a better availability of water and oxygenation for the roots, creating an excellent environment for the activation of enzymes responsible for the hydrolysis of reserve substances in the seeds and for the start of the germination process and emergence of the seedlings (Taiz and Zeiger, 2009).

The obtained data show that, the tested substrates presented higher values for seedlings emergence in relation to the results obtained by Kuhn et al. (2012) that observed an average of 30%. It is supposed that, the chicken manure had caused some compression in the superficial levels of the substrate in the presence of irrigation water, causing a reduction of penetration water in the substrate, occurring a reduction of the level of oxygenation in function of the crust. This fact may had contributed for the reduction of the germination process, providing a reduction on the emergence

Another hypothesis that needs to be analyzed is that, one increase of the substrate temperature near to the seed provided by the process fermentation of chicken

manure, may have affected the development of the germination process and emergence. According to what can be observed in the work of Andrade et al. (2006), that the increase of 5 degrees was enough for reducing the mean values of germination of *Dalbergia nigra*. On the other hand, the commercial substrate Macplant®, promoted the drainage of water through the contact with the poly tubes, standardizing its humidity.

Nascimento et al. (2003) observed that, little alterations in the chemical and physical composition or in the biological proprieties of the substrate can influence the germination of different vegetal species, as tomato and pepper, as well as their development. In the same way, the substrate can also cause physical impediment, through large particles, prejudicing the emergence of the seedlings, which have too small and delicate roots, causing poor vigor (Wagner-Junior et al., 2006).

In relation to the seedlings emergence, it was observed that, the beginning of the emergence started at 15 DAS, being this result similar to what was observed by Díaz et al. (2010). The behavior of all the evaluated substrate is observed in Figure 1. Alterations in the emergence

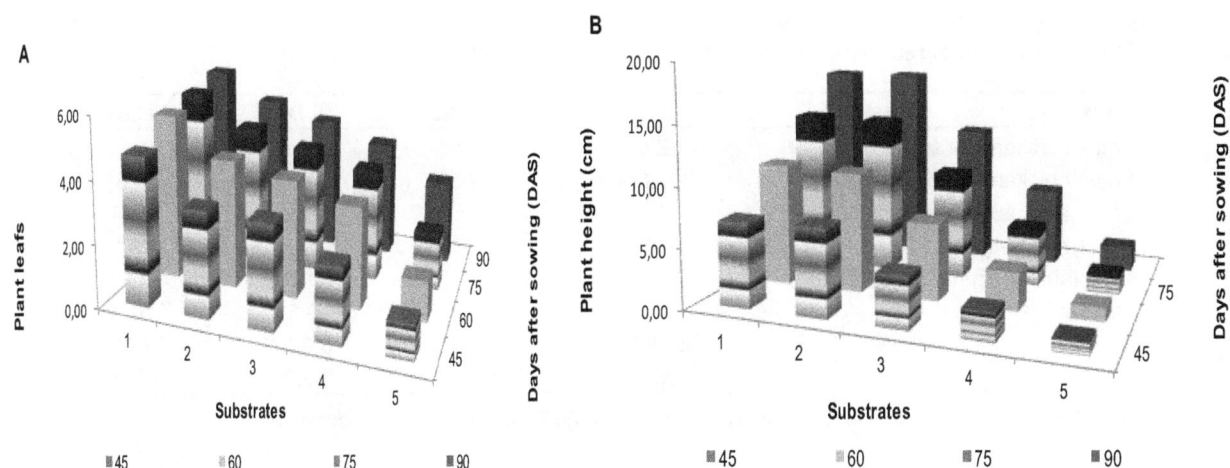

Figure 2. Leaves (A) plant height (B) of Cape gooseberry plants in response to different substrate compositions, Federal Technological University of Parana, Campus of Dois Vizinhos, 2009. Substrate 1: Soil + chicken manure (2:1 v/v) 2: Soil + chicken manure + sand (2:1:1 v/v/v); 3: Soil + humus + vermiculite (2:1:1 v/v/v); 4: Soil + humus (2:1 v/v); 5: Macplant® (commercial substrate).

Table 2. Evolution of the number of leaves per Cape gooseberry plant in response to different substrate combinations, Federal Technological University of Parana, Campus of Dois Vizinhos, 2009.

Substrate	45 DAS	60 DAS	75 DAS	90 DAS
Soil + chicken manure (2:1)	4.8^a	5.4^a	5.3^a	5.8^a
Soil + chicken manure + sand (2:1:1)	3.4^{ab}	4.1^{ab}	4.5^{ab}	4.9^{ab}
Soil + humus + vermiculite (2:1:1)	3.3^{ab}	3.7^b	4.0^b	4.4^{ab}
Soil + humus (2:1)	2.3^{bc}	3.2^b	3.6^b	3.8^{bc}
Macplant®(commercial substrate)	1.3^c	1.4^c	1.8^c	2.5^c
Means	2.99	3.56	3.89	4.25
Pr > Fc	0.0008	0.0001	0.0001	0.0014
CV (%)	24.87	18.72	16.13	17.68

Means followed by the same letter do not differ statistically by the Tukey test t 5% of probability.

percentage magnitude were evident, mainly for the substrates that contain humus in it composition, however, the less emergence percentage was observed with the commercial substrate.

Similar results were obtained by Oliveira et al. (2009) studying the germination of soursop (*Annona muricata*) seeds found that, the use of humus in substrate composition improved nutritional quality of the substrate

As can be observed in Figure 2, the seedlings that emerged in the substrates with soil + chicken manure and with soil + chicken manure + sand, presented higher initial development in relation to the humus substrates (Table 2).

The higher initial growth of seedlings was observed with the use of chicken manure in the mix with soil and with soil + sand. This can be due to the higher availability of nutrients, increasing the growth of seedlings in relation to other substrates (Figure 2). Lucena et al. (2009) also

observed that, the manure of chicken promoted the development of Pacara Earpod (*Enterolobium cotortosilicum*) seedlings with higher evidence when compared to obtained values for the control.

All the evaluated substrates, except the commercial substrate used in this study (Macplant®), have advantages related to the physical properties, which we can emphasize the greater availability of water and the drainage of the excess water.

Table 2 shows that, the seedlings cultivated in the substrate soil + chicken manure had higher number of leaves in relation to those cultivated with other substrates, since the first evaluation. Kuhn et al. (2012) indicates that, at 45 DAS the Cape gooseberry seedlings conducted in commercial substrate plus the mix with vermiculite presented average of 3.5 leaves per plant.

Contrary to our results, Neves et al. (2010) found that, the use of substrate with soil + chicken manure presented

Table 3. Plant height of Cape gooseberry plants in response to different substrate compositions, in the period of 45 to 90 DAS, Federal Technological University of Parana, Campus of Dois Vizinhos, 2009.

Substrate	45 DAS	60 DAS	75 DAS	90 DAS
Soil + chicken manure (2:1)	7.5^a	10.5^a	13.5^a	16.8^a
Soil + chicken manure + sand (2:1:1)	7.0^a	10.3^a	13.5^a	16.8^a
Soil + humus + vermiculite (2:1:1)	4.3^b	6.8^b	9.1^b	11.8^{ab}
Soil + humus (2:1)	2.5^{bc}	3.5^c	4.9^c	6.5^{bc}
Macplant®(commercial substrate)	1.0^c	1.0^d	1.6^c	2.3^{bc}
Means	4.45	6.4	8.53	10.8
Pr > Fc	0.0000	0.0000	0.0000	0.0000
CV (%)	20.72	16.88	22.22	27.05

Means followed by the same letter do not differ statistically by the Tukey test t 5% of probability.

Table 4. Parameters of grow of Cape gooseberry plants in response to different substrate compositions in the production of seedlings, Federal Technological University of Parana, Campus of Dois Vizinhos, 2009.

Substrate	Dry mass of root system (A) (g)	Dry mass of shoot system (B) (g)	Relation (A)/(B)
Soil + chicken manure (2:1)	0.147^a	0.287^a	1.93^{ab}
Soil + chicken manure + sand (2:1:1)	0.107^a	0.244^a	2.46^a
Soil + humus + vermiculite (2:1:1)	0.058^b	0.126^b	2.37^a
Soil + humus (2:1)	0.024^{bc}	0.045^{bc}	1.99^{ab}
Macplant®(commercial substrate)	0.006^c	0.006^c	1.19^b
Means	0.069	0.141	1.99
Pr > Fc	0.0000	0.0000	0.0059
CV (%)	30.55	30.84	20.2

Means followed by the same letter don't differ statistically by the Tukey test t 5% of probability.

lesser values in relation to other substrates in moringa (*Moringa oleifera* Lam) seedlings for the following variables: plant height, index of emergence velocity, dry mass of roots and dry mass of shoots.

In Table 3 can be observed that, the height of plants in substrates containing humus were statistically different, being inferior according to the Tukey test in all conducted evaluations, being only superior to the commercial substrate Macplant® in almost all evaluations.

As can be seen in Table 4, the variables dry mass of shoot system and root system presented higher values when substrates with presence of chicken manure were used, being the inferior means presented by the Macplant® substrate.

David et al. (2008) show that, the application of organic matter in chicken manure provides the development and production of dry matter in passion fruit plants. This can be result of the nutrient concentration, according to Severino et al. (2006), the chicken manure present in it composition mainly 2.95% of nitrogen, 3.97% of phosphorus, 1.10% of potassium, 4.71% of calcium and 6.93% of magnesium. However, Neves et al. (2010) observed lower values of dry mass for the substrate from chicken manure in relation to other organic materials.

Regarding to the inclusion of small fruits in the last years of Brazilian fruit production, studies are important in order to conduct the emergence of seedlings in adequate conditions, as indeed, when these conditions are obtained, the development of culture is improved.

Besides, in the southwest region of Parana, there is a significant production of many different byproducts from agricultural origin, being that the replacement of commercial substrates by compounds produced in the farms can decrease the production expenses, in this way increasing the economic gain of the production of fruits. These knowledge can also be used in the formulation of new substrates, with the purpose of using available raw materials and distribute them in other productive centers.

Conclusions

The use of substrates as soil + humus + vermiculite (2:1:1 v/v/v) and soil + humus (2:1 v/v) provided higher emergence of Cape gooseberry seedlings. The initial development of seedlings is higher in substrates containing chicken manure. The isolated use of the commercial substrate Macplant® must be avoided in the propagation of this culture, because it can affect the emergence and initial development of *P. peruviana* L. negatively.

REFERENCES

Alvino FO, Rayol BR (2007). Different substrate effects in the germination of Ochroma pyramidale (cav. ex lam.) urb. (bombacaceae). Ciên. Flor. 17(1):71-75.

Andrade ACS, Pereira TS, Fernandes MJ, Cruz APM, Carvalho ASR (2006). Germination substrate, temperature and post-seminal development of Dalbergia nigra seeds. Braz. Agric. Res. 41(3):517-523.

Andrade L (2008). *Physalis* or uchuva – Fruit from Colombia reach Brazil. Rural Mag. 38:11-12.

Brasil (2009). Ministry of Agriculture, Livestock and Food Supply. Rules for seed testing. Ministry of Agriculture, Livestock and Food Supply. Agriculture Defense Department. Brasília, DF: Mapa/ACS. P. 395.

Castro PRC, Kluge RA, Peres LEP (2005). Manual of plant physiology: Theory and practice. Piracicaba: Agronômica Ceres. P. 650.

Chaves AC, Schuch MW, Erig AC (2005). *In vitro* establishment and multiplication of *Physalis peruviana* L. Ciência e Agrotecnologia 29(6):1281-1287.

David MA, Mendonça V, Reis LL, Silva EA, Tosta MS, Freire PA (2008). Effect of single superphosphate and organic matter doses on the growth of yellow passion fruit seedlings. Pesquisa Agropecuária Trop. 38(3):147-152.

Dias MA, Lopes JC, Corrêa NB, Dias DCFS (2008). Pepper seed germination and seedling development due to substrate and water sheet. J. Seed Sci. 30(3):115-121.

Díaz LA, Fischer G, Pulido SP (2010). Coco peat as a substitute for peat moss in the production of Cape gooseberry (*Physalis peruviana* L.) seedlings. Colomb. J. Hortic. Sci. 4(2):153-162.

Ferraz ABF, Limberger RP, Bordignon S, von Poser GL, Henriques AT (2005). Essential oil composition of six Hypericum species from Southern Brazil. Flav Frag J 20: 335-339.

Inmet (2012). Weather Station A843 de Dois Vizinhos, PR. Available at <http://www.inmet.gov.br/sonabra/maps/pg_automaticas.php> Acesso em: 27 de Março de 2012.

Kuhn PR, Kulczynski SM, Bellé C, Koch F, Werner CJ (2012). Initial development of seedlings physalis (*Physalis peruviana*) through the seeds of fruits mature and green under different substrates. Encyclopedia Biosph. 8(15):1378-1385.

Lima CSL, Gonçalves MA, Tomaz ZFP, Rufato AR, Fachinello JC (2010). Periods replanting and training systems of Cape-gooseberry. Ciência Rural 40(12):2472-2479.

Lucena AMA, Costa FX, Silva H, Guerra HOC (2009). Germination of forest species on substrates fertilized with organic matter. Revista de Biologia e Ciências da Terra 4(2):208-212.

Mota RV (2006). Characterization of black-berry juice prepared in a domestic extractor. Food Sci. Technol. 26(2):303-308.

Muniz J, Kretzschmar AA, Rufato L, Pelizza TR, Marchi T, Duarte AE, Lima APF, Garanhani F (2011). Conduction systems for physalis production in southern Brazil. Braz. Mag. Fruit Cult. 33(3):830-838.

Nascimento WM, Sousa RB, Silva JBC, Carrijo AO (2003). Seed germination and stand establishment of vegetable crops in different substrates under tropical conditions. Acta Horticulturae 609:483-485.

Neves JMG, Silva HP, Duarte RF (2010). Use of alternative substrates for production of Seedlings moringa. Green Mag. 5(1):173-177.

OLIVEIRA DCSG, WERREN JH, VERHULST EC, GIEBEL JD,KAMPING A, BEUKEBOOM LW, VAN DE ZANDE L (2009).Identification and characterization of the doublesex gene of Nasonia. Insect Mol. Biol. 18: 315–324.

Rodrigues E, Rockenbach II, Cataneo C, Gonzaga LV, Chaves ES, Fett R (2009). Mineral sand essential fatty acid sof the exotic fruit *Physalis peruviana* L. Rev. Food Sci. Technol. 29(3):642-645.

Sas Institute (1999). SAS user's guide statistics: version 8.0 edition. Cary. P. 956.

Severino LS, Lima RLS, Beltrão NEM (2006). Chemical composition of eleven organic materials used in substrates for seedling production. Comunicado Técnico 278, Ministerio da Agricultura Pecuaria e Abastecimento. Campina Grande. P. 5.

Taiz L, Zeiger E (2009). Plant physiology. 5. ed. Massachusetts: Sinauer. P. 819.

Velasquez HJC, Giraldo OHB, Arango SSP (2007). Preliminary study of mechanical resistance to fracture and firmness force for uchuva (Physalis peruviana L) fruits. Revista Facultad Nacional de Agronomía -Medellín 60(1):3785-3796.

Wagner-Júnior A, Santos CEM, Silva JOC, Alexandre RS, Negreiros JRS, Pimentel LD, Álvares VS, Bruckner CH (2006). Influence of soaking water ph and substrates in the seeds germination and initial development of the sweet passion fruit. Curr. Agric. Sci. Technol. 12(2):231-236.

10

Harvest frequency effect on plant height, grass tiller production, plant cover and percentage dry matter production of some forage grasses and legumes in the derived savannah, Nigeria

C. C. Onyeonagu* and J. E. Asiegbu

Department of Crop Science University of Nigeria, Nsukka, Enugu State, Nigeria.

Four legumes, *Lablab purpureus*, *Stylosanthes hamata*, *Centrosema pascuorum* and *Stylosanthes guyanensis* and 4 grasses, *Sorghum almum*, *Panicum maximum*, *Chloris gayana* and *Andropogon gayanus* were investigated in a 2-year study at Nsukka, derived savannah, Nigeria. The response of these species to cutting management (4 and 8-weekly intervals) was evaluated. Increasing the interval between harvests increased (P<0.05) plant height and percentage dry matter production in the grass and legume species. Cutting treatment did not influence the extent of legume cover; however, the extent of grass cover was increased (P<0.05) by 30% when the interval between harvests was increased from 4 to 8 weeks. Weed cover in the grass plots was depressed (P<0.05) by 21% with increased interval of cut from 4 to 8 weeks. The tallest (P<0.05) plants among the legumes were obtained in *S. guianensis* when cutting was done at the interval of 8 weeks. Harvesting the grasses at the interval of 8 weeks produced the tallest plants in *A. gayanus* in 2007. The highest (P<0.05) tiller number per meter square was produced in both years when *A. gayanus* was harvested at 4-weekly interval. *S. guyanensis* and *S. hamata* suppressed (P<0.05) weed growth more than *L. purpureus* or *C. pascuorum*. Percentage dry matter production was lower (P<0.05) in *L. purpureus* compared with the other legumes. Grass cover remained relatively high with *P. maximum* and *A. gayanus*.

Key words: Cutting frequency, forage species, seasonal yield, dry matter content.

INTRODUCTION

Improved grasses and legumes have been recommended for intensive livestock production in Nigeria due to their high forage production and nutritive value (De Leeuw and Brinckman, 1974; Olubajo, 1974). Various grass species at present exist in Nigeria and the notable ones include *Andropogon gayanus*, *Panicum maximum*, *Chloris gayana* and *Sorghum almum* (Agishi, 1979). The legumes, which include; *Stylosanthes guyanensis*, *Lablab purpureus*, *Stylosanthes hamata* and *Centrosema* spp.

have proved very valuable (Onifade and Agishi, 1988).

The productivity, chemical composition and nutritive value of grasses and legumes found in Nigeria vary greatly according to species, the nature and fertility of the soil, water relations, season, disease control and the stage of growth at which the forage species are cut or grazed (Aregheore, 1996; Nuru, 1996). Cutting frequency has been shown to produce different effects on the quantity and quality of forage grasses and legumes at different seasons of the year depending on the species of forage (Njarui and Wandera, 2004; Enoh et al., 2005). The response to cutting of a forage plant depends upon its seasonal yield of carbohydrate storage, its growth

---

*Corresponding author. E-mail: onyeonagu@yahoo.com.

Table 1. Plant cover score of legume, grass and weed species.

Score	Percentage cover	Degree of cover
1	< 20	Very low
2	20 - 39	Low
3	40 - 59	Medium
4	60 - 79	High
5	80 - 100	Very high

habit and extent of inflorescence development (Dev, 2001). As pastures mature they are characterized by high content of fibre with a higher grade of lignification and low protein content (Enoh et al., 2005). Most improved grasses fed at early stages of maturity are more digestible and are eaten in larger quantities than at more mature stages (Mero, 1985). Legumes contain higher crude protein (CP) and minerals than grasses and increase total dry matter intake when used as supplements to low CP grass diets (Mero and Uden, 1990). Some of the available forage grass and legume species have not been evaluated under frequent cutting regimes in the derived savannah of Nigeria. The need thus arises to device better means of maximizing the performance of these pasture species without compromising either the dry matter production or quality. This paper reports the results of a study that was conducted to investigate the seasonal grass tiller production, percentage dry matter production, persistence and plant growth under two cutting regimes of four selected herbaceous legumes and four grasses in Nsukka, derived savanna of Nigeria.

MATERIALS AND METHODS

Experimental site

The experiment was conducted at the Teaching and Research Farm of the Department of Crop Science, Faculty of Agriculture, University of Nigeria, Nsukka. Nsukka is located at Latitude 06° 52' N and Longitude 07° 24' E, and on altitude of 447.2 m above sea level. Four herbaceous legumes (*L. purpureus, S. hamata, C. pascuorum,* and *S. guyanensis*) and 4 grasses (*S. almum, P. maximum, C. gayana* and *A. gayanus*) were evaluated in 2006 and 2007.

The design of the experiment was an 8 × 2 factorial in randomized complete block arrangement, with 3 replications. Treatments comprised eight forage species and two cutting frequencies (4 and 8 weeks). The marking of the field into blocks (2.2 × 36 m per block) and plots (2.25 × 2.2 m per plot) was done on 15 June 2006. Planting was done from 19 to 24 June 2006 with 1m² sampling area. The seeds of *L. purpureus* were planted by broadcast at the rate of 15 kg ha⁻¹ while 5.6 kg ha⁻¹ was the seed rate for the other 3 legume species. The seeds of *S. almum, C. gayana,* and *A. gayanus* were planted at the rate of 15, 5 and 5.6 kg ha⁻¹, respectively. *P. maximum* rooted cuttings were planted at 20 × 30 cm spacing. Basal application of 75 and 44 kg P ha⁻¹ as potassium chloride and single superphosphate, respectively, was made by broadcasting in 2006 and 2007.

Measurements

Cutting was done at uniform height of about 15 cm with shears. The harvest intervals of 4 and 8 weeks gave 2 and 1 samples, respectively in 2006 (that is 8 weeks from September 22 to November 17) and 6 and 3 samples, respectively in 2007 (that is 24 weeks from June 12 to November 27). Fresh samples of the grass and legume species weighing 100 g were put in paper envelops and dried in a forced air oven set at 80°C and weighed after attaining constant dry weights. These were used to calculate the dry matter content of the forage species. Plant height (cm) was taken using the mean of three readings taken at random from the sample area in each plot. Tiller counts for the grass species were made on each plot using a 25 cm square quadrant. The mean of three throws per plot was used to calculate tiller population per m². Plant scoring of the plots was done to determine the extent of cover by the forage species, weed species (that is, any plant other than the species planted) and bare ground cover. The scoring is shown in Table 1

Scoring was done using five point grading score as suggested by Snedecor and Cochran (1967) for subjective evaluation.

Statistical analysis

All data collected were statistically analysed using the procedure outlined by Steel and Torrie (1980) for factorial experiment in a randomized complete block design using GENSTAT (1995) statistical package. Separation of treatment means for statistical significance was done using the standard error of the difference between two means (s.e.d.). Square root transformation of the form

$\sqrt{x + 0.5}$, where x is the observation, was used whenever there is zero value. The 4- and 8-week intervals of cut each spanning the yearly harvest period were separately analysed and compared for effect of season as done by Omaliko (1980).

RESULTS

In 2006, cutting interval, species and their interactions showed no significant effects on the height of the legumes (Table 2). The longer interval of 8 weeks significantly (P<0.05) increased plant height by 46% over situations where cutting was frequent in 2007. *S. guyanensis* was significantly (P<0.05) taller than the other legume species, followed by *L. purpureus* and *S.hamata* which did not differ with each other statistically (P>0.05) but were significantly (P<0.05) taller than *C. pascuorum.* Cutting interval × species interactions showed no significant (P>0.05) effects on the height of the legumes in 2007.

Cutting the legumes at 8-weekly interval during the first harvest period of 2007, resulted in plants that were statistically (P<0.05) taller than those harvested every 4 weeks (Table 3). Plant height was not significantly affected by cutting treatment in the last two periods. The legume species showed significant (P<0.05) variability in plant height in all the harvest periods. During the first two harvest periods of 2007, *C. pascuorum* produced plants that were statistically (P<0.05) shorter compared with those from other legumes. There was significant cutting interval and species × cutting interval interaction effect on

Table 2. Plant heights of 4 legumes defoliated at 4 and 8-week intervals.

Species	Cutting intervals (weeks)		
	4	8	Mean
		2006	
Dolichos bean (*L. purpureus*)	70.4	59.9	65.1
Verano stylo (*S. hamata*)	31.6	51.9	41.7
Centro (*C. pascuorum*)	28.7	48.2	38.5
Cook stylo (*S. guyanensis*)	35.9	67.1	51.5
Mean	41.6	56.8	49.2
		2007	
Dolichos bean (*L. purpureus*)	30.6	43.2	36.9
Verano stylo (*S. hamata*)	26.7	41.7	34.2
Centro (*C. pascuorum*)	24.2	30.4	27.3
Cook stylo (*S. guyanensis*)	33.5	52.0	42.7
Mean	28.7	41.8	35.3
	2006		**2007**
s.e.d. between 2 cutting frequency means (C)	10.99		1.85
s.e.d. between 2 species means (S)	15.54		2.62
s.e.d. between 2 C × S means	21.98		3.70

Table 3. Plant heights of 4 legumes at various periods of the year 2007 for 4 and 8-weekly intervals.

Species	Cutting intervals (weeks)		
	4	8	Mean
	June12 to August 7		
Dolichos bean (*L. purpureus*)	40.9	51.3	46.1
Verano stylo (*S. hamata*)	30.4	53.7	42.1
Centro (*C. pascuorum*)	32.7	40.6	36.7
Cook stylo (*S. guyanensis*)	34.6	55.3	45.0
Mean	34.7	50.2	42.5
	August 7 to October 2		
Dolichos bean (*L. purpureus*)	20.3(4.6)[1]	18.7(3.8)	19.5(4.2)
Verano stylo (*S. hamata*)	26.9(5.2)	41.9(6.5)	34.4(5.9)
Centro (*C. pascuorum*)	20.6(4.5)	16.6(3.6)	18.6(4.0)
Cook stylo (*S. guyanensis*)	37.0(6.1)	57.3(7.6)	47.1(6.9)
Mean	26.2(5.1)	33.6(5.4)	29.9(5.2)
	October 2 to November 27		
Dolichos bean (*L. purpureus*)	0.0(0.7)	0.0(0.7)	0.0(0.7)
Verano stylo (*S. hamata*)	22.8(4.8)	29.4(5.5)	26.1(5.1)
Centro (*C. pascuorum*)	13.3(3.2)	10.2(2.9)	11.8(3.1)
Cook stylo (*S. guyanensis*)	28.7(5.4)	43.2(6.6)	36.0(6.0)
Mean	16.2(3.5)	20.7(3.9)	18.5(3.7)
	1st period	**2nd period**	**3rd period**
s.e.d. between 2 cutting means (C)	2.09	0.49	0.40
s.e.d. between 2 species means (S)	2.95	0.69	0.56
s.e.d. between 2 C x S means	4.18	0.98	0.79

Values in parentheses are square root transformed values to which s.e.d. are applicable.

plant height in all the harvest periods of 2007. *S. guyanensis* harvested at 8-weekly interval significantly (P<0.05) produced the tallest plants in all the harvest periods of 2007.

Table 4. Plant heights of four forage grasses defoliated at 4 and 8-weekly intervals.

Species	Cutting intervals (weeks)		
	4	8	Mean
2006			
Columbus grass (*S. almum*)	110.1	125.8	117.9
Guinea grass (*P. maximum*)	127.2	144.9	136.1
Rhodes grass (*C. gayana*)	119.0	138.6	128.8
Gamba grass (*A. gayanus*)	140.1	100.0	120.1
Mean	124.1	127.3	125.7
2007			
Columbus grass (*S. almum*)	78.4	119.4	98.9
Guinea grass (*P. maximum*)	60.9	121.3	91.1
Rhodes grass (*C. gayana*)	58.2	90.1	74.2
Gamba grass (*A. gayanus*)	68.4	130.7	99.5
Mean	66.5	115.4	90.9
	2006		**2007**
s.e.d. between 2 cutting frequency means (C)	15.51		3.92
s.e.d. between 2 species means (S)	21.93		5.54
s.e.d. between 2 C×S means	31.01		7.84

In 2006 harvest season, plant height remained statistically (P>0.05) similar among the grass species (Table 4). Cutting interval and species × cutting interval interactions did not show any significant (P>0.05) effects on grass height. Increasing the interval between harvests in 2007 from 4 to 8 weeks significantly (P<0.05) increased plant height by 74% on average. *A. gayanus* had significantly (P<0.05) taller plants than *C. gayana* but had similar height values with *S. almum* and *P. maximum*. Cutting *A. gayanus* at 8 weekly intervals significantly (P<0.05) increased plant height relative to the 4 week cutting interval.

Grass height was significantly (P<0.05) increased with the cutting interval of 8 weeks relative to the 4 week interval for all the harvest periods of 2007 (Table 5). Plant height varied statistically (P<0.05) among the grass species at the second and third periods of 2007. *A. gayanus* had statistically, similar plant height with *S. almum* for all the harvest periods.

Grass tiller population was not significantly (P>0.05) influenced by cutting frequency in both 2006 and 2007 (Table 6). In 2006, *A. gayanus* produced significantly (P<0.05) higher number of tillers relative to *P. maximum* but was statistically (P>0.05) similar to *S. almum* or *C.gayana.*. In 2007, *P. maximum* and *C. gayana* had significantly (P<0.05) lower number of tillers compared with *S. almum*. Cutting interval × species interaction showed significant effect on the number of grass tillers in 2006 and 2007. Harvesting *A. gayanus* at 4-weekly intervals resulted to number of tillers that were statistically higher than those of the other grass species.

Cutting at the 4 weeks interval significantly (P<0.05)

increased grass tiller number relative to the 8 week cutting interval during the first two periods of 2007 (Table 7). Cutting interval did not influence tiller number at the third period of the year. *A. gayanus* significantly (P<0.05) produced higher number of tillers than *P.maxmum* and *C. gayana* in the first two periods but had values that were statistically similar with *S. almum* in all the harvest periods.

The extent of legume and weed cover and bare ground area in plots before application of cutting management in 2006 showed legume cover to be mostly very high (Table 8). The extent of weed cover and bare ground area were relatively very low. Cutting interval did not significantly (P>0.05) influence the extent of legume cover in 2007. Species comparison showed that *S.guyanensis* significantly (P<0.05) covered more of the plot than the other legumes. *S. hamata* also significantly covered the plot more than *L. purpureus* and *C. pascuorum* which did not differ with each other. The highest plot cover was obtained when *S. guyanensis* was harvested either at the 4- or 8-weekly interval, while harvesting at either the 4- or 8-weekly interval produced the least (P<0.05) plot cover in *L. purpureus*. Weed cover and bare ground area where not influenced by cutting management in 2007. *S. guyanensis* and *S. hamata* had significantly (P<0.05) lower weed covers compared with *L. purpureus* and *C. pascuorum*, which did not differ with each other.

Except for *S. almum* the other three grasses had similarly high grass cover in the establishment year and there were no clear effects of cutting frequency (Table 9). However, by the end of the first harvest year, grass cover was on average depressed (P<0.05) by 24% although

Table 5. Plant heights of four forage grasses at various periods of the year 2007 for 4 and 8-weekly intervals.

Species	Cutting intervals (weeks)		
	4	8	Mean
June12 to August 7			
Columbus grass (*S. almum*)	65.5	69.0	67.3
Guinea grass (*P. maximum*)	65.2	109.6	87.4
Rhodes grass (*C. gayana*)	63.5	101.2	82.3
Gamba grass (*A. gayanus*)	65.6	123.3	94.5
Mean	64.9	100.8	82.9
August 7 to October 2			
Columbus grass (*S. almum*)	61.3	138.6	99.9
Guinea grass (*P. maximum*)	64.2	161.1	112.7
Rhodes grass (*C. gayana*)	60.9	85.5	73.2
Gamba grass (*A. gayanus*)	64.2	124.8	94.5
Mean	62.7	127.5	95.1
October 2 to November 27			
Columbus grass (*S. almum*)	108.3	150.7	129.5
Guinea grass (*P. maximum*)	53.4	93.1	73.3
Rhodes grass (*C. gayana*)	50.3	83.4	66.9
Gamba grass (*A. gayanus*)	77.1	143.0	110.0
Mean	72.3	117.5	94.9
	1st period	2nd period	3rd period
s.e.d. between 2 cutting frequency means (C)	10.76	7.27	6.45
s.e.d. between 2 species means (S)	15.22	10.28	9.12
s.e.d. between 2 C × S means	21.52	14.54	12.90

Table 6. Tiller population per square metre of four forage grasses defoliated at 4 and 8-weekly intervals.

Species	Cutting intervals (weeks)		
	4	8	Mean
2006			
Columbus grass (*S. almum*)	534.9	432.0	483.5
Guinea grass (*P. maximum*)	354.7	428.3	391.5
Rhodes grass (*C. gayana*)	764.8	439.2	602.0
Gamba grass (*A. gayanus*)	1034.1	748.3	891.2
Mean	672.1	511.9	592.0
2007			
Columbus grass (*S. almum*)	796.3	890.0	843.1
Guinea grass (*P. maximum*)	538.9	404.8	471.9
Rhodes grass (*C. gayana*)	443.0	197.0	320.0
Gamba grass (*A. gayanus*)	895.6	675.6	785.6
Mean	668.5	541.8	605.1
	2006		**2007**
s.e.d. between 2 cutting frequency means (C)	98.51		65.09
s.e.d. between 2 species means (S)	139.31		92.05
s.e.d. between 2 C×S means	197.02		130.18

remained relatively high with *P. maximum* and *A. gayanus*. In 2006, the dry matter content of legume herbage was significantly (P<0.05) increased by 19% when the interval between cuts was increased from 4 to 8

Table 7. Tiller population per square metre of four forage grasses at various periods of the year 2007 for 4 and 8-weekly intervals.

Species	Cutting intervals (weeks)		
	4	8	Mean
	June12 to August 7		
Columbus grass (*S. almum*)	500.3	535.5	517.9
Guinea grass (*P. maximum*)	392.0	325.3	358.7
Rhodes grass (*C. gayana*)	661.6	163.2	412.4
Gamba grass (*A. gayanus*)	941.3	346.7	644.0
Mean	623.8	342.7	483.2
	August 7 to October 2		
Columbus grass (*S. almum*)	944.8	408.5	676.7
Guinea grass (*P. maximum*)	706.7	348.3	527.5
Rhodes grass (*C. gayana*)	488.0	234.1	361.1
Gamba grass (*A. gayanus*)	939.2	515.7	727.5
Mean	769.7	376.7	573.2
	October 2 to November 27		
Columbus grass (*S. almum*)	943.7	1725.9	1334.8
Guinea grass (*P. maximum*)	518.1	540.8	529.5
Rhodes grass (*C. gayana*)	179.5	193.6	186.5
Gamba grass (*A. gayanus*)	806.4	1164.3	985.3
Mean	611.9	906.1	759.0
	1st period	2nd period	3rd period
s.e.d. between 2 cutting frequency means (C)	68.57	45.66	144.42
s.e.d. between 2 species means (S)	96.98	64.57	204.24
s.e.d. between 2 CxS means	137.15	91.32	288.84

weeks (Table 10). *S. hamata* had significantly higher dry matter content than *L. purpureus* and *S. guyanensis* but did not differ statistically with *C. pascuorum*. In 2007, cutting management did not influence the percentage dry matter of legumes. Species comparison showed that *L. purpureum* had significantly (P<0.05) the least percentage dry matter compared with the other legumes. *S. hamata* though similar to *S. guyanensis*, produced percentage dry matter that was statistically higher than that of *C. pascuorum*. The legume dry matter content was not significantly affected by species × cutting interval interactions in both years.

Whether for 4- or 8-week interval of cuts, legume dry matter content in 2007 was greater later in the season than earlier (Table 11). Increasing the interval between harvests from 4 to 8 weeks in 2006 significantly (P<0.05) increased the percentage dry matter of grass species by 29% (Table 12). Grass dry matter content did not vary statistically among the grass species in 2006. In 2007, the dry matter content of grass increased significantly (P<0.05) by 21% with increase in interval between cuts from 4 to 8 weeks. *S. almum* produced significantly (P<0.05) higher dry matter content than *C. gayana* though similar to *P. maximum* or *A. gayanus*. Percentage dry matter was not significantly affected by cutting interval × species interaction in any of the years. Whether

for 4- or 8-week interval of cuts, grass dry matter content was greater late in the season than earlier (Table 13).

DISCUSSION

The outstanding performance of the two *Stylosanthes* species in this study area over the other legumes in plant height, plot cover, weed control and percentage dry matter production agreed with the report by de Andrade et al. (2004). This has been attributed to the ability of the *Stylosanthes* species to adapt to low fertility soils of the tropics (Hall and Glatzle, 2004). Among the grasses, *A. gayanus* and *P. maximum* adapted better in this study area than the other tested grass species with better performance in plant height, plot cover and weed control. The outstanding performance of *A. gayanus* and *P. maximum* in this study area could be attributed to their ability to thrive well in acid and infertile soils of the tropics (Grof, 1981; Okeagu, 1991). *A. gayanus* have been shown to be relatively free of major pest and disease problems, resistant to drought and fire and generally adapted to savannah environment (Grof, 1981).

The observed decrease in the heights of grasses and legumes with increase in cutting frequency agrees with the report by Adams et al. (1991) that frequent defoliation

Table 8. Legume and weed covers and bare ground area of four forage legumes defoliated at 4 and 8-weekly intervals.

Species	Cutting intervals (weeks)			Cutting intervals (weeks)		
	4	8	Mean	4	8	Mean
	Initial assessment (10 - 7 - 06)			Final assessment in 2007 (27 - 11 - 07)		
	Legume cover			Legume cover		
Dolichos bean (*L. purpureus*)	5.0	5.0	5.0	1.0	1.0	1.0
Verano stylo (*S. hamata*)	5.0	5.0	5.0	5.0	3.7	4.3
Centro (*C. pascuorum*)	5.0	5.0	5.0	1.3	1.3	1.3
Cook stylo (*S. guyanensis*)	4.7	4.7	4.7	5.0	5.0	5.0
Mean	4.9	4.9	4.9	3.1	2.8	2.9
	Weed cover			Weed cover		
Dolichos bean (*L. purpureus*)	1.3	1.3	1.3	4.7	4.7	4.7
Verano stylo (*S. hamata*)	1.3	1.3	1.3	2.0	2.7	2.3
Centro (*C. pascuorum*)	1.0	1.0	1.0	4.3	4.7	4.5
Cook stylo (*S. guyanensis*)	1.3	1.3	1.3	1.0	1.3	1.2
Mean	1.2	1.2	1.2	3.0	3.3	3.2
	Bare ground area			Bare ground area		
Dolichos bean (*L. purpureus*)	1.0	1.0	1.0	1.7	1.7	1.7
Verano stylo (*S. hamata*)	1.0	1.0	1.0	1.0	1.0	1.0
Centro (*C. pascuorum*)	1.0	1.0	1.0	1.3	1.3	1.3
Cook stylo (*S. guyanensis*)	1.0	1.0	1.0	1.0	1.0	1.0
Mean	1.0	1.0	1.0	1.2	1.2	1.2
	2006			2007		
	Legume	Weed	Bare ground	Legume	Weed	Bare ground
s.e.d. between 2 cutting frequency means (C)	0.11	0.19	0.00	0.15	0.34	0.21
s.e.d. between 2 species means (S)	0.15	0.27	0.00	0.21	0.48	0.30
s.e.d. between 2 C × S means	0.22	0.38	0.00	0.30	0.68	0.42

Table 9. Grass and weed covers and bare ground area of four forage grasses defoliated at 4 and 8-weekly intervals.

Species	Cutting intervals (weeks)			Cutting intervals (weeks)		
	4	8	Mean	4	8	Mean
	Initial assessment (10 - 7- 06)			Final assessment in 2007 (27 - 11 - 07)		
	Grass cover			Grass cover		
Columbus grass (*S. almum*)	3.7	4.0	3.8	1.7	4.3	3.0
Guinea grass (*P. maximum*)	5.0	5.0	5.0	4.0	4.7	4.3
Rhodes grass (*C. gayana*)	5.0	4.7	4.8	2.3	2.0	2.2
Gamba grass (*A. gayanus*)	4.7	5.0	4.8	4.0	4.7	4.3
Mean	4.6	4.7	4.6	3.0	3.9	3.5
	Weed cover			Weed cover		
Columbus grass (*S. almum*)	2.3	2.0	2.2	5.0	3.7	4.3
Guinea grass (*P. maximum*)	1.0	1.0	1.0	2.7	2.0	2.3
Rhodes grass (*C. gayana*)	1.3	1.3	1.3	4.7	4.7	4.7
Gamba grass (*A. gayanus*)	1.3	1.0	1.2	2.7	2.0	2.7
Mean	1.5	1.3	1.4	3.9	3.1	3.5
	Bare ground area			Bare ground area		
Columbus grass (*S. almum*)	1.7	1.3	1.5	1.0	1.0	1.0
Guinea grass (*P. maximum*)	1.0	1.0	1.0	1.7	1.0	1.3
Rhodes grass (*C. gayana*)	1.0	1.0	1.0	1.0	1.0	1.0
Gamba grass (*A. gayanus*)	1.0	1.0	1.0	1.3	1.0	1.2
Mean	1.2	1.1	1.1	1.2	1.0	1.1

Table 9. Contd.

	2006			2007		
	Grass	Weed	Bare ground	Grass	Weed	Bare ground
s.e.d. between 2 cutting frequency means (C)	0.25	0.18	0.11	0.37	0.30	0.11
s.e.d. between 2 species means (S)	0.35	0.25	0.16	0.52	0.42	0.16
s.e.d. between 2 CxS means	0.50	0.35	0.23	0.73	0.60	0.23

Table 10. Dry matter contents (%) of four forage legumes defoliated at 4 and 8-weekly intervals.

Species	Cutting intervals (weeks)		
	4	8	Mean
	2006		
Dolichos bean (*L. purpureus*)	16.6	18.2	17.4
Verano stylo (*S. hamata*)	30.3	35.6	32.9
Centro (*C. pascuorum*)	29.6	35.3	32.4
Cook stylo (*S. guyanensis*)	25.2	31.5	28.3
Mean	25.4	30.1	27.8
	2007		
Dolichos bean (*L. purpureus*)	11.5	12.1	11.8
Verano stylo (*S. hamata*)	25.1	29.0	27.0
Centro (*C. pascuorum*)	18.7	19.0	18.8
Cook stylo (*S. guyanensis*)	22.7	26.6	24.7
Mean	19.5	21.7	20.6
	2006		**2007**
s.e.d. between 2 cutting frequency means (C)	1.03		1.73
s.e.d. between 2 species means (S)	1.45		2.45
s.e.d. between 2 C × S means	2.06		3.47

of Himalayan grasslands by large number of cattle reduced the ability of the grasses to replenish leaf area, set seeds and store food reserves in their roots, thereby reducing plant growth. Frequent grazing was simulated by the frequent harvest of 4-week interval in this present study. Poor performance of forage species due to excessive removal of photosynthetic tissues and reduction in root growth and available soil N and other nutrients as a result of accelerated desiccation of surface soil have been associated with frequent cutting regime (Sheley et al., 2002; Donaghy and Fulkerson, 2002).

Grass tiller number per meter square increased with frequent cutting interval as found by Onyeonagu and Asiegbu (2005) with *P. maximum*. They obtained a higher number of tillers with the 3-weekly interval (655 tillers per m^3) compared with the 9 weeks interval of cuts (521 tillers per m^2). Chapman and Lemaire (1993) reported similar increase in tiller population of grass species with frequent defoliation. The significant increase in the percentage dry matter production in the grass and legume species obtained in this present study agree with the report by

Wilman et al. (1976) on perennial ryegrass. The increasing proportion of the stem fraction with increasing interval of cuts (Asiegbu and Onyeonagu, 2008) would help to account for increasing dry matter percentage with increasing interval of cut. The observed significant effect of cutting interval on the grass, and weed cover was also reported by Asiegbu and Onyeonagu (2008). They indicated that long cutting intervals produced better competitive ability in the desired pasture species over the weed species and that this could account for the better plot cover by the desired species than the weeds.

Conclusion

Among the legumes, the tallest plants were produced with *S. guianensis* when the 8 weeks cutting interval was used. For the grasses, harvesting at 8-weekly interval gave the tallest plants in *A. gayanus*. The highest tiller number per meter square was obtained in *A. gayanus* with the frequent cutting interval of 4 weeks. *S. guianensis*

Table 11. Dry matter contents (%) of four forage legumes at various periods of the year 2007 for 4 and 8-weekly intervals.

Species	Cutting intervals (weeks)		
	4	8	Mean
	June12 – August 7		
Dolichos bean (*L. purpureus*)	10.0	10.7	10.4
Verano stylo (*S. hamata*)	21.6	21.3	21.5
Centro (*C. pascuorum*)	17.0	13.9	15.4
Cook stylo (*S. guyanensis*)	16.8	20.5	18.7
Mean	16.4	16.6	16.5
	August 7 – October 2		
Dolichos bean (*L. purpureus*)	6.7(2.0)	9.5(2.8)	8.1(2.4)
Verano stylo (*S. hamata*)	22.2(4.8)	28.9(5.4)	25.6(5.1)
Centro (*C. pascuorum*)	17.9(4.3)	8.9(2.2)	13.4(3.2)
Cook stylo (*S. guyanensis*)	20.7(4.6)	28.1(5.3)	24.4(5.0)
Mean	16.9(3.9)	18.9(3.9)	17.9(3.9)
	October 2 – November 27		
Dolichos bean (*L. purpureus*)	11.5	12.1	11.8
Verano stylo (*S. hamata*)	25.1	29.0	27.0
Centro (*C. pascuorum*)	18.7	19.0	18.8
Cook stylo (*S. guyanensis*)	22.7	26.6	24.7
Mean	19.5	21.7	20.6
	1st period	2nd period	3rd period
s.e.d. between 2 cutting means (C)	1.46	0.58	1.73
s.e.d. between 2 species means (S)	2.07	0.82	2.45
s.e.d. between 2 C × S means	2.92	1.16	3.47

Numbers in parentheses are square root transformed values to which s.e.d. are applicable.

Table 12. Dry matter contents (%) of four forage grasses defoliated at 4 and 8-weekly intervals.

Species	Cutting intervals (weeks)		
	4	8	Mean
	2006		
Columbus grass (*S. almum*)	28.1	33.5	30.8
Guinea grass (*P. maximum*)	25.4	27.4	26.4
Rhodes grass (*C. gayana*)	28.3	39.9	34.1
Gamba grass (*A. gayanus*)	26.0	37.5	31.7
Mean	26.9	34.6	30.7
	2007		
Columbus grass (*S. almum*)	23.6	27.8	25.7
Guinea grass (*P. maximum*)	22.0	27.1	24.6
Rhodes grass (*C. gayana*)	19.1	22.6	20.8
Gamba grass (*A. gayanus*)	20.2	24.9	22.5
Mean	21.2	25.6	23.4
	2006		2007
s.e.d. between 2 cutting frequency means (C)	1.90		1.06
s.e.d. between 2 species means (S)	2.69		1.50
s.e.d. between 2 C x S means	3.80		2.12

Table 13. Dry matter contents (%) of four forage grasses at various periods of the year 2007 for 4 and 8-weekly intervals.

Species	Cutting intervals (weeks)		
	4	8	Mean
June12 – August 7			
Columbus grass (*S. almum*)	19.1(4.4)	24.1(4.9)	21.6(4.7)
Guinea grass (*P. maximum*)	19.5(4.5)	12.6(3.2)	16.1(3.8)
Rhodes grass (*C. gayana*)	14.3(3.8)	13.4(3.7)	13.9(3.8)
Gamba grass (*A. gayanus*)	17.7(4.2)	20.4(4.5)	19.1(4.4)
Mean	17.7(4.2)	17.6(4.1)	17.6(4.2)
August 7 – October 2			
Columbus grass (*S. almum*)	22.1	26.9	24.5
Guinea grass (*P. maximum*)	21.5	24.3	22.9
Rhodes grass (*C. gayana*)	19.1	24.3	21.7
Gamba grass (*A. gayanus*)	18.8	24.9	21.9
Mean	20.4	25.1	22.7
October 2 – November 27			
Columbus grass (*S. almum*)	23.6	27.8	25.7
Guinea grass (*P. maximum*)	22.0	27.1	24.6
Rhodes grass (*C. gayana*)	19.1	22.6	20.8
Gamba grass (*A. gayanus*)	20.2	24.9	22.5
Mean	21.2	25.6	23.4
	1st period	2nd period	3rd period
s.e.d. between 2 cutting frequency means (C)	0.40	1.64	1.06
s.e.d. between 2 species means (S)	0.56	2.33	1.50
s.e.d. between 2 CxS means	0.80	3.29	2.12

Numbers in parentheses are square root transformed values to which s.e.d. are applicable.

and *S. hamata* had better plot cover and suppressed weed growth than the other tested legumes. *A. gayanus* and *P. maximum* had better plot cover and suppressed weed growth than the other grass species. Harvesting those species at 8 weekly intervals provides a good management option for higher plant growth and better percentage dry matter production.

REFERENCES

Adams BW, Ehlert G, Robertson A (1991). Grazing systems for public grazing lands. Range notes No 10 Alberta Forestry lands and Wildlife, Public Lands, Division, Leth-Bridge, Alberta. pp. 1-8.

Agishi EC (1979). The performance of young heifers grazing buffel grass-verano pastures. Annual Report, National Animal Production Research Institute (NAPRI), Shika. pp. 90-91.

Aregheore EM (1996). Natural grassland and ruminant interactions in the dry season in Delta State, Nigeria. World Rev. Anim. Prod. 31(1-2):74-79.

Asiegbu JE, Onyeonagu CC (2008). Effect of cutting frequency and nitrogen application on herbage yield and nitrogen content of a degraded *Panicum maximum* pasture. Nig. J. Anim. Prod. 35(1):114-127.

Chapman DF, Lemaire G (1993). Morphogenetic and structural determinants of plant regrowth after defoliation. In: Baker, M.J. (Ed.), Grassland for Our World. SIR Publishing, New Zealand. pp. 55-64.

de Andrade RP, Karia CT. Ramos AKB (2004). *Stylosanthes* as a forage legume at its centre of diversity. In: Chakraborty, S. (ed.) High-yielding anthracnose resistant *Stylosanthes* for agricultural systems. ACIAR Monograph 3:268.

De Leeuw PN, Brinckman WL (1974). Pasture and rangeland improvement in Northern Guinea and Sudan Zone of Nigeria. In: Loosli JK, Oyenuga VA, Babatunde GM eds. Animal Production in the Tropics. Proceedings of the International Symposium of Animal Production in the tropics, held at the University of Ibadan, Nigeria 26–29 March 1973. pp. 124-151.

Dev I (2001). Problems and prospects of forage production and utilization of Indian Himalaya. ENVIS Bull. Himalayan Ecol. Dev. 9(2):1-13.

Donaghy DJ, Fulkerson WJ (2002). The impact of defoliation frequency and nitrogen fertilizer application in spring on summer survival of perennial ryegrass under grazing in subtropical Australia. Grass Forage Sci. 57(4):351-359.

Enoh MB, Kijora C, Peters KJ, Yonkeu S (2005). Effect of stage of harvest on DM yield, nutrient content, in vitro and in situ parameters and their relationship of native and Brachiaria grasses in the Adamawa Plateau of Cameroon. Livestock Res. Rural Dev. 17(1):1-9.

GENSTAT (1995) Genstat Release 7.2DE, Discovery Third Edition, Lawes Agricultural Trust Rothamsted Experimental station.

Grof B (1981). The performance of *Andropogon gayanus*-legume associations in Colombia. J. Agric. Sci. (Camb.) 96:233-237.

Hall TJ, Glatzle A (2004). Cattle production from *Stylosanthes* pastures. In: Chakraborty, S, (ed.) High-yielding anthracnose resistant *Stylosanthes* for agricultural systems. ACIAR Monograph 3:268.

Mero RN (1985). The effect of supplementing Rhodes grass (*Chloris gayana*) with *Siratro* and *Macroptilium atropurpureum* on dry matter digestibility and voluntary intake. M.Sc Thesis. Sokoine University of Agriculture, Tanzania.

Mero RN, Uden P (1990). Effect of supplementing mature grass hay with dried *Leucaena* leaves on organic matter digestibility and voluntary intake by sheep. Anim. Feed Sci. Technol. 31:1-8.

Njarui DMG, Wandera FP (2004). Effect of cutting frequency on productivity of five selected herbaceous legumes and five grasses in semi-arid tropical Kenya. Tropical Grasslands 38:158-166.

Nuru S (1996). Agricultural development in the age of sustainability: livestock production. In: Sustaining the future. Economic, social and environmental change in sub-Saharan Africa (Edited by George Benneh, William B. Morgan and Juha, I. Uitto). The United Nations University, 1996.

Okeagu MU (1991). The Agronomy of newly introduced grass pastures in the northern Guinea Savanna Zone of Nigeria. 3. *Panicum maximum* (Jacq). World Rev. Anim. Prod. Italy 26(1):87-90.

Olubajo FO (1974). Pasture research at the University of Ibadan. In Lossli, J.K., Oyenuga, V.A. and Babatunde, G.M. eds. Animal Production in the Tropics. Proceedings of the International Symposium of Animal Production in the tropics, held at the University of Ibadan, Ibadan, Nigeria 26-29 March 1973. pp. 67-78.

Omaliko CPE (1980). Influence of initial cutting date and cutting frequency on yield and quality of star, elephant and guinea grass. Grasslands. Forage Sci. 35:139-145.

Onifade OS, Agishi EC (1988). A review of forage production and utilization in Nigeria savanna. In: Utilization of research results on forage and agricultural by-products materials as animal feed resources (B.H. Dzowela, A.N. Said, Aarat Wendem Ageneh and J.A Kategile Eds.). Proceedings of the first joint workshop held in Lilongwe, Malawi, by the Pasture Network for Eastern and Southern Africa (PANESA) and African Research Network for agricultural by-products (ARNAB). International Livestock Centre for Africa (ILCA). 5-9 December, 1988. pp. 114-125.

Onyeonagu CC, Asiegbu JE (2005). Effect of cutting management and N-fertilizer application on plant height, tiller production and percentage dry matter in a run-down *Panicum maximum* pasture. J. Agric. Food, Environ. Exten. 4(2):28-33.

Sheley R, Goodwin K, Rinella M (2002). Mowing to manage noxious weeds. Mont Guide Fact Sheet no. 200104/Agriculture. Montana State University Extension Service.

Snedecor GW, Cochram WG (1967). Statistical methods 6th edition, Lowa State University Press America. p. 246.

Steel GD, Torrie JA (1980). Principles and procedure of statistics: A biometrical approach, 2nd edition, McGraw – Hill Book Company, Inc. New York 6331. p. 31.

Wilman D, Koocheki, A, Lwoga, AB (1976). The effect of interval between harvests and nitrogen application on the proportion and yield of crop fractions and on the digestibility and digestible yield and digestible yield and nitrogen content and yield of two perennial ryegrass varieties in the second harvestyear. J. Agric. Sci (Cambridge). 87:59-74.

11

Genetic variability of morphological and yield traits in Dolichos bean (*Lablab purpureus* L.)

A. M. Parmar[1], A. P. Singh[1], N. P. S. Dhillon[1] and M. Jamwal[2]

[1]Department of Vegetable crops, Punjab Agricultural University, Ludhiana-141004, Punjab, India.
[2]Division of Fruit Science, Sher-e- Kashmir University of Agricultural Sciences and Technology of Jammu, Chatha, Jammu-180 009, (J&K), India.

Thirty genotypes of Dolichos bean (*Lablab purpureus*) were evaluated to study the genetic variability on yield, yield-contributing and related characters. The highest and lowest coefficient of variation was observed for single podded clusters per plant and protein content respectively. Little or no difference between the phenotypic and genotypic coefficients of variability in the expression of various horticultural traits studied for protein content, days to 50% flowering, days to first pod set, pod length, pod weight, weight of 10 green pods and days to maturity. Phenotypic coefficient of variation was high for the single podded clusters per plant. Days to 50% flowering, pod length, width of pod, weight of 10 green pods accounted for the higher heritability and higher genetic advance. Significant positive phenotypic correlations were observed between yield other yield components including days to first pod set, days to 50% flowering, number of pods per plant, weight of 10 green pods and length of pod. The genotypes used in the study are of diverse nature and can be used in the breeding programme for development of superior genotypes in Dolichos bean.

Key words: Dolichos bean, correlation, genetic advance, genetic variability, heritability.

INTRODUCTION

Dolichos bean (*Lablab purpureus*L.) 2n=2x=22, 24 is an important leguminous vegetable crop grown throughout India and is commonly known as Sem. It is potentially a herbaceous perennial but cultivated as an annual with bushy, erect or climbing races. Sem is primarily grown for green pods and is rich in protein (3.8%, green pod basis). The dry seeds are also used for various vegetable preparations and foliage of the crop provides hay, silage and green manures (Bose et al., 1993). It is photo-sensitive and both short day and long day types areavailable (Anonymous, 1961).India is the centre of diversity of Dolichos and large numbers of indigenous strains are available in northern India. Although this crop

has originated in India but very little work has been done for the genetic improvement of yield and quality. A great range of variation exists for the plant and pod characters amongst the accessions grown all over the country. The success of any breeding programme in general and improvement of specific trait through selection in particular, totally depends upon the genetic variability present in the available germplasm of a particular crop. Since, many of the plant characters are governed by polygenes and greatly influenced by environmental conditions; the progress of breeding is, however, conditioned by the magnitude, nature and interrelationship of genotypic and non-genotypic variation.

This suggests a redundancy need to partition the overall variability into heritable and non-heritable components. For the success of the crop improvement programme, the characters for which variability is present, it should be highly heritable as progress due to selection depends on heritability, selection intensity and genetic advance of the character. Heritability and genetic advance estimates for different targeted traits help the breeder to apply appropriate breeding methodology in the crop improvement programme. One of the main thrust in any crop improvement programme is to enhance yield. Yield is a complex trait and is dependent on many other ancillary characters which are mostly inherited quantitatively. Meager information is available for genetic variability in dolichos bean addressing the morphological and yield traits. Hence, an attempt was made with specific objective to examine the genetic parameters of variability to identify major characters for achieving higher yield.

MATERIALS AND METHODS

The experimental material for the present investigation comprised 30 dolichos bean genotypes viz. PD 1, PD 2, PD 3, PD 4, PD 5, PD 6, PD 7, PD 8, PD 9, PD 10, PD 11, PD 12, PD 13, PD 14, PD 15, PD 16, PD 17, Pushpa, PD 19, Pusa Sem 2, PD 21, PD 22, PD 23, PD 24, Pusa Sem 3, PD 26, PD 27, PD 28, PD 29 and PD 30. These were evaluated in randomized complete block design with three replications during (kharif) seasons of two consecutive years 2002 to 2003 at the Vegetable Experimental Area, Punjab Agricultural University, Ludhiana. The sowing was done on ridges with spacing of 1.25 m and 45 cm, ridge to ridge and plant to plant respectively. The recommended package of practices was followed to raise a good crop. Observations were recorded on five randomly taken competitive plants for eleven characters, viz. days to 50% flowering, days to first pod set, days to maturity, number of pods per cluster, pod length, width of pod, weight of 10 green pods, yield per plant, number of single podded clusters per plant, number of pods per plant and protein content.

Analysis of variance was performed following the standard procedures. The phenotypic and genotypic coefficients of variation (PCV, GCV) were computed as described by Burton and Devane (1953). Heritability in broad sense and genetic advance (% of mean) were calculated according to Allard (1960). Estimates of genotypic and phenotypic correlation were obtained using the formulae given by Al-Jibouri et al. (1958).

RESULTS AND DISCUSSION

The 30 genotypes involved in the study varied significantly among themselves for all the horticultural traits studied (Table 1) as revealed by analysis of variance over the years. The minimum number of days to 50% flowering was found in Pushpa (39.33) and the maximum number of days to 50% flowering was taken by genotype PD 21 (147.30). Genotype PD 1 and PD-17 had taken 91.66 days to first pod set whereas PD 21 and PD 22 had taken 140.33 and 108 days to first pod set respectively. The minimum number of days taken to first

maturity was found in the cultivar Pushpa (57.33) and the maximum number of days to maturity of pods among genotypes was recorded in PD-21 (155.66). The genotypes PD-24 (497.66) and PD-12 (487.00) followed by Pusa Sem-2 (460.66) gave higher number of pods per plant and the lower number was obtained from PD-21 (46.00) followed by Pushpa (63.33) and PD-28 (105.00). Pods of genotype PD-21 had maximum pod width (30.03 mm) and the lowest width was observed in PD-9 (14.96). The highest mean value for 10 pod weight was recorded for PD-2 (64.33 g) and the lowest mean value for the 10 pod weight was recorded in PD-29 (27.66 g). Yield/plant is the most important horticultural trait of a crop. The genotype PD-10 out yielded all the cultivars for marketable fruit where production of 2.03 kg pods per plant was recorded. The lowest marketable yield was recorded from PD-21 (0.260 kg). Similar pattern of variability in germplasm evaluation of different sizes for various horticultural traits in dolichos bean have earlier been reported by Borah and Shadeque (1992).

The highest coefficient of variation was observed for single podded clusters per plant (33.78), suggesting the highest variability in the material which can be exploited for further improvement (Table 2). The lowest coefficient of variation was observed in the protein content (2.62) followed by days to maturity (3.91), days to 50% flowering (4.01) and days to first pod set (4.83). The experimental material exhibited a wide range of phenotypic variability ranged from 7.30 to 89.98 (Table 2). Phenotypic coefficient of variation was high for the single podded clusters per plant (89.98) and the lowest phenotypic coefficient of variation was observed in protein content (7.30). In case of genotypic variability, the experimental material exhibited wide ranging from 7.12 (protein content) to 83.40 (single podded clusters per plant). A close proximity in the phenotypic and genotypic coefficients of variability (Table 2) was observed indicating a little influence of environment in the expression of various horticultural traits studied for protein content, days to 50% flowering, days to first pod set, pod length, pod weight, weight of 10 green pods and days to maturity. This suggests that selection for improvement of these characters is possible and effective on the phenotypic basis. Similar findings pertaining to different traits including weight of 10 green pods and days to maturity in dolichos bean is found Borah and Shadeque (1992) and Ali et al. (2005).

Heritability is the transmissibility of characteristics from parents to offsprings. It is of fundamental importance in practicability of selection, because it acts as predictive instrument in expressing the reliability of phenotypic value as guide to breeding value. Heritability is a useful indicator of the progress that can be expected as a result of exercising selection on the pertinent population. In the present investigation, heritability (%) ranged from 59.02 to 95.09 (Table 2). High heritability estimates were obtained for protein content (95.09), days to 50%

Table 1. Mean values for pod yield and related horticultural traits of 30 dolichos bean genotypes.

S/N	Genotype	Days to 50% flowering	Days to first pod set	Days to maturity	No. of pods per cluster	Pod length (cm)	Width of pod (mm)	Wt. of 10 green pods (g)	Yield per plant (kg)	No. of single podded clusters per plant	No. of pods per plant	Protein content (%)
1	PD 1	103.00	91.66	106.00	8.07	9.82	26.00	61.33	1.10	2.33	168.33	18.23
2	PD 2	120.00	103.00	119.33	5.13	9.39	28.53	64.33	0.63	4.33	116.00	21.00
3	PD 3	93.67	95.00	107.00	5.07	9.21	23.23	43.67	0.80	1.33	178.67	20.60
4	PD 4	101.00	98.00	110.33	7.20	9.90	21.30	41.00	0.89	0.00	220.67	21.90
5	PD 5	106.00	101.00	113.33	6.70	5.18	18.33	35.00	0.66	2.00	188.00	21.50
6	PD 6	95.33	95.67	109.67	8.70	6.57	21.93	59.00	1.26	0.00	220.33	19.73
7	PD 7	94.67	95.33	107.00	7.40	9.33	25.43	55.67	1.83	3.00	362.00	17.70
8	PD 8	95.00	94.67	106.33	5.53	8.70	25.33	48.00	1.26	3.67	253.33	18.10
9	PD 9	91.00	95.67	106.33	6.40	8.23	14.96	38.00	0.88	4.66	237.67	18.77
10	PD 10	105.00	106.00	118.00	5.63	10.48	22.07	53.00	2.03	1.00	382.67	18.43
11	PD 11	110.30	106.00	112.67	3.66	8.66	22.13	55.67	0.91	3.00	166.00	22.03
12	PD 12	93.67	95.33	106.33	8.27	9.11	17.80	29.33	1.38	3.67	487.00	20.30
13	PD 13	95.67	96.67	110.00	7.10	9.71	22.67	52.67	1.37	0.33	267.33	19.73
14	PD 14	106.67	101.67	116.33	6.25	10.17	19.63	56.67	0.76	0.33	136.00	21.17
15	PD 15	95.00	102.00	108.00	8.43	8.16	18.16	40.00	1.03	0.00	302.00	18.60
16	PD 16	103.67	102.00	105.67	6.63	7.13	18.40	37.67	0.58	0.00	160.00	20.60
17	PD 17	93.33	91.66	101.33	7.33	8.50	15.03	30.00	0.76	4.00	249.00	19.70
18	Pushpa	39.33	49.00	57.33	6.97	6.57	16.50	34.67	0.26	0.00	63.33	19.10
19	PD 19	100.33	97.33	112.67	8.30	9.52	20.13	58.00	0.88	1.00	164.00	20.10
20	Pusa Sem 2	101.33	101.33	110.67	12.93	10.00	19.70	40.33	1.79	2.33	460.66	20.10
21	PD 21	147.33	140.33	155.66	5.40	5.43	30.03	56.00	0.26	7.33	46.67	22.13
22	PD 22	105.00	108.00	117.33	5.90	10.23	21.42	49.33	1.78	2.67	357.33	19.16
23	PD 23	96.33	99.00	111.33	6.80	9.03	18.43	42.33	0.76	0.00	185.00	20.60
24	PD 24	98.00	100.67	106.67	11.23	6.57	23.78	31.67	1.60	2.00	497.66	16.26
25	Pusa Sem 3	93.67	97.00	106.00	11.26	8.30	22.57	49.00	1.70	4.33	371.67	20.30
26	PD 26	91.67	92.00	102.33	6.00	9.94	17.63	32.00	0.42	2.00	143.00	18.57
27	PD 27	94.33	93.00	106.33	6.07	7.68	15.05	52.67	1.33	0.00	250.00	19.43
28	PD 28	102.67	102.67	114.33	5.10	9.40	20.57	53.00	0.52	3.67	105.00	18.47
29	PD 29	99.33	99.67	111.33	7.27	5.70	15.13	27.66	0.54	4.33	177.33	21.83
30	PD 30	88.00	92.33	105.00	5.70	9.31	15.17	33.33	0.43	2.33	131.67	20.50
	CD (P=0.05)	6.47	7.72	6.99	2.35	0.90	2.45	6.68	0.53	1.20	117.33	0.52

Table 2. Range, mean, phenotypic coefficient of variation (PCV), genotypic coefficient of variation (GCV), heritability, expected genetic gain and genetic advances (GA) in dolichos bean.

Characters	General mean	Range	GCV	PCV	CV	h² (%)	Expected genetic gain	GA % of mean
Days to 50% flowering	98.68	39.33-147.33	15.76	16.26	4.01	93.91	31.05	31.46
Days to first pod set	97.86	49.00-140.33	12.70	13.59	4.83	87.38	23.94	24.46
Days to maturity	109.34	57.33-155.66	12.32	12.92	3.91	90.83	196.15	24.18
Number of pods per cluster	7.01	3.66-12.93	24.60	32.02	20.50	59.02	2.73	38.93
Length of pod (cm)	8.53	5.18-10.48	17.23	18.79	7.50	84.06	20.13	32.54
Width of pod (mm)	20.57	14.96-30.03	19.08	20.43	7.29	87.27	7.55	36.72
Weight of 10 green pods (g)	45.36	27.66-64.33	23.31	24.99	9.02	86.98	20.31	44.79
Pod yield per plant (kg)	1.01	0.26-2.03	45.26	55.42	31.97	66.71	0.77	76.16
Number of single podded clusters per plant	2.19	0.00-7.33	83.40	89.98	33.78	85.91	3.49	159.24
Number of pods per plant	234.84	63.33-497.66	48.05	56.93	30.55	71.21	196.15	83.52
Protein content	19.84	16.26-22.13	7.12	7.30	2.62	95.09	2.84	25.32

flowering (93.91), days to maturity (90.83), days to first pod set (87.38), width of pod (87.27), weight of 10 green pods (86.98), single podded cluster per plant (85.91) and pod length (84.06). High heritability showed the possibility of effective selection based on the phenotypic expression. High heritability estimates of these characters were also recorded by Nayar (1982), Borah and Shadeque (1992), Basu et al. (1999), Rai et al. (2008) and Pandiyan et al. (2006). Low heritability for number of pods per cluster (59.02) was observed. The highest genetic advance as percentage of mean was observed in single podded clusters per plant (159.24), number of pods per plant (83.52), yield per plant (76.16) and weight of 10 green pods (44.78). These results are in conformity with Nayar (1982). Low genetic advance was observed for days to first pod set (24.46) and days to maturity (24.18) and was also reported by Das et al. (1987). Days to 50% flowering, pod length, width of pod, weight of 10 green pods accounted for the higher heritability and higher genetic advance which suggests the role of additive gene action in the expression of these characters. So, selection will be effective for the improvement of these characters. High heritability coupled with moderate genetic advance was expressed by days to first pod set and days to maturity. So, these characters can be partially improved by selection. Similar results have been reported by Singh et al. (2011), Ali et al. (2005) and Das et al. (1987) supports the results of the present investigations. Pod yield of a crop is a complex character and is the ultimate product of action and interaction of various component characters.

Correlation of yield with other characters should be studied because sometimes the selection on the basis of yield may not be effective due to low heritability. If high correlation between easily measurable characters and yield is established, it would facilitate selection work. The correlation coefficient among the nine characters estimated from the data recorded on 30 genotypes of dolichos bean is presented in Table 3. Phenotypic correlations reflect observed relationship between traits arising from the combined effects of genotypes and environment, whereas genotypic correlations estimate the association between traits resulting from linkage or pleiotropy as well as physiological/metabolic constraints. Significant positive phenotypic correlations were observed between yield and number of pods per plant (0.459), weight of 10 green pods (0.409) and length of pod (0.375) were also reported by Lal et al. (2005) and Singh et al. (2011). In contrast, the correlation between yield and protein content of pods (-0.332) was significantly negative. Therefore, selection for yield and its positively correlated characters should result in correlated response for increased yield, but protein content in pods would be reduced. This positive correlation between yield and its contributing characters show simple, indirect selection criteria in the development of high yielding cultivars. Genotypic correlations between yield and other traits were slightly larger in magnitude and similar in direction to their corresponding phenotypic correlations. These results indicated that superior

Table 3. Estimation of correlation coefficients at the phenotypic and genotypic levels.

S/N	Characters	Path	Days to 50% flowering	Days to first pod set	No. of pods per cluster	Single podded clusters per plant	No. of pods per plant	Length of pod (cm)	Width of pod (mm)	Weight of 10 green pods (g)	Days to maturity	Pod yield per plant (kg)	Protein content
1	Days to 50% flowering	rp		0.915**	-0.156	-0.002	0.513**	0.101**	0.221	0.418*	-0.054	0.936**	0.359**
		rg		0.961	-0.209	-0.022	0.579	0.447	0.019	0.476	-0.075	0.979	0.379
2	Days to first pod set	rp			-0.070	-0.019	0.0419**	0.254	0.093	0.391**	0.065	0.964**	0.266*
		rg			-0.113	-0.072	0.0477	0.311	0.096	0.468	0.057	0.985	0.294
3	No. of pods per cluster	rp				-0.047	-0.034	-0.197	.0352**	-0.090	0.478**	-0.129	-0.267*
		rg				-0.061	-0.0833	-0.258	0.567	-0.0151	0.755	-0.214	-0.326
4	Single podded clusters per plant	rp					0.075	0.283*	0.327*	-0.170	0.193	0.009	-0.174
		rg					0.1060	0.307	0.364	-0.194	0.212	-0.020	-0.188
5	No. of pods per plant	rp						0.591**	0.192	0.296**	-0.003	0.459**	-0.047
		rg						0.712	0.260	0.325	-0.008	0.525	-0.050
6	Length of pod (cm)	rp							0.211	-0.002	-0.184	0.375	0.018
		rg							0.235	0.006	-0.235	0.421	0.031
7	Width of pod (mm)	rp								-0.025	0.883**	0.067	-0.386**
		rg								-0.081	0.861	0.045	-0.479
8	Weight of 10 green pods (g)	rp									-0.044	0.409**	-0.359**
		rg									-0.004	0.475	-0.128
9	Days to maturity	rp										-0.014	-0.379**
		rg										-0.048	-0.455
10	Pod yield per plant (kg)	rp											0.332**
		rg											0.360
11	Protein content	rp											
		rg											

$*P=0.05.$

yielding ability in dolichos bean was associated with enhanced expression for number of pods per plant and selection for them should result in a correlated response for increased yield. Similar results were obtained by Nandi et al. (1997) and Rai et al. (2008).

Among the genotypes evaluated, PD 10 was the highest yielder (2.06 kg) of pod per plant but recorded the second best due to unattractive pod characteristics (extra-large and white coloured pod). Pushpa produced earlier marketable yield and was bushy in nature, PD 21 and PD11 had maximum protein contents, while Pusa Sem 3, Pusa Sem 2 and PD-24 had medium sized green coloured pods. All these characters were not present in a single genotype, so attempts should be made to develop ideotype with all the desirable characters in a single plant through proper hybridization and selection. High heritability estimates were obtained for protein content, days to 50% flowering, days to maturity, days to first pod set, width of pod, weight of 10 green pods, single podded clusters per plant and pod length. High heritability showed that selection based on phenotype for these characters would be effective. The highest genetic advance as percentage of mean was observed in single podded clusters per plant, number of pods per plant, yield per plant and weight of 10 green pods.

Days to 50% flowering, pod length, width of pods and weight of 10 green pods accounted for the higher heritability and genetic advance which suggests the role of additive gene action. So, selection could be effective for the improvement of these characters. Significant positive phenotypic correlations were observed between yield and other yield components viz. days to first pod set, days to 50% flowering, number of pods per plant, weight of 10 green pods and length of pod. In contrast, the correlation between yield and protein content of pods was significantly negative; therefore, selection of positively correlated characters with yield would be easy to practice for obtaining high breeding types. Early selection based on days to 50% flowering will result in high marketable yield and is thus desirable character for early life cycle selection.

In summary, there was adequate genetic variability within the germplasm evaluated for the improvement of all the characters studied viz. pod yield, growth and related traits, which can be utilized for further improvement through selection.

REFERENCES

Ali F, Sikdar B, Roy AK, Joarder OI (2005). Correlation and genetic variation of different genotypes of Lablab bean, *Lablab purpureus* (L.) Sweet. Bangl. J. Bot. 34(2):125-28.

Al-Jibouri HA, Millar PA, Robinson HP (1958). Genotypic and environmental variances and covariances in upland cotton crosses in interspecific origin. Agron. J. 50:633-7.

Allard RW (1960). Principles of Plant Breeding, John Wiley and Sons Inc New York.

Anonymous (1961). Agriculture and Horticulture seeds, FAO, United Nations, Rome. pp. 531.

Basu AK, Pal D, Sasmala SC, Samanta SK (1999). Genetic analysis for embryo weight, cotyledon weight and seed protein in lablab bean. Vegetable Sci. 26:37-40.

Bose TK, Som MG, Kabir J (1993). Vegetable Crops, Naya Prakash Kolkata, India. p. 612.

Borah P and Shadeque A (1992). Studies on genetic variability of common Dolichos bean. Indian J. Hort. 49:270-273.

Burton GW, Devane CH (1953). Estimating heritability in tall Fescue (*Festiucaarundinaceal*) from replicated clonal material. Agron. J. 45:514-18.

Das AK, Hazra P, Som MG (1987). Genetic variability and heritability studies in dolichos bean (*Dolichos lablab*). *Vegetable Sci.* 14:173-89.

Lal H, Rai M, Verma A, Vishwanath (2005). Analysis of genetic divergence of Dolichos Bean (*Lablab purpureus*) genotypes. Vegetable Sci. 32(2):129-132.

Nandi A, Tripathy P, Lenka D (1997). Correlation, co-heritability and path analysis studied in *Dolichos* bean. ACIAR-Food-Legume-Newslett. 25:1-2.

Nayar KM (1982). Studies on genetic divergence and behavior of few intervarietal crosses in field bean (*Lablab purpureus*).Mysore J. Agric. Sci. 16:486.

Pandiyan M, Subhalakshmi, Jebaraj S (2006). Genetic variability in greengram(Vignaradiata (L.)Wilezek). Int. J. Plant Sci. 1(1):72-75.

Rai N, Singh PK, Verma A, Lal H, Yadav DS, Rai M (2008). Multivariate characterization of Indian bean [*Lablab purpureus* (L.)Sweet] genotypes. Indian J. Plant Gen. Res. 21(1):42-45.

Singh PK, Rai N, Lal H, Bhardwaj DR, Singh R, Singh AP (2011). Correlation, path and cluster analysis in hyacinth bean (*Lablab purpureus* L. sweet). J Agric. Tech. 7(4):1117-1124.

12

The effects of organic fortification with pulse sprout extracts from horse gram and cowpea on the seedling quality characteristics of rice variety Co. 43

Jayanthi M., R. Umarani and V. Vijayalakshmi

Department of Seed Science and Technology, Tamil Nadu Agricultural University, Coimbatore-641 003, India.

Rice is the staple food for over half of the Indian population ways and means to fortify seeds organically for better seed vigour has become important and emphasized. Seed fortification is one of the important seed invigouration treatment. The study was conducted to determine the effects of seed fortification with pulse sprout extract on seedling quality characteristics of rice seeds. Laboratory experiment was conducted with the treatments which include fortification of rice seeds with water soaking, 1, 2, 3, 4 and 5% of horse gram sprout extract and cowpea sprout extract. Seed fortification was done in seeds with $12 \pm 1\%$ moisture content by soaking for 12 h. Later the soaked seeds were dried in shade for one day followed by sun drying to reach the original moisture content. The seedling quality characteristics were radical protrusion (%), days to maximum germination (days), speed of germination, germination (%), shoot length (cm), root length (cm), vigour index and dry matter production (g seedlings^{-10}). Among all the treatments used, 2% horse gram sprout extract enhanced comparatively better seedling quality characteristics. Hence, seed fortification with 2% horse gram sprout extract could be recommended for rice as a pre-sowing seed invigorative treatment.

Key words: Rice seeds, germination, vigour index, pulse sprout extract, horse gram, cowpea.

INTRODUCTION

Rice is the major staple food for more than two billion people in Asia and one third of the calorific intake of nearly one billion people of Africa and Latin America. The uninterrupted and disproportionate use of chemical fertilizers over a longer period of time has resulted in deterioration of soil quality and reduced yield. To maintain long – term food production there is a need for sustainable agricultural practices. This is one of the aims of organic farming and consumers are prepared to pay higher prices for certified organic products. In many developing countries agriculture is still largely based on low inputs, because farmers cannot afford the high costs of chemical fertilizers and pesticides. For such farmers, organic farming can provide a better economic alternative because the advantages are two fold; (i) the inputs are of lesser cost and (ii) the produces fetch higher price. Sprouting improves nutritional quality of seeds reported by Deshpande et al. (2002). Germination unfolds, and enzymes trigger elaborate biochemical changes. Proteins break into amino acids. Water-soluble vitamins such as B complex and vitamin C are created. Fats and carbohydrates are converted into simple sugars. Weight increases as the seed absorbs water and minerals (Vidal-Valverde et al., 2002). Kareem et al. (1989) reported thatseeds of rice Cv.IR 36 and IR 42 treated with 2.5% neem kernel extract gave more vigourous seedlings than untreated seeds. Joseph and Nair (1989) concluded that seed hardening in rice with ten percent cowdung solution registered its superiority in early germination, root and shoot growth and vigour index. Hence it was hypothesized that application of the nutrient extract from the sprouted pulses in the form of seed fortification will enable better crop growth and productivity of rice. Horse gram and cowpea are the low cost and easily available pulses in Tamil Nadu. Seed fortification treatment mainly supplies nutrient to the seed to germinate into vigourous seedlings. The objective of this research wok was to determine the effects of seed fortification with pulse sprout extract on seed germination and seedling quality characteristics in rice seeds.

MATERIALS AND METHODS

The study was conducted on rice variety CO 43. Seeds of rice obtained from Paddy Breeding Station, Tamil Nadu Agricultural University, Coimbatore was used for the study. The seeds were manually cleaned before conducting the studies. The Experiment was conducted at Department of Seed Science and Technology, Tamil Nadu Agricultural University, Coimbatore. Horse gram and cowpea pulse sprout extract were used for seed fortification. Horse gram and cowpea seeds (50 g each) were soaked overnight (16 h) at room temperature and incubated in a wet cloth for 12 h to enable sprouting. 100 g of sprouts were obtained from 50 g of dry seeds. Later, 100 g of sprouts of horse gram and cowpea were separately ground in a mixer-grinder by using 100 ml of ice water refrigerated at 5°C. The ground substance was squeezed through cloth bag and 100 ml extract of 100% concentration was obtained.

The rice seeds variety Co 43 were soaked in water and also varied concentrations such as 1, 2, 3, 4 and 5% of both horse gram and cowpea sprout extracts for 12 h. Later the soaked seeds were dried in shade for one day followed by sun drying to reach the original moisture content, 12 ± 1. Untreated seeds were maintained as control. Fortified seeds were subjected to germination test with four replicates of 100 seeds in between paper towels. The test conditions were 25 ± 2°C and 95 ± 5% RH, illuminated with fluorescent light (750 – 1250 Lux). The seeds were checked daily upto 14 days for protrusion of radicle. The number of normal seedlings were counted after 14 days and expressed as germination percentage. Observation were also made on radical protrusion (%), days to maximum germination (days) (Mauromicale and Cavallaro, 1995), speed of germination (Maguire, 1962), germination (%) (ISTA,1999), shoot length (cm) (the distance between collar region to the tip of the primary leaf), root length (cm) (the distance between collar region to the tip of the primary root), vigour index (Abdul-Baki and Anderson,1973) and dry matter production (g seedlings^{-10}), for which seedlings were dried in a hot air oven maintained at 85°C for 48 h and cooling in a desiccator for 30 min and afterwards weighing in an electronic digital balance. All the data collected were subjected to analysis of variance (ANOVA) using the Factorial Completely Randomized Design and means were separated using the least significant differences method (LSD) at 5% level of significance (Panse and Sukatme, 1985) only when a significant "F" test was obtained. Wherever necessary, the percent values were transformed to angular (Arc-sine) values before analysis.

RESULTS AND DISCUSSION

Highly significant variations were observed in the evaluated seedling quality parameters obtained from the experiments (Table 1). The results showed that horse gram 2% pulse sprout extract effected the maximum radical protrusion of 71% whereas the control was 60%. The seeds fortified with horse gram 2% pulse sprout extract required 4 days for maximum germination, while the control seeds required 5.3 days. With respect to the speed of germination, the seed treated with horse gram 2 and 3% pulse sprout extract and 4 and 5% cowpea pulse sprout extract resulted in the maximum speed of germination of 5.8, whereas the control was 5.2. As shown in Table 1, the germination percentage of 80 was the highest and this was with the seeds fortified with 2% horse gram extract, whereas that of the control seeds was 60%. In Table 2, the minimum shoot length of the control seeds was 9.2 cm and this was followed by horse gram 1% extract of 9.3 cm. The horsegram 2% pulse sprout extract induced the maximum shoot length of 11.4 cm. Among the treatments 5% horse gram extract induced the minimum root length 16.9 cm and the control brought about a value of 17.8 cm. The maximum root length obtained with the use of horsegram 2% extract was 19.8 cm. The control seeds yielded the minimum dry matter production of 0.12 g seedlings^{-10} whereas the maximum dry matter production obtained by the use of 2 and 3% horsegram pulse sprout extract was 0.19 g seedlings^{-10}. Among the evaluated seedling vigour characters and computed vigour index values of all the concentrations, the seeds fortified with 2% horse gram pulse sprout extract yielded the highest value of 2496 while the control seeds brought about the lowest vigour index of 1620.

In this experiment it was found that the seed fortified with horse gram extract recorded higher values for all the parameters compared to the same concentrations of cowpea extract. This might be because bioactive substances present in horse gram extracts are more optimum for rice seeds than cow pea extracts. Kadam and Salunkhe (1985) reported that horse gram is an excellent source of iron and molybdenum. Pre-sowing seed treatments with nutrients produce physiological effects on seed and thereby improve its emergence and productivity (Natarajan, 2003). Sprouting has been identified as an inexpensive and effective technology for improving the quality of legumes, by enhancing their digestability, increasing the content of amino acids.(Chang and Harrold,1988) and reducing the levels of antinutrients (Vidal valvarde et al., 2002).The results revealed that irrespective of the pulse sprout all the treatments recorded better performance compared to control. This may be due to nutritional quality increased in pulse sprouts which is used for seed fortification. Augustin and Klein (1989) reported that the content of phosphorous, potassium, zinc and copper increased

Table 1. Effect of squeezed extract of sprouted horsegram and cowpea at lower concentration on seed germination and seedling vigour of rice seeds.

Treatment and concentration	Radicle protrusion (%)	Days to maximum germination (Days)	Speed of germination	Germination (%)
Control	60(52.53)[c]	5.3	5.2	60(50.76)[d]
Water	63(50.76)[bc]	4.6	5.5	62(51.94)[cd]
Horse gram 1%	65(53.73)[b]	4.3	5.7	76(60.66)[ab]
Horse gram 2%	71(57.41)[a]	4.0	5.8	80(63.43)[a]
Horse gram 3%	68(55.55)[ab]	4.0	5.8	72(58.05)[b]
Horse gram 4%	65(53.73)[b]	4.0	5.6	68(55.55)[bc]
Horse gram 5%	55(47.87)[d]	4.3	5.2	64(53.13)[c]
Cowpea 1%	60(50.76)[c]	5.3	5.7	62(51.9)[cd]
Cowpea 2%	60(50.76)[c]	5.3	5.7	64(53.13)[c]
Cowpea 3%	62(51.94)[bc]	4.6	5.7	64(53.13)[c]
Cowpea 4%	62(51.94)[bc]	4.3	5.8	72(58.05)[b]
Cowpea 5%	62(51.94)[bc]	4.6	5.8	68(55.55)[bc]
Mean	52(43.56)	4.46	5.62	55 (47.87)
SEd	4.158	0.503	0.217	6.198
C.D.(0.05)				
T	0.64	0.083	0.096	0.69
C	0.72	0.09	0.19	0.77
T × C	1.44	0.18	1.44	1.55

Figures in parenthesis are arcsine values.

Table 2. Effect of squeezed extract of sprouted horsegram and cowpea at lower concentration on seed germination and seedling vigour of rice seeds.

Treatments concentrations	Shoot length (cm)	Root length (cm)	Dry matter production (g seedlings^{-10})	Vigour index
Control	9.2	17.8	0.12	1620
Water	10.7	17.3	0.14	1736
Horse gram 1%	9.3	18.5	0.16	2113
Horse gram 2%	11.4	19.8	0.19	2496
Horse gram 3%	11.1	19.0	0.19	2167
Horse gram 4%	10.2	18.0	0.17	1918
Horse gram 5%	9.6	16.9	0.14	1696
Cowpea 1%	9.7	17.1	0.14	1686
Cowpea 2%	10.7	17.5	0.14	1779
Cowpea 3%	10.7	17.1	0.15	1736
Cowpea 4%	11.1	18.0	0.15	1804
Cowpea 5%	10.9	17.6	0.15	1710
Mean	**10.38**	**17.9**	**0.153**	**1918**
SEd	0.757	0.857	0.021	259.752
C.D.(0.05)				
T	0.17	0.31	0.002	31.80
C	0.19	0.34	0.002	35.55
T×C	0.39	0.69	0.005	71.11

significantly as result of germination in various legumes. He also observed that germination of a wide array of legumes significantly improved the thiamine content between 7.2 and 147.7%. Similar results was reported by Martin-Cabrejas et al. (2007) that the remarkable increase in sodium content increases the nutritional qualities of sprouted legumes. Seed fortification with micronutrients such as, manganese sulphate, ferrous sulphate, zinc sulphate, ammonium chloride, ammonium molybdate, potassium chloride, potassium dihydrogen phosphate, magnesium sulphate, borax and ammonium sulphate in three concentrations of 0.5, 0.75 and 1.0 percent have been proved to be better than control in terms of speed of germination, germination percent, root length (cm), shoot length (cm), dry matter production (g seedlings^{-10}) as well as vigour index (Marimuthu, 2007). Grzywnowicz-Gazda (1982) reported that soaking of spring barley seeds in borax, manganese sulphate, Ammonium molybdate, zinc sulphate, Ferrous sulphate and magnesium sulphate either individually or in mixture for 24 h increased the trace element concentration in the grain. The seed treatment was also found to increase the germination capacity, growth and dry matter production of barley seedlings when compared with untreated control. Nitrogen containing compound might have stimulated the germination by increasing the seed cytokinin content, occurring naturally in seeds, which interacted with growth inhibitors and enhance the metabolic process, leading to higher germination (Khan, 1980). FaShui et al. (1996) noticed an increase in germination rate, seed vigour, seedling fresh weight, seedling height and root length when maize seeds were soaked in solution of 0.3% $CaCl_2$, 0.1% $ZnSO_4$ either individually or in combination. Seed fortification with molybdenum as a Sodium molybdate, Zinc sulphate and Manganese sulphate at 0.1, 0.2 and 0.2% respectively, brought maximum seed physiological quality attributes in laboratory (Natesan, 2006). The fortification with zinc sulphate was found to increase the level of vitamins, biotins and thiamins and its coenzymes (Srimathi and Malarkodi, 2000). The overall performance of the treatments underscore that manganese sulphate (1.0%), ferrous sulphate (0.75%) and ammonium molybdate (0.75%) can well serve as seed fortification agent to increase the seed vigour of rice seeds. During seed fortification, the first phase of germination ends with completion of imbibition process and hence the time taken from sowing to emergence is much reduced (Hegarty, 1970). Fortification of seed increased the field germination of corn by promoting embryo growth (Zubenko, 1959). The improvement in field emergence due to fortification could also be ascribed to activation of cells, which results in the enhancement of mitochondrial activity leading to the formation of more high energy compounds and vital biomolecules, which are made available during the early phase of germination observed by (Dharmalingam et al., 1988). These initial changes culminate in enlargement of the latent embryo. The probable reason for higher germination in fortified treatment could be due to greater hydration of colloids, higher viscosity and elasticity of protoplasm, increase in bound water content, lower water deficit, more efficient root system (May et al., 1962) and increased metabolic activity (Joseph and Nair, 1989). It might also be due to enhanced metabolic activity resulted

in early germination as stated by Kamalam and Nair (1991). Thus it is obvious that the presence of bioacitve substances in sprouted horse gram and cowpea extracts such as, amino acid, vitamins and minerals could have resulted in fortification of rice seeds as corroborated by earlier reports. The seed vigour extended due to fortification of rice seeds with pulse sprout extracts had resulted in better seedling growth as reflected in germination percentage, root length and shoot length. The increase in dry matter production due to the treatments in the present study is in accordance with the report of Periyathambi and Palaniyappan (1981), Selvaraju (1992) in sorghum and Rangasamy et al. (1993) in agricultural crops.

The seeds fortified with horsegram 2% extract recorded the maximum increase over the control to the tune of 9.2, 11, 33, 58 and 54% for radical protrusion percentage, root length (cm), germination percentage, dry matter production (g seedling^{-10}) and vigour index, respectively. It was followed by horsegram 3% extract. With respect to cowpea pulse sprout, the maximum increment of 3, 1, 13, 25, and 11, respectively was recorded in cowpea 4% extract which was followed by 5% extract. Thus the study on seed fortification revealed that with respect to horse gram, 2% horse gram sprout extract was comparatively the most appropriate concentration followed by 3 %. Similarly, among the cowpea extract concentrations, 4% concentration was also the most appropriate followed by 5% concentration.

REFERENCES

Abdul-Baki AA, Anderson JD (1973). Vigor determination in soybean seed by multiple criteria. Crop Sci. 13:630-633.

Augustin J, Klein B (1989). Nutrient composition of raw, cooked, canned and sprouted legumes. Legumes, chemistry, technology and human nutrition, pp. 187–217.

Chang KC, Harrold RL (1988). Changes in selected biochemical components in vitro protein digestibility and amino acids in two bean cultivars during germination. J. Food Sci. 53:783–787.

Deshpande SS, Salunkhe DK, Oyewole OB, Azam-Ali S, Battcock M, Bressani R (2002). Fermented grain legumes, seeds and nuts. A global perspective. FAO, Agricultural Services Bulletin, P. 142.

Dharmalingam C, Paramasivam K, Sivasubramanian V (1988). Seed hardening to overcome adversity. The Hindu, November 16 (Wednesday).

FaShui H, ChengCang M, XuMing W, GuangMing J (1996). Effect of ca^{2+} and zn^{2+} on seed vigour and some enzyme activities during seed germination of maize. Plant Physiol. Comm. 32(2):100-112.

Grzywnowicz-Gazda Z (1982). Response of spring barley to fertilizing with microelements and magnesium through the seeds. Acta agraria et silvestria Series agrarian 1(21):65-78.

Hegarty TW (1970). The possibilities of increasing field establishment by seed hardening. Hort. Res. 10:59-64.

ISTA. (1999). Seed Sci. and Technol. (Supplement Rules), 27:25-30.

Joseph K, Nair NR (1989). Effect of seed hardening on germination and seedling vigour in paddy. Seed Res. 17(2):188-190.

Kadam SS, Salunkhe DK (1985). Nutritional composition , processing and utilization of horse gram and moth bean. Food Sci. Nutr. 22(1):1-26.

Kamalam J, Nair NR (1991). Studies on the effect of soil moisture content and seed soaking on the germination of legumes. Legume Res. 14(3):153-154.

Kareem AA, Sexena RC, Boncodin MEM, Krishnasamy V, Seshu DV (1989). Neem as seed treatment for rice before sowing ; Effects on two homopterous for insects and seedling vigour. J. Econ. Ent. 82(4):1219-1223.

Khan AA (1980). The physiology and biochemistry of seed dormancy and germination. North Holland Publishing Company, Amsterdam, New York, Pages, P. 127.

Maguire JD (1962). Speed of germination – aid in selection and evaluation for seedling emergence and vigour. Crop Sci. 2:176-177.

Marimuthu A (2007). Comprehensive seed enhancement technique for direct sown rice (Oryzasativa.L.)cv CO 43. M.Sc. (Ag.) Thesis, Tamil Nadu Agricultural University, Coimbatore, pp. 6-17.

Martin-Cabrejas MA, Diaz MF, Aguilera Y, Benitez V, Molla E (2007). Influence of germination on the soluble carbohydrates and dietary fibre fractions in non-conventional legumes. Food Chem. 107:1045-1052.

Mauromicale G, Cavallaro V (1995). Effects of seed osmopriming on germination of tomato at different water potential. Seed Sci. Technol. 23:393-403.

May LH, Milthorpe EJ, Milthorpe FL (1962). Pre-sowing hardening of plants to drought. Field Crop Abstr. 15:93-98.

Natarajan K (2003). Ph.D. Thesis, Dept. of Seed Sci. Technol. Tamil Nadu Agric. Uni. Coimbatore, India.

Natesan P (2006). Designing integrated seed treatment in blackgram (Vignamungo(L).Hepper) for irrigated and rainfed ecosystem. Ph.D. Thesis, Tamil Nadu Agricultural University, Coimbatore, pp. 5-19.

Panse VG, Sukhatme PV (1985). Statistical methods for Agricultural workers. ICAR, Publication, New Delhi, pp. 327-340.

Periyathambi C, Palaniyappan SP (1981). Effect of premonsoon sowing, depth of seed placement and seed hardening on total drymatterproduction, growth and yield components of rainfed sorghum. Madras Agric. J. 68(2):100-104.

Rangasamy A, Purushothaman S, Devasenapathy P (1993). Seed hardening in relation to seedling quality characters of crops. Madras Agric. J. 80(9):535-537.

Selvaraju K (1992). Studies on certain aspects of seed management practices for seed production under moisture stress condition in sorghum cv. Co 26 (Sorghum bicolor(L.)Moench). M.Sc.(Ag) Thesis, Tamil Nadu Agricultural university,Coimbatore, pp. 122-140.

Srimathi PK, Malarkodi K (2000). Influence of nutrient pelleting on seed yield and soybean cv. Co-1. Indian Agric. 44(1-2):79-82.

Vidal-Valverde C, Frías J, Sierra I, Blazquez I, Lambien F, Kuo YH (2002). New functional legume food by germination. Effect on the nutritive value of beans, lentils and peas. Eur. Food Res. Technol. 215:472–476.

Zubenko VKH (1959). The effect of preplanting hardening of seeds against drought on the productivity of corn in late plantings. Plant Physiol. 6:341-343.

13

Effects of methanol and some micro-macronutrients foliar applications on maize (*Zea mays* L.) maternal plants on subsequent generation yield and reserved mineral nutrients of the seed

M. Yarnia[1], E. Khalilvand Behrouzyar[1]*, F.R. Khoii[1], M. Mogaddam[2] and M. N. Safarzadeh Vishkaii[3]

[1]Department of Crop Production and Plant Breeding, Tabriz Branch, Islamic Azad University, Tabriz, Iran.
[2]Department of Crop Production and Plant Breeding, Tabriz University. Iran.
[3]Department of Crop Production and Plant Breeding, Rasht Branch, Islamic Azad University, Rasht, Iran.

Tests were done to investigate effects of foliar applications in various combinations on maternal corn plants for reserved mineral nutrients and seed yield in a subsequent generation. The experiment was done as a factorial based on randomized complete block design (RCBD) with three replications during the growing season of 2009 to 2011. Treatments tested in the investigation were as follows: Four growth stages; 8 to 10 leaf, tasseling, seed-filling and at all stages and seven foliar applications of methanol, Zn, B, Mg, N, Mn, a mixture of all combinations and a separate plot as the control. Data analysis showed a significant effect of the combination type of foliar application in different stages on reserves of N, Mg, Zn, Mn, B, and seed yield of a subsequent generation. Detailed results of the study showed that foliar application with a mixture of all combinations in all stages had the highest reserve of N in seeds (1/45%). Results also proved that Mg foliar application at all stages had the highest effect on Mg reserve in seeds (0/172 mg kg^{-1}) and Zn foliar application at the tasseling stage had the highest effect on reserved Zn of seed (44 mg kg^{-1}). Results showed that Mn foliar application at the 8 to 10 leaf stage had the highest effect on reserved Mn of seed (12 mg kg^{-1}) and B foliar application at all stages had the highest effect on reserved B of seed (9/1 mg kg^{-1}). In conclusion a mixture of all combinations at all stages had the highest (1309 g m^{-2}) and at the tasseling stage had the lowest (713 g m^{-2}) seed yield in the subsequent generation.

Key words: Foliar application, methanol, micro-macronutrients, *Zea mays*.

INTRODUCTION

Maize (*Zea mays* L.) is an important crop worldwide (Graham, 2008). Seed development can be affected by many factors (biotic and/or environmental) that have a negative effect on seed quality and adverse consequences for crop yield (Welch, 1995). Genotype, seed size and weight and environmental stresses such as water deficit, temperature extremes, pathogens, deficient nutrient supply, mineral toxicity, salinity, soil acidity and anaerobiosis are all factors that directly affect growth and nourishment of a maternal plant. These adverse conditions have a direct impact on seed development, seed nutrient reserves and ultimate seed quality (Welch, 1986). As environmental conditions affect seed quality during seed formation this also affects seedling establishment at the next growing season (Zakaria et al., 2009).

A shortage of nitrogen (N) restricts the growth of all

---

*Corresponding author. E-mail: e.khalilvand@iaut.ac.ir.

Abbreviations: NFA, Nitrogen **foliar** application; **MgFA,** magnesium foliar application; **ZnFA,** zinc foliar application; **MnFA,** manganese **foliar** application; **BFA,** boron foliar application; **MFA,** methanol foliar application.

plant organs such as roots, stems, leaves, flowers and fruit (including seeds) (Barker and Bryson, 2007). N-Foliar application serves to increase reproductive structures that can increase seed yield (Welch, 1995). As Sawan et al. (1989) reported increasing N in cotton plants from 108 to 216 kg ha^{-1} increased seed viability in terms of germination velocity and total germination as well as seedling vigor.

Magnesium (Mg) has major physiological and molecular roles in plants; it is a component of the chlorophyll molecule, a co-factor for many enzyme processes associated with phosphorylation and the hydrolysis of various compounds, as well as a structural stabilizer for various nucleotides (Merhaut, 2007). Current known reports on research in to the effect of Mg on seed vigor are insufficient. Welch's (1986) studies on barley (*Hordeum vulgare* L.) showed that seedlings grown from a maternal plant, which were fed insufficient amounts of Mg had less chlorophyll in comparison to those grown from sufficiently fed maternal plants.

Zinc (Zn) can be readily transported from vegetative tissue into reproductive tissue according to a plant's capacity. However, the transformation from vegetative tissue into reproductive tissue decreases when a zinc supply is inadequate (Welch, 1995). An increased zinc content of bread wheat grain from 355 ng to 1465 ng grain^{-1} leads to an increased subsequent yield (Yilmaz et al., 1998). Gangloff et al. (2002) found that an application of zinc sulphate in maize plants increased dry matter and zinc accumulation in leaf and grain.

Manganese (Mn) is involved in many biochemical functions. It primarily acts as an activator of enzymes, involved in respiration, amino acid and lignin synthesis and hormone concentrations (Humphries et al., 2007). Mn foliar application (25 mg Mn L^{-1} as MnEDTA) on cotton (*Gossypium barbadense* L.) grown on Mn-poor soil increased seed yield, seed weight, seed viability and seedling vigor in terms of length of hypocotyls, radicle and fresh and dry weights of seedling (Sawan et al., 1993).

Deficiency of boron (B) can cause reductions in crop yield, impair crop quality, or have both of these effects (Gupta, 2007). B deficiency in a maternal plant causes the cotyledon leaves to become yellow, serrated pointed with distinguishing colors such as yellow or tan, as they reach the reproductive stage. Moreover it causes a high proportion of cells in B-deficient seeds to become empty or to collapse (Welch, 1986). However, during seed development the need for B is higher than it is at the growth period (Marschner, 1995). A shortage of B in the maternal base leads to a decrease in viability of the produced seed. In black gram (*Vigna mungo* L.) the concentration of 6 mg kg^{-1} dry weight of seed is considered as the critical level for normal seed viability (Bell et al., 1989).

Methanol spray is used to increase CO_2 fixation in crops (Nadali et al., 2010). Foliar-spray applications of aqueous methanol are reported to increase yield, accelerate maturity and reduce effects of drought stress (Ramirez et al., 2006; Downie et al., 2004). Mirakhori et al. (2009) demonstrated that 21% (v/v) methanol spray poses the greatest impact on yield, and other physiological traits in soybean. Foliar application is the best way to nourish plants that grow in soil with poor quality due to adverse pH (Ishii et al., 2002). In the dry and semi-dry areas of Iran, the absorption of micronutrients is low due to a high pH level of the soil. In order to use chemical fertilizers efficiently, it is essential that fertilizer is applied by foliar-applications. The aim of this study was to investigate the effects of different combinations of foliar applications on corn maternal plants on yield and reserved mineral nutrients in seeds produced from hybrid corn seeds.

MATERIALS AND METHODS

This research was done at the Agricultural Station of Tabriz Islamic Azad University during the two farming years of 2009 to 2011. The station is located 5 km from Tabriz-Iran at 46° and 17′E and 38° and 5′N, at an altitude of 136 meters above sea level. The experiment was conducted as a factorial in a Randomized Complete Block Design in three replications. The experimental factors were; (A) Four growth stages (a_1: 8 to 10 leaf; a_2: tasseling; a_3: seed-filling, and a_4: all stages); (B) the 7 foliar applications were (b_1: Methanol, b_2: N (Urea), b_3: Mg sulfate ($MgSO_4 \cdot H_2O$), b_4: Zn sulfate ($ZnSo_4$), b_5: boric acid (H_3BO_3), b_6: Mn sulfate ($MnSO_4.H_2O$), and b_7: a mixture of all combinations. In order to perform the orthogonal contrast between treatments and the control, there was an untreated control plot applied as a separate plot. The soil used for tests included on average 68% sand, 18% silt and 14% clay. Taking into account the triangular texture of the soil, the experimental texture of the soil was sandy loam. The pH of the soil was low-average alkaline (7/8 to 8/9), which makes it difficult for a plant to absorb micronutrients such as Fe, Mn, Cu, B and Zn (Table 1).

Plant material

B73 was used as the maternal plant and Mo17 as the paternal plant. In each plot, a row of paternal plants was planted around 3 rows of maternal plants (Beck, 2004). As soon as a tassel appeared in the maternal plants, they were cut. The distance between the site of the experiment and other farms was at least 400 m.

Experimental material

The first growing season

In order to prepare the field for planting the corn to produce hybrid seeds from a maternal base, the first growing season (that is, 2009 to 2010), was spent plowing, harrowing, making furrows and plotting. Then pre-planting fertilization was carried out using urea fertilizer at the rate of 150 kg ha^{-1} in furrows. Each plot consisted of 5 rows, 75 cm row spacing and 25 cm plant intervals. Seeds were planted at the depth of 5 to 7 cm in the water strain of each furrow. In order to pollinate from paternal bases, tassels were cut as soon as they appeared in the maternal bases to extract productive seeds from maternal plants. Hybrid seeds were planted at the end of the first growing season.

Mineral treatment: Mineral concentrations of foliar applications were determined in ratios of 5 to 1000. Considering the size of the

Table 1. Soil physical and chemical analysis.

Clay (%)	Silt (%)	Sand (%)	K (ava) (ppm)	B (ava) (ppm)	Zn (ava) (ppm)	N (%)	Ec × 10³
14	18	68	600	0/93	0/4	0/133	1/57

plots, applications were made with a hand sprayer to facilitate effective foliar application in terms of precision and delicacy. Spraying was done thoroughly until foliar dropped from the plants. Furthermore, Tween80 was used as a surfactant to enable the leaves to absorb nutrient minerals. The control plots were water sprayed consistently to avoid the effects of foliar application used for experimental plots.

Methanol treatment: 10% volume- methanol was used for the MFA. Amounts of 1 g amino acid Glycine and 1 mg Tetrahydrofolate were added to 1 L of methanol.

Measurement of mineral concentration: Mineral concentrations were determined in all treatments. In order to determine exact amounts of extracts of Mg, Mn, B and Zn, the dry ashing procedure was used and then mixed with hydrochloric acid. Dry ashing was done according to the following procedure; a 2 g sample of seeds was put inside a porcelain crucible and then placed in a muffle furnace. The temperature of the furnace was slowly increased to the ashing temperature of up to 550°C for over 2 h and kept for 6 h. After cooling, 10 cc hydrochloric (2 mol) was added. This time the temperature was increased up to 80°C. As soon as white steam was observed, the content of the porcelain crucible was filtered through a Whatman No. 40 filter paper. In this way, 100 cc extract was obtained (Karla, 1998). The digestion method was used to produce an extract for measuring N; the digestion method was used, which is performed inside a volumetric flask with sulfuric acid 96%, salicylic acid and hydrogen peroxide 30%. In this method a 0.3 g sample of seeds was put in the volumetric flask. Then the digestion mixture, including 100 cc of sulfuric acid, 6 g salicylic acid and 18 cc de-ionized water were added and were shaken to mix well (Waling et al., 1989). After it had cooled down, 5 drops of hydrogen peroxide were added and heated to 280°c for 5 to 10 min until white steam emerged. When the sample became colorless and cool, 10 cc of water was added (to make 100 cc). Having made the extract, the atomic spectrophotometer absorption was used (Hanlon, 1998).

The second growing season

To prepare the field for planting, hybrid seeds were collected from the mother plant. And preparations were made during the second growing season (that is, 2010-2011), such as plowing, harrowing, making furrows and plotting. Each plot consisted of 5 rows, 75 cm row spacing and 25 cm plant intervals. Plots were irrigated according to the condition of the soil and climate. Field irrigation was done once a week up to the end of growth.

Statistical analysis

Data was checked for analysis of variance, means comparison (based on Least Significant Difference Test) and the correlation between different treatments by MStatc and SPSS17 packages.

RESULTS AND DISCUSSION

Results for analysis of variance showed a significant effect for the combination of types of foliar application at different stages on reserves of N, Mg, Zn, Mn, B ($p<0.01$) and subsequent generation seed yield (Table 2). Furthermore, analysis of variance between treatments and the control showed significant difference (Table 5).

Seed N concentration

Data on comparison of means showed that foliar application with a mixture of all combinations at all stages (1/45% dry matter of seed) had the highest N concentration. Furthermore, N concentration in foliar application treatments with NFA in seed-filling, MgFA in tasseling, ZnFA in seed-filling, BFA at all stages, and MFA at all stages showed no significant difference. The orthogonal contrast of treatments and the control showed significant difference (Table 3). Based on these results, the reserved N showed a 26% increase due to foliar application with all mixtures at all growth stages compared with the control. The results of this study showed that BFA at all stages influenced N absorption, which in turn increased the concentration of N in the seed. Protein and soluble nitrogenous compounds are decreased in boron-deficient plants (Gupta, 2007). Boron deficiency did not substantially affect relative amino acid composition (Dugger, 1983), but it did enhance the proportion of inorganic nitrogen, particularly nitrate, in plant tissues and translocation fluids (Shelp, 1990). A number of researchers have reported increases in nitrate concentrations as well as corresponding decreases in nitrate reductase activity in sugar beet, tomato, sunflower and corn plants due to boron deficiency (Gupta, 2007; Kastori and Petrovic, 1989). The positive effect of nitrogen fertilizer in the mother plant can contribute to its role in delaying the aging cycle and providing enough time to obtain photosynthetic matter and consequently more weight and higher N seed reserve (Delouch, 1980). ZnFA at the seed-filling stage led to increased N accumulation. The role of zinc and magnesium in protein synthesis has been known about for a long time (Storey, 2007; Marschner, 1995). Protein synthesis resumes when zinc is resupplied because zinc is a structural component of ribosomes and responsible for their structural integrity (Storey, 2007). As the major portion of nitrogen in plants is in the form of proteins, it is clear that with an increase of protein synthesis, nitrogen will be stored in the seeds

Table 2. The analysis of variance of measured traits in experiment.

S.O.V	df	N	Mg	Zn	Mn	B	Seed yield
Rep	2	0/460**	0/011**	28/048**	67/58**	14/07**	46011*
FAS	3	0/030**	0/0001**	39/76**	22/22**	77/29**	43179*
FA	6	0/064**	0/001**	69/51**	8/836**	15/34**	105256**
FAS ×FA	18	0/063**	0/0004**	102/008**	25/06**	29/05**	81328**
Error	54	0/005	0/000002	0/048	0/025	0/528	10587
CV		5/79	1/04	0/72	7/28	19/34	9/85

*,**, Significant at 5 and1% respectively; FAS, foliar application stage; FA, foliar application.

Table 3. Orthogonal contrast of between control Vs other treatments.

	N	Mg	Zn	Mn	B	Yield
Ms	0/032*	0/0004**	49/003**	23/843**	20/352**	1378/558*

*, **, Significant at 5 and1%, respectively.

and will lead to an increase of N concentration. Mikkelsen (2000) argued that an increase of Zn affects the production of N and its absorption. Gupta and Singh (1985) reported that an increase in nitrogen absorption from Zn application can be attributed to an increase in shoot biomass. It has also been reported that nitrogen absorption in corn increased from 53/4 mg in the control plant to 206/2 mg due to a treatment of 2/5 mg Zn kg^{-1} soil. Abou-Hussein and Faiyad (1995) reported that application of 60 kg ha Zn and 8 kg ha B resulted in increased N concentration in plants from 1/2 to 2%. A study conducted by Vahedi (2011) showed that the use of B leads to an increase in seed N concentration. Foliar application of methanol can increase the activity of nitrate reductase in leaves (Zbieć et al., 2003). Therefore, MFA may be the cause of an increase of nitrogen assimilation in leaves that consequently promotes N accumulation in seeds. In addition, MgFA at all stages including the seed-filling stage (1/02%) had the lowest reserve of N in seed (Table 4 and Figure 1). Results of this study are incompatible with those of Choudhury and Khanif (2001) reporting that magnesium treatment increased plant accumulation of nitrogen, applied as urea, in rice (*Oryza sativa* L.).

Seed Mg concentration

Data from comparison of means showed that MgFA at all stages (0/172 mg kg^{-1}) had the highest and MgFA at the seed-filling stage and MFA at the 8 to 10 leaf stage (0/121 mg kg^{-1}) had the lowest Mg concentration in seed (Table 4 and Figure 2). The orthogonal contrast of treatments and the control regarding the Mg reserve showed significant difference (Table 3). Based on these results, Mg reserved in the seed showed an average increase of 33% due to MgFA at all growth stages

compared with the control. Magnesium is a physiologically mobile within a plant. Therefore, magnesium can be reallocated from other plant parts and transported through the phloem to actively growing sinks (Merhaut, 2007). In this research, foliar application of maternal plants with Mg and the transfer of this element to seeds as an active sink has increased the concentration of seed-Mg. Nitrogen may either inhibit or promote magnesium accumulation in plants, depending on the particular form of nitrogen: with ammonium, magnesium uptake is suppressed and with nitrate, magnesium uptake is increased (Lasa et al., 2000). According to these results, NFA has increased the Mg-seed reserve. Based on results of correlations (Table 6), Mg-seed concentration had a positive and significant correlation with Mn-seed concentration (r = 0.363**). In hydroponically grown poinsettia, Mg concentrations in leaves increased as the proportion of nitrate-nitrogen to ammonium-nitrogen increased, even though all treatments received the same amounts of total nitrogen (Scoggins and Mills, 1998). In cauliflower (*Brassica oleracea* var. botrytis L.), increasing nitrate-nitrogen fertilization from 90 to 270 kg ha^{-1} increased yield as a response to increased Mg fertilization rates (22/5 to 90 kg ha^{-1}) (Batal et al., 1997).

Seed Zn concentration

Comparing means showed that ZnFA at the tasseling stage (44 mg kg^{-1}) had the highest effect on Zn seed reserve, whereas MFA at the tasseling stage (23/5 mg kg^{-1}) showed the least effect (Table 4 and Figure 3). The orthogonal contrast between treatments and the control showed a significant difference in terms of seed Zn concentration (Table 3). Based on these results, Zn reserve in the seed showed an average 83% increase

Table 4. Mean comparison of interaction between foliar application and growth stage based on LSD.

FAS	FA	N (%)	Mg (mg kg^{-1})	Zn (mg kg^{-1})	Mn (mg kg^{-1})	Mn (mg kg^{-1})	B (mg kg^{-1})	Yield (g m^{-2})
	Methanol	1.350^{a-d}	0.1210^p	33.00^g	7.000^j	7.000^j	0.8000^i	1125/19^{a-e}
	Zn	1.150^{e-jk}	0.1427^i	24.00^V	10.00^d	10.00^d	1.800^j	777/51^{ijk}
	B	1.250^{b-f}	0.1407^j	26.17^{qr}	4.000^m	4.000^m	4.600^{e-h}	1067/9^{b-h}
8-10 leaves	Mg	1.227^{c-g}	0.1400^j	26.50^q	6.500^k	6.500^k	0.8000^i	885/63^{g-k}
	N	1.200^{d-j}	0.1527^{cd}	29.50^m	9.500^e	9.500^e	4.600^{e-h}	1242/3^{abc}
	Mn	1.093^{h-k}	0.1507^{fgh}	25.00^T	12.00^a	12.00^a	0.8000^i	1120/4^{a-e}
	Mix	1.097^{f-k}	0.1517^{def}	33.50^f	3.000o	3.000o	1.107^i	1301/1^a
	Methanol	1.383^{ab}	0.1303^{no}	23.50^w	6.000^l	6.000^l	6.600^{cd}	1279/6^{ab}
	Zn	1.200^{d-j}	0.1313^{mn}	44.00^a	2.333^p	2.333^p	3.500^h	1002/7^{d-i}
	B	1.220^{c-h}	0.1300^o	31.50^j	3.500^n	3.500^n	3.700^{gh}	881/3^{g-k}
Tasseling	Mg	1.440^a	0.1333^{kl}	25.50s	7.500^i	7.500^i	7.200^{bc}	848^{h-k}
	N	1.210^{c-i}	0.1503^{gh}	25.50s	6.500^k	6.500^k	4.600^{e-h}	1051/5^{c-h}
	Mn	1.150^{e-k}	0.1537^c	30.50^k	4.000^m	4.000^m	1.800^i	1069/1^{b-h}
	Mix	1.300^{a-e}	0.1620^b	38.00^c	8.000^h	8.000^h	5.700^{cde}	713/9^k
	Methanol	1.210^{c-i}	0.1513^{efg}	30.00^l	8.500^g	8.500^g	0.8000^i	1106/6^{a-g}
	Zn	1.420^a	0.1500^h	35.00^e	7.000^j	7.000^j	0.9000^i	1138/4^{a-e}
	B	1.120^{f-k}	0.1323^{lm}	27.00^p	9.000^f	9.000^f	0.9000^i	1051/54^{c-h}
Seed-filling	Mg	1.020^k	0.1210p	30.50^k	11.00^b	11.00^b	3.700^{gh}	869/2^{h-k}
	N	1.400^{ab}	0.1337^k	32.00^i	4.000^m	4.000^m	4.600^{e-h}	1136/3^{a-e}
	Mn	1.207^{c-i}	0.1407^j	37.00^d	6.500^k	6.500^k	4.100^{fgh}	1169/1^{a-e}
	Mix	1.060^{jk}	0.1410^j	26.00r	8.000^h	8.000^h	0.7000^i	1114/5^{a-f}
	Methanol	1.440^a	0.1300o	42.50^b	10.50^c	10.50^c	6.500^{cd}	1218/4^{a-d}
	Zn	1.070^{h-k}	0.1317^m	32.50^h	9.000^f	9.000^f	6.400^{cd}	1112/4^{a-f}
	B	1.410^a	0.1317^m	28.00^n	6.500^k	6.500^k	9.100^a	1064/7^{b-h}
All stages	Mg	1.020^k	0.1723^a	27.50^o	3.000^o	3.000^o	5.300^{def}	967/25^{e-j}
	N	1.360^{abc}	0.1520^{de}	24.50^u	3.500^n	3.500^n	8.700^{ab}	889/34^{f-k}
	Mn	1.050^{jk}	0.1513^{efg}	30.00^l	8.500^g	8.500^g	0.8000^i	746/77^{jk}
	Mix	1.450^a	0.1423^i	24.50^u	7.500^i	7.500^i	5.100^{d-g}	1309/1^a
Control	-	1.15	0.129	24	4	4	1.1	982.34
LSD%	-	0.154	0.001	0.126	0.0912	0.0912	1.584	42.55

Within each column, means with the same lower case letter superscript are not significantly different (P > 0.05).

Table 5. Two-way analysis of variance of measured traits in experiment.

S.O.V	df	N	Mg	Zn	Mn	B	Seed yield
Rep	2	0/447**	0/011**	29/011**	71/391**	13/425**	1710/59*
Treat	28	0/058**	0/1×10^{-4} **	86/483**	21/241**	20/280**	2880/99**
Error	56	0/005	0/2×10^{-5}	0/154	0/248	0/22	367/878
CV		5/97	2/47	1/31	7/36	19/71	9/77

*, **, Significant at 5 and 1%, respectively.

due to ZnFA at the tasseling stage compared with the control. While zinc can be transported from vegetative tissue into reproductive tissue according to plant capacity (Welch, 1995) therefore, because of ZnFA and its transfer to seeds, the zn-seed concentration will increase. These results indicate that MFA at all stages had a positive

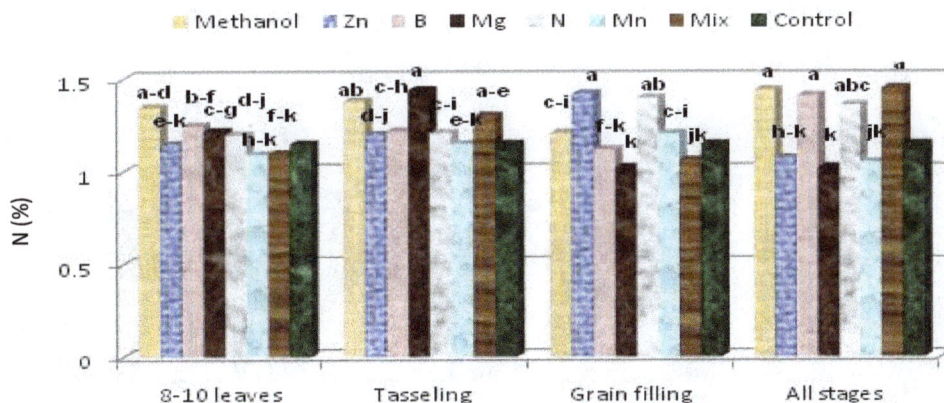

Figure 1. Effect of combination type of foliar application in the different stages on the reserve of N. (LSD5%=0.154)

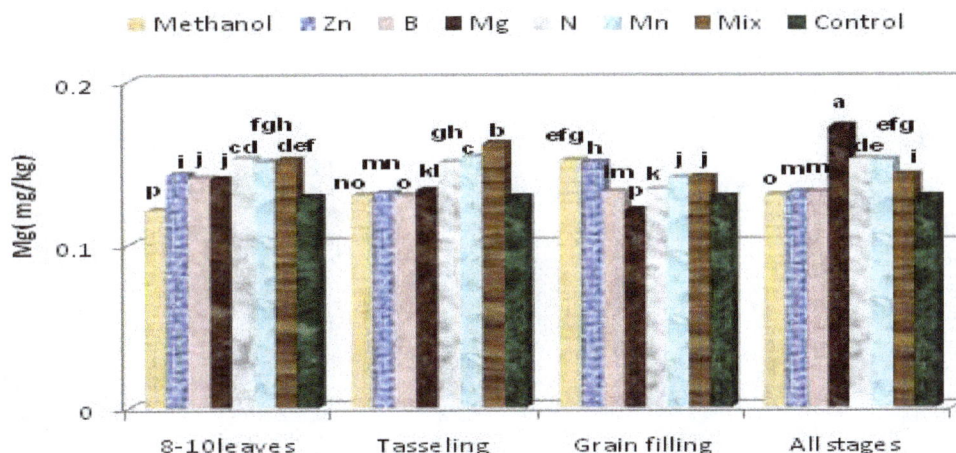

Figure 2. Effect of combination type of foliar application in the different stages on the reserve of Mg. (LSD5%=0.001)

Table 6. Correlation coefficients between measured traits.

	N	Mg	Zn	Mn	B	Seed yield
N	1	0.099	0.052	0.046	0.193*	0.308**
Mg		1	0.042	0.363**	0.098	-0.065
Zn			1	-0.008	0.202*	0.174
Mn				1	-0.048	0.205*
B					1	-0.072
Seed yield						1

*, ** Significant at 5 and 1% respectively.

positive effect and at the 8 to 10 leaf stage it had a negative effect on absorption and accumulation of Zn in the seed. Yilmaz et al. (1998) reported that fertilizer application can increase grain Zn concentration up to three or four-fold. Gangloff et al. (2002) determined that in maize plants and application of zinc sulphate increased dry matter and zinc accumulation in leaf and grain.

Similar results have been reported by Maralian (2009).

Seed Mn concentration

Comparison of means showed that MnFA at the 8 to 10 leaf stage (12 mg kg^{-1}) had the highest and ZnFA at the

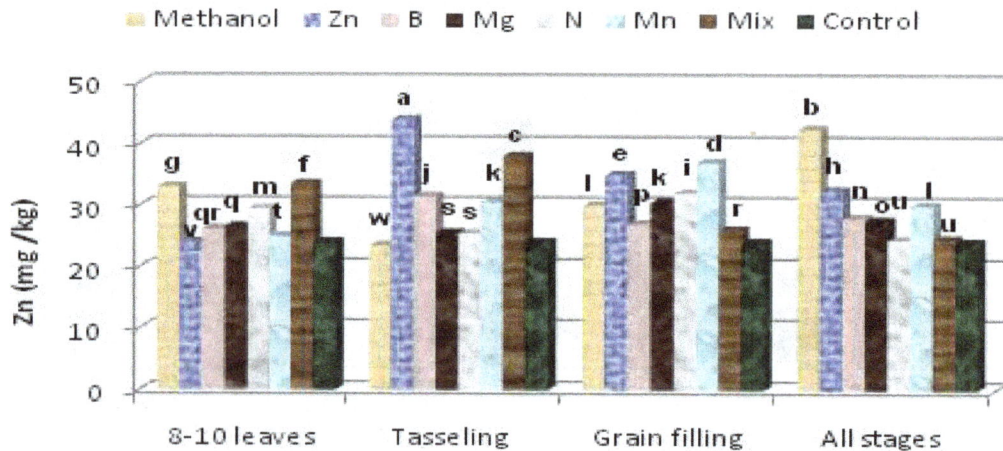

Figure 3. Effect of combination type of foliar application in the different stages on the reserve of Zn. (LSD5%=0.126).

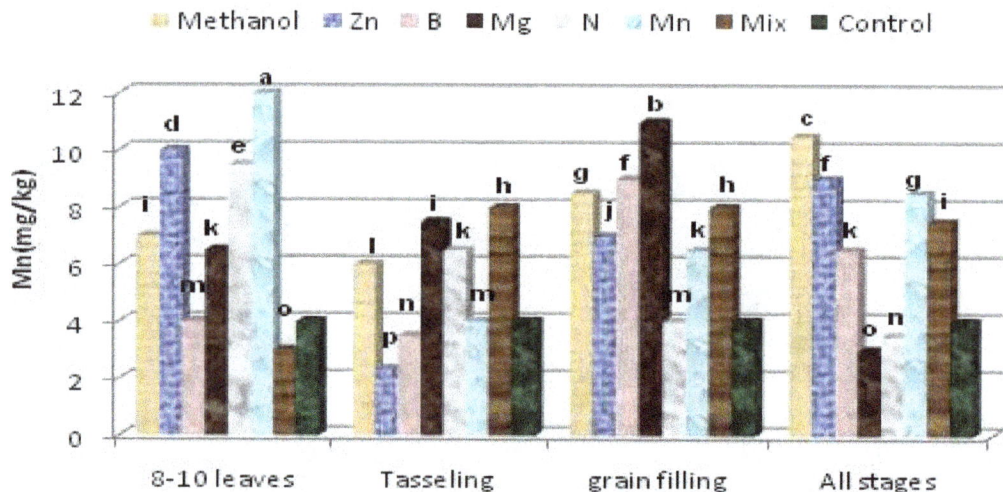

Figure 4. Effect of combination type of foliar application in the different stages on the reserve of Mn. (LSD5%=0.0912).

tasseling stage (2/33 mg kg^{-1}) had the lowest effect on the Mn concentration (Table 4 and Figure 4). The orthogonal contrast between treatments and the control showed a significant difference in terms of seed Mn concentration (Table 3). Mn reserved in the seed showed a 3-fold increase due to MnFA at the 8 to 10-leaf stage compared with the control. Likewise, there was a synergic difference between MnFA and MgFA. In other words, MnFA at the tassel-appearing stage increased the seed concentration of Mg but MgFA at the seed-filling stage increased the concentration of Mn in the seed. This point might be because of the capacity of Mg and Mn as a binding site on the plant's biological membrane (White and Cantor, 1971). A report given by Vahedi (2011) indicated that in soybean plants the highest concentration of Mn was observed in the Zn treatment. In addition, the use of B and Zn increased the Mn reserve in the seed.

Seed B concentration

Comparison of means showed that the BFA at all stages (9/1 mg kg^{-1}) had the highest effect on the seed B concentration from the treatment - mixture of all combinations - and MnFA at all stages (0/7 mg and 0/8 mg kg^{-1}), the MgFA at the 8 to 10 leaf stage (0/8 mg kg^{-1}), the MFA both at the tasseling stage (0/8 mg kg^{-1}), and at all stages (0/8 mg kg^{-1}). The ZnFA and BFA in the seed-filling stage (0/9 and 0/9 mg kg^{-1}) had the lowest effect on seed B reserves (Table 4 and Figure 5). The orthogonal contrast between the treatments and the control showed significant difference in terms of seed B concentration (Table 3). According to these results, the B reserve in the seeds showed 8-fold increase due to BFA at all stages compared with the control. The results of the effect of BFA on N concentration showed that the minerals had a

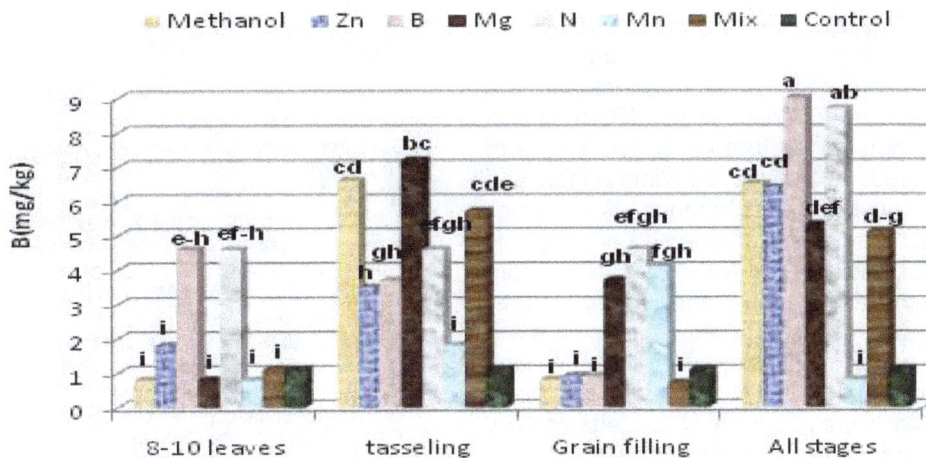

Figure 5. Effect of combination type of foliar application in the different stages on the reserve of B. (LSD5%= 1.584).

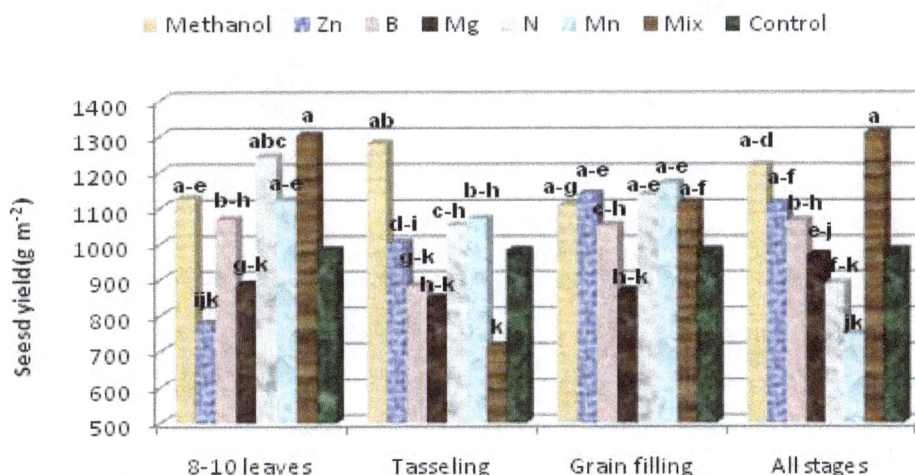

Figure 6. Effect of combination type of foliar application in the different stages on the seed yield. (LSD5%= 42.55).

synergic effect on each other. Boron deficiency in tobacco (*Nicotiana tabacum* L.) resulted in a decrease in leaf N concentration and reduced nitrate reductase activity (Camacho and Gonzales, 1999). Another report Rajaie et al. (2009) determined that Zn application in lime plants contributed to the amount of B in the plant's root. Mikkelsen (2000) in a similar statement refers to the effect of Zn in increasing levels of B and N absorption. A study on the effects of Zn and B on corn showed that high amounts of Zn and B in the soil help to increase seed yield and N absorption. Accordingly Zn-seed concentration had a positive and significant correlation with B-seed concentration (r = 0.202*) (Table 6).

Seed yield

Comparison of means showed that a mixture of all

combinations at all stages had the highest effect on seed yield of a subsequent generation (1309 gm^{-2}) (33% increase compared with the control); at the tasseling stage (713 gm^{-2}) it had the lowest effect. This indicates that there was no significant difference between those treatments that had foliar application with a mixture of all combinations at all stages, MFA application at the 8 to 10- leaf stage, the tasseling stage, the seed-filling stage and at all stages, the NFA at the 8 to 10 -leaf stage and the seed-filling stage, the ZnFA at the seed-filling stage and at all stages and a mixture of all combinations at the seed-filling stage (Table 4 and Figure 6). The orthogonal contrast comparing treatments and the control in terms of seed yield showed significant difference (Table 3). This could be interpreted as a result of the positive effect of MFA on seed yield at all four stages. Increasing the amounts of micronutrient minerals stored in seeds and grains of staple food crops increases the yield potential of these crops when they are sown in micronutrient-poor soil. The effect of breeding for micronutrient-dense staple seeds and grains on crop yield has been addressed in a number of recent reviews (Graham et al., 1999). Briefly these reviews have determined that increasing stores of micronutrients in seeds increases seedling vigor and viability, which enhances the performance of seedlings when seeds are planted in a micronutrient-poor soil. This improved seed vigor allows for the production of more and longer roots under micronutrient-deficient conditions, allowing seedlings to scavenge more soil volume for micronutrients and water in early growth, an advantage that can lead to improved yield compared with those seeds with low-micronutrient stores when affected by micronutrient deficiency stress during growth (Welch, 2002). Perry and Harrison (1973) showed that if mother plants undergo high temperature stress, physiological disorders would be induced in seeds that together with delayed germination decrease seedling growth and emergence and produce low yield in the field. Environmental conditions, especially those of soil nitrogen affect a seed's nitrogen content and can increase or decrease yield and yield components (Oskouie and Divsalar, 2011). Yilmaz et al. (1998) showed that an increase in zinc content of bread wheat grain from 355 ng to 1465 ng grain^{-1} leads to increased yields at a subsequent harvest. According to the results of correlations (Table 6), seed yield had a positive and significant correlation with N-seed concentration (r = 0.205**).

Conclusions

As KSC 704 maternal plant treated with foliar applications of Mg and B at all stages, Zn foliar application at tasseling and MN foliar application at 8 to 10 leaf stage, Mg, B, Zn and MN-seed reserves were more than 0/043, 8, 20 and 8 mg kg^{-1} in comparison with the control respectively. Increasing mineral nutrient stores in seeds serves to increase reserves of nutrients in seeds of the subsequent generation. Many countries where micronutrient deficiency is a problem for humans, there are also large areas of micronutrient-poor or deficient soil. Improving seed vigour in terms of its ability to improve micronutrient reserves will be very beneficial to agricultural production in these countries (Welch, 2002).

REFERENCES

Abou-Hussein EA, Faiyad MN (1996). The combined effect of poudrette, zinc and cobalt on corn growth and nutrients uptake in alluvial soils. Egyptian J. Soil Sci. 36 (1-4):47-58.

Barker AV, Bryson GM (2007). "Nitrogen", In: Handbook of Plant Nutrition (eds. Barker AV and Pilbeam DJ). Taylor and Francis Group, Boca Raton, FL. pp. 21-50.

Batal KM, Granberry DM, Mullinix JR (1997). Nitrogen, magnesium, and boron applications affect cauliflower yield, curd mass and hollow stem disorder. HortScience 32 (1):75–78.

Beck DL (2004). Hybrid corn seed production. In: Corn: Origin, History, Technology, and Production (eds. Smith CW, Betran J, Runge ECA). John Wiley & Sons, Hoboken, New Jersey. pp. 565-627.

Bell RW, McLay L, Plaskett D, dell B, Loneragan JF (1989). Germination and vigour of black gram (Vigna mungo L. Hepper). Seed from plants grown with and without boron. Austr. J. Agric. Res. 40:273-279.

Camacho CJJ, Gonzales FA (1999). Boron deficiency causes a drastic decrease in nitrate content and nitrate reductase activity, and increases the content of carbohydrates in leaves from tobacco plants. Planta 209:528–536.

Choudhury TMA, Khanif YM (2001). Evaluation of effects of nitrogen and magnesium fertilization on rice yield and fertilizer nitrogen efficiency using 15N tracer technique. J. Plant Nutr. 24(6):855–871.

Downie A, Miyazaki S, Bohnert H, John P, Coleman J, Parry M, Haslam R (2004). Expression profiling of the response of Arabidopsis thaliana to methanol stimulation. Phytochemistry 65:2305–2316.

Delouch JC (1980). Environmental effects on seed development and seed quality. Hortic. Sci.15:775-780

Dugger WM (1983). Boron in plant metabolism, In: Lauchli A and Bieleski RI (eds). Encyclopedia of Plant Physiology, new series, New York: Springer. pp. 626–650.

Gangloff WJ, Westfall DG, Peterson GA and Mortvedt JJ (2002). Relative availability coefficients of organic and inorganic Zn fertilizers. J. Plant Nutr. 25:259-273.

Graham RD (2008). Micronutrient deficiencies in crops and their global significance. In: micronutrient deficiencies in Global Crop Production (Ed Alloway BJ). Springer. pp. 41-61..

Graham RD, Senadhira D, Beebe S, Iglesias C, Monasterio I (1999). Breeding for micronutrient density in edible portions of staple food crops: conventional approaches. Field Crops Res. 60(1):57–80.

Gupta UC (2007). Boron", In: Handbook of Plant Nutrition (eds. Barker AV, Pilbeam DJ). Taylor and Francis Group, Boca Raton, FL. pp. 241-276.

Gupta VK, Singh B (1985). Residual effect of zinc and magnesium on maize crop. J. Indian Soc. Soil Sci. 33(1):204 -207.

Hanlon EA (1998). Elemental determination by atomic absorption spectrophotometery. In: handbook of reference methods for plant analysis (eds. kalra YP). CRC Press: Boca Raton. pp. 157-165.

Humphries MJ, Stangoulis JCR, Graham RD (2007). "Manganese", pp. 351-375. In: Handbook of Plant Nutrition (eds. Barker AV, Pilbeam DJ). Taylor and Francis Group, Boca Raton. FL.

Ishii T, Matsunaga T, Iwai H, Satoh S, Taoshita J (2002). Germanium dose not substitute for boron in cross-linking of rhamnogalacturonan II in pumpkin cell walls. Plant Physiol. 130 (4):1967-1973.

Karla Y (1998). Handbook of reference methods for plant analysis. Soil and Plant Analysis Council, Inc. CRC Press. p. 320.

Kastori R, Petrovic N (1989). Effect of boron on nitrate reductase activity in young sunflower plants. J. Plant Nutr. 12(5):621–632.

Lasa B, Frechilla S, Aleu M, González-Moro B, Lamsfus C, Aparicio-Tejo PM (2000). Effects of low and high levels of magnesium on the response of sunflower plants grown with ammonium and nitrate. Plant Soil 225:167–174.

Maralian H (2009). Effect of foliar application of Zn and Fe on wheat yield and quality. Afr. J. Biotechnol. 8 (24):6795-6798.

Marschner H (1995). Mineral Nutrition of Higher Plants. 2nd ed. Academic Press. London. pp. 301-306.

Merhaut DJ (2007). "Magnesium", pp. 145-183. In: Handbook of Plant Nutrition (eds. Barker AV, Pilbeam DJ). Taylor and Francis Group, Boca Raton. FL.

Mikkelsen RL (2000). Nutrient management for organic farming: A case study. J. Nat. Res. Life Sci. Educ. 29:88–92.

Mirakhori M, Paknejad F, Moradi F, Ardakani M, Zahedi H. Nazeri P (2009). Effect of Drought Stress and Methanol on Yield and Yield Components of Soybean Max (L 17). Am. J. Biochem. Biotechnol.5(4):162-169.

Nadali I, Paknejad F, Moradi F, Vazan S, Tookalo M, Jami Al-Ahmadi M, Pazoki A (2010). Effects of methanol on sugar beet (Beta vulgaris). Austr. J. Crop Sci. 4 (6):398-401.

Oskouie B, Divsalar M (2011). The effect of mother plant nitrogen on seed vigor and germination in rapeseed. ARPN J. Agric. Biol. Sci. 6(5):49-56.

Perry DA, Harrison JG (1973). Causes and development of bollow heart in pea seed. Annals Appl. Boil. 73(1):95-101.

Rajaie M, Ejraie AK, Owliaie H, Tavakoli AR (2009). Effect of zinc and boron interaction on growth and mineral composition of lemon seedlings in a calcareous soil. Int. J. Plant Prod. 3(1):39-50.

Ramirez I, Dorta F, Espinoza V, Jimenez E, Mercado A, Pen A-Cortes H (2006). Effects of foliar and root applications of methanol on the growth of Arabidopsis, Tobacco and Tomato plants. J. Plant Growth Regul. 25:30–44.

Sawan ZM, Maddah El Din MS, Gregg BR (1989). Effect of nitrogen fertilization and foliar application of calcium and micro-elements on cotton seed yield, viability and seedling vigor. Seed Sci. Technol. 17:421-431.

Sawan ZM, Maddah El Din MS, Gregg BR (1993). Cotton seed yield, viability and seedling vigor as affected by plant density, growth retardants, copper and manganese. Seed Sci. Technol. 21:417-431.

Scoggins HL, Mills HA (1998). Poinsettia growth, tissue nutrient concentration, and nutrient uptake as influenced by nitrogen form and stage of growth. J. Plant Nutr.. 21 (1):191–198.

Shelp BJ (1990). The influence of nutrition on nitrogen partitioning in broccoli plants. Commun. Soil Sci. Plant Anal. 21:49–60.

Storey JB (2007). Zinc", pp: 411-437. In: Handbook of Plant Nutrition (eds. Barker AV, Pilbeam DJ). Taylor and Francis Group, Boca Raton. FL.

Vahedi A (2011). Study of the effects of micronutrient application on the absorption of macro- and micronutrients in the Soybean Cultivar Telar in the North of Iran. J. Am. Sci. 7 (6):1252-1257.

Waling I, Van Vark W, Houba VJG, Van der lee JJ (1989). Soil and plant Analysis, a series of syllabi part7. Plant Analysis procedures. Wageningen Agricultural University.

White JP, Cantor CR (1971). Role of magnesium in binding of tetracycline to Escherichia coli ribosomes. J. Mol. Biol. 58:397–400.

Welch RM (1986). Effects of nutrient deficiencies on seed production and quality. Advan. Plant Nutr. 2:205-247.

Welch RM (1995). Micronutrient nutrition of plants. CRC Critical Rev. Plant Sci. 14:49-82.

Welch RM (2002). Breeding Strategies for Biofortified Staple Plant Foods to Reduce Micronutrient Malnutrition Globall. J. Nutr. 132(3):495S-499S.

Yilmaz A, Ekiz H, Gultekin I, Torun B, Karanlik S, Cakmak I (1998). effect of seed zinc content on grain yield and zinc concentration of wheat grown in zinc-deficient calcareous soils. J. Plant Nutr. 21:2257-2264.

Zakaria MS, Ashraf HF, Serag EY (2009). Direct and residual effects of nitrogen fertilization, foliar application of potassium and plant growth retardant on Egyptian cotton growth, seed yield, seed viability and seedling vigor. Acta Ecologica Sinica 29:116-123.

Zbieć I, Karczmarczyk S, Podsiadło C (2003). Response of some cultivated plants to methanol as compared to supplemental irrigation. Electronic J. Polish Agric. Univer. Agron. 6(1):1-7.

14

Exploiting genotype x environment interaction in maize breeding in Zimbabwe

Casper Nyaradzai Kamutando[1,2], Dean Muungani[2], Doreen Rudo Masvodza[3] and Edmore Gasura[1]

[1]Department of Crop Science,University of Zimbabwe,P.O.Box MP167,Mt Pleasant,Harare,Zimbabwe.
[2]Agriseeds (PvtLtd,5 Wimbledon Drive,P.O.Box,6766,Eastlea,Harare,Zimbabwe.
[3]Department of Biosciences, Bindura University of Science Education, P. Bag 1020, Bindura, Zimbabwe.

Agriseeds Company produces several hybrids yearly. These hybrids need to be evaluated for yield stability before release. In this study, fifty-eight newly developed hybrids were planted at five sites and evaluated for grain yield and other traits. The objective was to assess the stability of Agriseeds hybrids in Zimbabwe and to identify strategies of minimizing evaluation cost of hybrids in multi-locations. Across site, analysis of variance indicated significant differences (p<0.001) in grain yield, days to silking, days to anthesis and anthesis-silking interval on genotypes, environments and genotype × environment interactions (GEI). Stable hybrids were 10A3WH04 (6.7 tha^{-1}) and 10A3WH24 (6.7 tha^{-1}) while hybrid 10A3WH03 (6.5 tha^{-1}) showed specific adaptability. Since all the evaluation sites fell into one mega-environment, a few representative sites with a few replications will be ideal to capture much of the variance due to GEI. Furthermore, Agriseeds should not establish separate breeding programmes for these environments. Rather, suitable culling and discriminating environments must be captured in few sites to be utilized and these sites are Harare, Gwebi and any one of Kadoma, Matopos and Shamwa.

Key words: *Zea mays* L., Genotype × environment interactions, grain yield, bi-plots, stability analysis.

INTRODUCTION

The huge demand for maize (*Zea mays* L.) as food and feed in Zimbabwe has resulted in the rise of the private seed companies. Agriseeds is one of the private companies aimed at developing and marketing improved maize seeds in Zimbabwe. This company develops several new hybrids every year. These hybrids need to be evaluated for the presence of genotype × environment interactions (GEI) in grain yield and other agronomic traits (Mohammadi and Haghparast, 2010; Tiawari et al., 2011). The number of materials evaluated and the number of test environments required in multi-location trials affects the cost of plant breeding, particularly to young emerging companies such as Agriseeds. Reduction in the number of test sites requires a thorough

understanding of the genotype and GEI (Bernardo, 2002).

Southern Africa has been divided into mega-environments by CIMMYT based on maize regional trials data (Setimela et al., 2005). Zimbabwe was also found to have diverse agro-ecological environments and has been divided into natural regions based on their potential in crop production (Rukuni et al., 2006). Natural regions 2a and 2b normally experience adequate rainfall, followed by natural regions 3 and 4, where rainfall distribution and amount vary from season to season. However, maize is grown in all agro-ecological regions of the country which are highly variable in terms of soil characteristics, rainfall and temperature during the growing season (Muungani et al., 2007; Rukuni et al., 2006).

Breeding programmes are intended to develop new varieties with superior agronomic performance compared to those in current production by farmers. Prior to release of the new varieties, they are evaluated in yield trials at several locations for two or more seasons in multi-environmental trials (METs). The variety trials provide important information that enables selection and recommendation of crop cultivars (Yan and Tinker, 2006; Yang et al., 2009). Comparisons are made with the performance of the commonly grown commercial varieties (checks). Genotype x environment interactions (GEI) are a major challenge when identifying superior genotypes using MET data because it slows down the selection process and makes genotype recommendations difficult (Hassanpanah, 2009). A genotype is defined as an individual's genetic makeup while an environment refers to a set of non-genetic factors that affect the phenotypic value associated with a cultivar (Fan et al., 2007). Crop varieties show wide fluctuations in their yielding ability when grown over varied environments or agro-climatic zones (Fan et al., 2007). Each genotype may have a specific environment for its maximum performance, but successful new varieties must show high performance for yield and other essential agronomic traits, and their superiority should be reliable over wide range of environments (Fan et al., 2007). Plant breeders desire stable cultivars with good performance under all conditions within the targeted production region (Caliskan et al., 2007).

The regression models have been used often by plant breeders to assess yield stability (Finlay and Wilkinson, 1963). Yield stability is a measure of the ability of a genotype to maintain relative performance across a wide range of environments. In general stable genotypes should perform more or less the same across environments. An appropriate stable cultivar is capable of utilizing resources that are available in high yield environments, while maintaining above average in all other environments (Finlay and Wilkinson, 1963). Furthermore, biplots have also been developed and they display the genotype + genotype x environment interaction (GGE) of a MET data. The GGE refers to the genotype main effect (G) plus the genotype x environment interaction (GEI), which are the two sources of variation (Yan et al., 2001). Yang et al. (2009) described a biplot as a descriptive statistical tool. The biplots allow the researcher to concentrate on the part of the MET data that is most useful to cultivar selection (Kang, 2003; Yan and Tinker, 2006).

Currently, the stability of the recently developed Agriseeds hybrids is unknown, yet this is crucial in cultivar recommendation in specific or general environments. Furthermore, the logical number of test environments needed for early and advanced generation testing for Agriseeds hybrids is unknown because of the poor understanding of GEI patterns. However, this is essential in reducing the cost of cultivar evaluation in multiple locations. The objective of this study was to assess the stability of Agriseeds hybrids across major production environments in Zimbabwe and to identify strategies of minimizing the evaluation cost of hybrids in multi-locations.

MATERIALS AND METHODS

Fifty-eight experimental maize hybrids from Agriseeds (Pvt) Ltd together with 12 commercial check hybrid varieties from various seed companies in Zimbabwe were evaluated at five sites during the 2011-2012 summer season. The sites represent the major maize growing agro-ecological regions in Zimbabwe (Table 1). The experiments were grown using an α-lattice (0,1) design with three replications. Two row-plots of 4 m length, with an inter-row spacing of 0.75 m and an intra-row spacing of 0.25 m were used. Basal fertilizer (N-7, P_2O_5-14, K_2O-7) was broadcasted at 400 kg ha^{-1} and disced into the soil before planting. All sites received two applications of 200 kg ha^{-1} of ammonium nitrate as top dressing. The first and second applications were at four and eight weeks after crop emergence, respectively. All the sites were rain fed and hand weeding was done to control weeds. Data was recorded for grain yield (GY) (shelled grain weight per plot adjusted to 12.5% grain moisture and converted to tons per hectare), anthesis date (MF) (number of days after planting when 50% of the plants shed pollen), silking date (FF) (number of days after planting when 50% of the plants extrude silks) and anthesis-silking interval (ASI) (the difference between silking date and anthesis date, FF – MF).

Data analyses

Individual site and across site analysis of variance for all the agronomic traits were done using Genstat Software version 13 (Genstat, 2010) and the appropriate denominators were used for the F-test. The variance components due to error, genotypes and genotype x environment interaction were calculated and used to estimate the broad sense coefficient of genetic determination (fixed parent equivalent of broad sense heritability). The means of genotypes per site were ranked to assess the importance of cross-over genotype x environment interactions. Stability analysis for yield was done based on the Finlay and Wilkinson (1963) regression model. The genotype + genotype x environment interaction (GGE) scatter plots (Yan and Tinker, 2006) were generated using Genstat Version 13 (Genstat, 2010) to identify genotypes adapted to specific environments, the most discriminating and suitable culling environments. Decisions on the number of testing sites and number of replications per site were calculated by making replication and environment the subject of the formula in the following equation as stated by Bernardo (2002) as $(V_E/re)+ (V_{GE}/e)$, where, 5% LSD is the least significant difference, V_E is the error variance, V_{GE} is the genotype x environment interaction variance, r is the number of replications, and e is the number of environments used in the experiment.

RESULTS

The hybrids showed significant differences (p<0.001) in for grain yield (GY), number of days to silking (FF), number of days to anthesis (MF) and anthesis-silking interval (ASI) (Table 2). The five sites used in the experiment, that is, Harare, Gwebi, Shamva, Kadoma and Matopos were significantly different (p<0.001) in term

Table 1. Description of the evaluation sites for the Agriseeds experimental hybrid trials.

Trial site	GIS position	[a]Soil type	[b]Altitude (masl)	[b]Mean rainfall (mm)	[c]Natural region
Agriseeds Research Station, Harare	30°56'E and 17°44'S	Red clay	1400	750-1000	**2a**
Gwebi Variety Testing Station, Gwebi	31°32'E and 17°41'S	Red clay	1448	750-1000	**2a**
Panmure Experiment Station, Shamva	31°47'E and 17°35'S	Red clay	881	650-800	**2b**
Cotton Research Station, Kadoma	29°53'E and 18°19'S	Sandy loamy soil	1149	650-800	**3**
Matopos Experiment Station, Matopos	28°28'E and 20°24'S	Red clay	1138	450-650	**4**

Source: [a]Nyamapfene (1991); [b]Rukuni et al. (2006). [c]Natural region 2 is subdivided into a and b based on various agro-ecological factors

Table 2. Across sites analysis of variance mean squares, variance components and broad sense coefficient of genetic variation values.

Source	DF	Grain yield	Silking date	Anthesis date	Anthesis-silking interval
Site	4	1813.555***	5062.94***	6848.84***	549.287***
Rep /site	10	10.584***	81.49***	60.68***	8.188***
Block(Rep/Site)	195	1.6477***	8.046***	7.057***	1.776***
Hybrid	69	7.094***	56.75***	53.35***	6.577***
Site*Hybrid	276	2.151***	8.99***	7.91***	2.373***
Error	495	1.118	4.349	4.114	1.131
Total	1049				
Error variance component		1.118	4.349	4.114	1.131
GxE variance component		0.344	1.547	1.265	0.414
Genotype variance component		0.329	3.184	3.029	0.28
Broad sense heritability					
Single plot basis (%)		18.4	35.1	36.0	15.4
Across environments basis (%)		69.7	84.2	85.2	63.9

*** Significant at 0.1% probability level.

terms of their average performance for all traits studied (Table 2). Harare site had an average yield of 10.5 tha[-1], followed by Gwebi (7.1 tha[-1]), Shamva (5 tha[-1]), Kadoma (4.1 tha[-1]) and Matopos (3 tha[-1]). There were significant interactions (p<0.001) between the sites (environments) and the hybrids (genotypes) for all the traits measured in the study.

The genotype x environment interaction (GEI) variance component for grain yield was higher than the genotype and error variance components. However, the error term was higher than the genotype and GEI variance components. The broad sense coefficients of genetic variation were low (less than 40%) on single plot basis but high on across environments basis (above 63%) for all traits measured.

Selected genotypes yield ranks across environments

Genotype ranks across environments were non-consistent (Table 3). Genotypes that had high means for yield in Harare changed over the other four sites. The changes were subjectively high with few hybrids that remained in the top 10 in some environments.

Grain yield stability

The existence of genotype x environment interaction (GEI) raised the need to identify stable and high yielding genotypes. The Finlay and Wilkinson (1963) regression model showed that hybrids such as 10A3WH04 (6.70 tha[-1]) and 10A3WH24 (6.70 tha[-1]) as well as a check from Seed Co (SC 533) had mean yield greater than average mean, 5.95 tha[-1], and they showed average stability based on the regression coefficient (b=1) (Table 4).

Experimental hybrids such as 10A3WH03 (6.50 tha[-1]) and some check hybrids from Seed Co (SC 727 and SC 637) and Pannar Seeds (Pan 5M-35) were high yielding but had below average stability (b>1). Hybrids such as 10A3WH02 (5.87 tha[-1]) and 10A3WH14 (5.20 th[-1]) had below average yields and below average stability (b>1). Hybrids such as10A3WH29(4.8 tha[-1]) and 10AH37 (5.0 tha[-1]) as well as checks that include SC 403, Pan 4M-21 and ZS 259 gave below average yield and had above average stability (b<1). Checks like SC 727 (7.24 tha[-1]), SC 637 (7.21 tha[-1]) and Pan 5M-35 (6.43 tha[-1]) are high yielding checks, but have below average stability. Pan 4M-19 (4.94 tha[-1]) and Pan 7M-97 (5.71 tha[-1]) are low yielding check hybrids and they have average stability.

Table 3. Genotype means of the top 10 yielding hybrids (based on Harare site) and their rank changes across other sites.

Genotype name	Harare		Gwebi		Shamva		Matopos		Kadoma	
	Grain yield	Rank	Grain yield	Rank	Grain yield	Rank	Grain yield	Rank	Grain yield	Rank
SC727	16.7	1	4.747	67	7.818	1	2.688	58	5.246	7
10A3WH41	13.474	2	7.773	25	5.239	31	3.221	25	4.796	16
10AH09	13.039	3	8.144	15	5.091	36	3.056	35	5.177	8
09A3WH07	12.63	4	8.652	9	6.297	5	3.695	5	5.142	9
10A3WH03	12.263	5	8.929	4	5.939	11	3.214	26	3.236	58
10A3WH10	12.132	6	8.082	16	6.007	10	3.446	13	5.101	12
10A3WH06	11.999	7	7.109	34	5.552	22	3.222	24	5.537	3
Pan5M-35	11.921	8	7.311	31	4.552	54	3.036	38	4.627	20
10AH03	11.906	9	7.725	27	5.534	23	2.871	46	4.627	15
10AH05	11.874	10	8.37	11	6.107	7	3.232	22	5.348	6

On the other hand, checks like Pan 4M-21 (4.79 tha^{-1}), SC 403 (4.57 tha^{-1}) SC411 (5.32 tha^{-1}) and ZS 259 (4.84 tha^{-1}) showed to be low yielding, but have above average stability.

Genotypes for specific environments

A significant cross-over genotype x environment interactions raised the need to identify hybrids that performed better in specific environments. The genotype + genotype x environment interaction (GGE) scatter plot showed that most of the hybrids, such as 10A3WH20 and 10A3WH10 were found to be suitable to all sites. However, 10A3WH10 and SC 727 performed better in high potential areas like Harare. Most of the hybrids including 10A3WH03 and 10A3WH06 were all found to be concentrated close to Matopos and Kadoma, which are low potential areas (Figure 1). Some experimental hybrids, including some checks (ZS 259 and SC 403) were not specific to any environment. The environments were also grouped into one mega-environment (Figure 1).

Discriminating and culling environments

Since there was one mega-environment, the better testing environments had to be found. The genotype + genotype x environment (GGE) scatter plot showed the most discriminating and suitable culling environments to be Harare, Gwebi and any one of either Kadoma, Shamva or Matopos (Figure 3).

Decision on the number of testing sites and number of replications per site

Agriseeds requires detecting critical differences of 0.8, 1.0 and 1.5 tha^{-1} among the varieties under testing. Based on the equation by Bernardo (2002), the number of environments required are 8, 5 and 2 when there are three replications, respectively. However, when there are two replications the number of sites will increase to 14, 8 and 3, respectively, based on the same critical distances to be detected.

DISCUSSION

Existence of genetic variability among the hybrids for grain yield, anthesis-silking interval ASI and days to maturity raises possibilities of identifying high yielding hybrids with desirable physiological maturity periods and suitable pollen-silk synchronization under diverse environments. Grain yield has been singled out as the most important trait in cereals. Late maturing varieties are needed to achieve high yield in high potential environments, where there is a low risk of occurrence of drought. However, early maturing varieties are desirable in low potential drought prone areas, since they have the capacity to escape late season drought (Banziger et al., 2004). The significant ASI and its high broad sense coefficient of genetic variation (fixed parent equivalent to broad sense heritability calculated from individuals selected from a random mating population) suggest genetic differences in synchronization and therefore selection of hybrids that exhibit good pollen-silk synchronization under drought is possible. A ASI and/or negative ASI is desirable for hybrids to be grown in drought prone areas such as Kadoma and Matopos. Shorter ASI improves the pollen-silk synchronization, a major trait that is affected under drought. The need for a shorter ASI to achieve high grain yield has been observed by Bassetti and Westigate (1993) and Anderson et al. (2004), where the potential number of florets that could become grains was limited by the receptivity of the silks. Asychrony is correlated with reduced number of grains per plant and grain yield in maize (Edmeades et al., 1993). Bolanos and Edmeades (1993) noted a yield decline by 90% as ASI increases from -0.4 to 10 days. To this regard, ASI has been widely

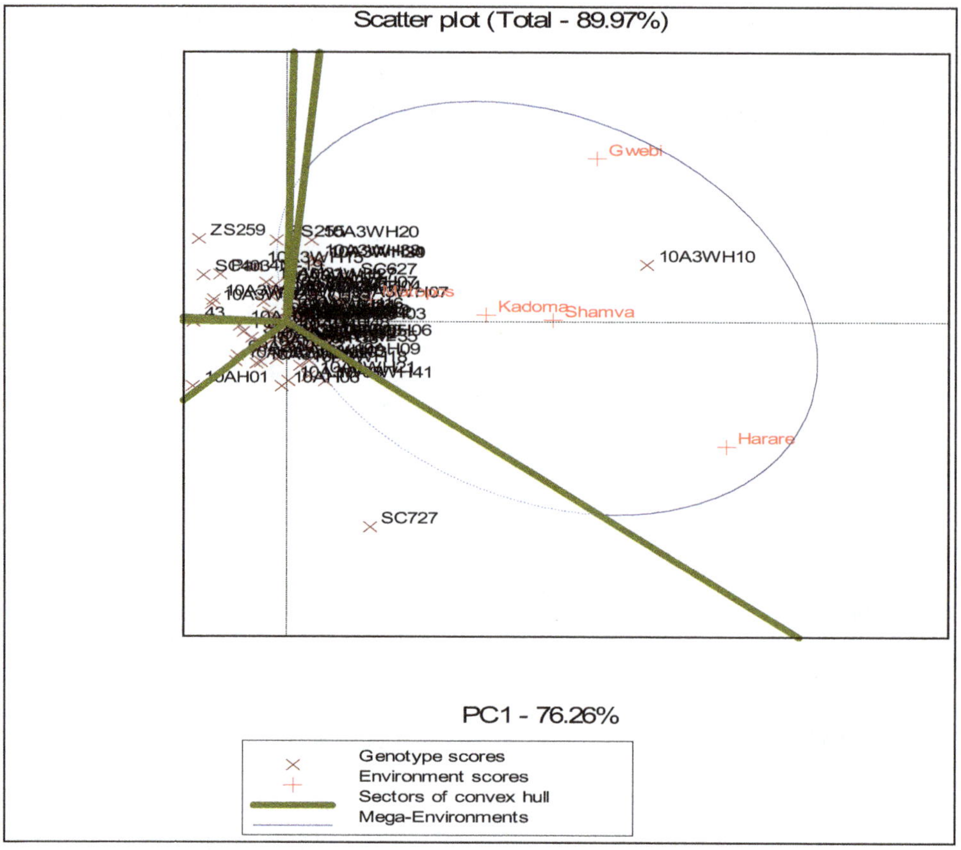

Figure 1. The GGE scatter plot showing all sites to be in one mega-environment.

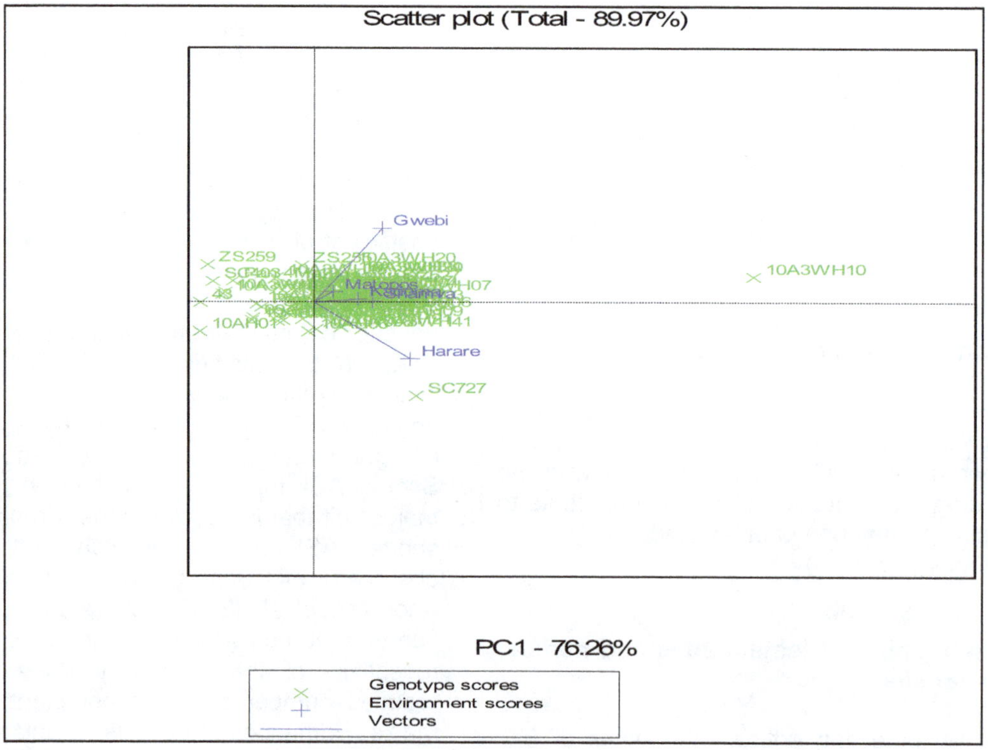

Figure 2. Three groups of discriminating and culling sites of Agriseeds hybrid trials.

Table 4. Genotypes stability parameters for grain yield across sites based on the Finlay and Wilkinson (1963) stability model.

Hybrid name	Yield (tha^{-1})	b-value	p-value	Hybrid name	Yield (tha^{-1})	b-value	p-value
10A3WH29	4.76	0.53	0.003	Pan 53	6.44	1.00	0.006
SC403	4.57	0.55	0.006	10A3WH25	6.08	1.00	0.001
ZS259	4.48	0.59	0.077	10A3WH40	5.78	1.01	0.002
10A3WH15	5.62	0.68	0.016	10A3WH18	6.26	1.01	0.012
Pan 4M-21	4.79	0.72	0.009	10A3WH08	6.05	1.02	0.011
10A3WH28	4.52	0.84	0.004	10A3WH19	5.49	1.02	0.001
10AH01	4.09	0.84	0.024	SC533	6.34	1.02	0.001
10A3WH33	5.69	0.84	0.001	10AH12	5.26	1.03	0.013
10AH31	5.71	0.84	0.003	10AH49	6.05	1.04	0.001
SC411	5.37	0.85	0.001	Pan 7M-97	5.71	1.05	0.002
10AH37	5.01	0.87	0.005	10AH48	6.53	1.07	0.001
09A3WH05	5.93	0.88	0.001	10A3WH06	6.77	1.09	0.001
ZS255	5.95	0.88	0.026	10AH02	5.32	1.09	0.006
10A3WH11	5.88	0.89	0.002	09A3WH07	7.11	1.09	0.003
10A3WH34	6.18	0.90	0.001	Pan 4M-19	4.94	1.09	0.001
10A3WH20	6.64	0.90	0.024	10A3WH42	6.25	1.09	0.001
10A3WH27	0.03	0.90	0.004	10A3WH14	5.24	1.11	0.002
10A3WH37	5.78	0.91	0.001	10A3WH30	6.59	1.11	0.008
10A3WH35	5.80	0.91	0.004	10A3WH26	6.15	1.12	0.001
10A3WH38	6.64	0.92	0.015	10A3WH05	5.74	1.12	0.003
10A3WH23	6.27	0.93	0.002	10A3WH10	7.03	1.12	0.001
10A3WH32	5.48	0.93	0.005	10AH06	5.80	1.13	0.005
09A3WH10	5.99	0.93	0.005	10A3WH39	5.99	1.13	0.004
10A3WH17	6.16	0.94	0.001	10A3WH16	6.01	1.14	0.003
10A3WH12	6.15	0.94	0.002	10AH05	6.82	1.14	0.001
10A3WH31	5.52	0.94	0.001	10AH03	6.38	1.15	0.002
10A3WH22	6.47	0.95	0.001	10A3WH02	5.87	1.16	0.001
10A3WH36	5.86	0.96	1.163	SC637	7.21	1.18	0.004
10A3WH09	6.13	0.96	0.001	Pan 5M-35	6.43	1.20	0.001
10AH34	5.00	0.96	0.001	10A3WH07	6.26	1.22	0.005
10A3WH24	6.74	0.96	0.007	10A3WH03	6.50	1.23	0.001
10AH42	5.67	0.97	0.001	10A3WH21	6.19	1.27	0.002
09A3WH11	5.74	0.97	0.002	10A3WH41	6.58	1.31	0.003
10A3WH13	4.94	0.99	0.008	10AH09	6.96	1.31	0.001
10A3WH04	6.67	1.00	0.001	SC727	7.24	1.55	0.058

used in indirect selection of higher grain yield under drought conditions (Banziger et al., 2004). The yield differences that were observed across sites also confirm the site potential as evidenced by their natural regions (NR) (Rukuni et al., 2006).

The best site was Harare (NR2a) followed by Gwebi (NR2a), Shamva (NR2b), Kadoma (NR3) and Matopos (NR4). Natural region 2a (Harare and Gwebi) is associated with highest rainfall followed by NR2b (Shamva).

Although Gwebi and Harare are in the same natural region, Gwebi has the least yield because this site is associated with high incidences of diseases such as maize streak virus (MSV) and leaf blight. Although NR3 is better than NR4, these areas are prone to drought, with the highest frequencies of the occurrences of the mid-season and late season drought spells that greatly impact on maize yield.

The NR4 is not good for maize production although farmers insist to grow this crop. Harare and Gwebi are rich sites (yielded above the mean), Shamva is an average site (yielded about the mean) while Kadoma and Matopos are poor sites (yielded below the mean).

This classification of sites is also backed up by their geographical classification into various natural regions (Nyamapfene, 1991; Rukuni et al., 2006)

Significance of the genotype x environment interaction, variance components and broad sense coefficient of genetic determination of the traits studied

Genotype x environment interaction (GEI) has been widely reported to impede the speed at which desirable cultivars are made (Caliskan et al., 2007). In this study, the cross-over interactions were common and the GEI variance components for grain yield were larger than the error and the genotype variance components. The large contribution of GEI to grain yield makes is difficult for breeding and selection of better maize varieties. For example, it reduces the heritability and gain in selection and it confuses early generation selection. The high broad sense coefficient of genetic determination across environments raises the possibility of identifying the suitable genotypes across environments. To increase heritability, more number of sites and replications will be required as evidenced by lower heritability estimates at single plot basis but higher estimates as the number of replications and sites are increased. In other studies, broad sense heritability for maize was also found to be low at single plot basis but high at across site basis (Hallauer and Miranda, 1988). Heritability can be improved by increasing the number of sites and replications per site but this has the consequences of increasing the cost of research and this is detrimental to a small company like Agriseeds. To this regard, logical decisions must be made to attempt to reduce the number of sites and replications per site. Fehr (1987) recommended the use of few replications and then increase the number of sites. Based on Bernardo (2002), if the number of replications are three and two, then the number of sites required to achieve a critical distance of 1.0 tha^{-1} were found to be five (5) and eight (8), respectively. Hence, for Agriseeds to reduce their cost of research, they have to use more replications per site and reduce the number of sites which are in the same mega-environment, in order to detect the same difference in yield.

Changes in the genotype ranks across environments suggest the existence of cross-over genotype by environment interactions. Cross-over interactions has been reported to be the major worry for breeders as it results in changes of cultivars ranks across environments. Changes in cultivar ranks across environments make it difficult to recommend a single best genotype for all environments based on evaluations from a single site (Fehr, 1987).

Causes of genotype x environment interactions (GEI)

Grain yield, a quantitative trait, has been widely reported to be due to the interaction of many genes with small effects. The effect of the environment on quantitative

traits has been widely reported to be significant. Grain yield formation is influenced by the duration and rate of grain filling (Lee and Tollennaar, 2007). Furthermore, the variability for 1000 kernel weight and kernel number per ear are predictive of the genotype's sink capacity, hence grain yield (Lee and Tollenaar, 2007; Wang et al., 1999). Grain filling follows three stages, that is, lag phase (rapid cell division and differentiation), linear phase (rapid dry matter accumulation) and the final phase (Lee and Tollenaar, 2007). Extremes in temperature and low amounts of rainfall affects these critical stages, thereby resulting in GEI across the five sites used in the study, since they differ in rainfall distribution and temperature (Rukuni et al., 2006). In deed in this study, the natural regions of Zimbabwe were observed to experience different rainfall amount and pattern. This makes it critical to select genotypes that have stable ASI in order to achieve some levels of drought tolerance.

Grain yield stability

Grain yield is the most important trait because it is the one that gives an economic benefit to the consumers. Hence, good hybrids should give high yield and should be stable across different environments in which they are grown as alluded by Finlay and Wilkinson (1963). Fan et al. (2007) pointed genotype by environment interactions (GEI) as the basic cause of differences between genotypes in their yield stability. Hybrids such as 10A3WH04 (6.67 tha^{-1}) and 10A3WH24 (6.70 tha^{-1}) were high yielding and they showed above average stability and can be taken for further evaluations to estimate the genotype x years interactions and genotypes x environments x years interactions. Although some varieties showed high yielding capabilities, their stability was poor. On the other hand, some hybrids like 10A3WH29 (4.80 tha^{-1}) and 10AH37 (5.0 tha^{-1}) gave above average stability but low yield. This is the scenario exhibited by most of the hybrids, that is, they tend to be stable, yet they produce uneconomic yields. These varieties are generally undesirable to the farmer who wants higher yields in order to get higher returns per each dollar invested. Hence, breeders select higher yielding hybrids and discard low yielding ones, regardless of their stability performance abilities across different environments. Furthermore, hybrids like 10A3WH14 (5.20 tha^{-1}) showed that low yields and below average stability are considered as poor performing cultivars and they have to be discarded as they are of little benefit to the farmers.

Genotypes for specific environments, discriminating and culling environments

Generally, crop varieties show wide fluctuations in their

yielding abilities when grown over varied environments or agro-climatic zones (Fan et al., 2007). Based on the genotype + genotype x environment (GGE) scatter plot, hybrids such as 10A3WH20 and 10A3WH10 were found to be suitable to all the sites. This means that these varieties have above average stability. The scatter plot also demonstrated that the most yielding varieties like SC 727 and 10A3WH10 favors the highest potential environments, that is, Harare and Gwebi. Surprisingly, most hybrids, for example 10A3WH06 were found to favor low potential environments. These genotypes are highly stable under adverse environmental conditions, hence can be recommended for low potential areas like natural regions 3 and 4 of Zimbabwe. The existence of one mega-environment suggest that there is no need to initiate separate breeding programmes for Agriseeds, however, discriminating and culling environments are needed. These proved to be Harare, Gwebi and any one among Shamva, Kadoma and Matopos. Although, sites fell in one mega-environment, the site means and the natural region classification system show that it will be critical to evaluate in all these sites. The existence of one mega-environment is not supported by CIMMYT where Southern Africa was partitioned into various mega-environments (Setimela et al., 2005). However, the existence of one mega-environment shows that it is not essential to have separate breeding programmes for various environments for Agriseeds. The existence of one mega-environment also shows that cross-over interactions could be occurring within a few varieties and thus, selection of stable genotypes is needed. The genotype + genotype x environment interaction (GGE) scatter plot showed the most discriminating and suitable culling environments to be Harare, Gwebi and any one of Kadoma, Shamva and Matopos. These environments are the most discriminating and are good as the testing environments for both early generation testing and advanced testing.

CONCLUSIONS AND RECOMMENDATIONS

Genotypes x environment interaction (GEI) effects are huge in grain yield abilities of Agriseeds materials, thus necessitating the need to do multi-locational trials. Although our data showed the testing sites in Zimbabwe to represent one mega-environment for maize, evaluation of maize in all these sites is critical since the sites represent various natural regions of Zimbabwe. Furthermore, the data was based on one year that might need further investigations. Further increasing the number of testing sites will not improve the breeding and selection efficiency, but would rather increase the cost of the breeding programmes. Gwebi and Harare are good culling sites and could be used in early generation evaluation of the breeding materials. The better hybrids for various sites are 10A3WH04 (6.7 tha^{-1}) and

10A3WH24 (6.7 tha^{-1}), since they are high yielding and have above average stability. The hybrid 10A3WH10 (7 tha^{-1}) is suited for the high potential areas such as (Gwebi and Harare) while hybrids such as 10A3WH03 (6.5 tha^{-1}) and 10A3WH06 (6.8 tha^{-1}) are suited for low potential areas (Kadoma and Matopos). Further evaluation of stable and specific genotypes is required before release of the hybrids, in order to determine their genotype x years and genotype x environments x years interactions. However, an increase in the number of locations will reduce the number of testing years in attempting to derive desirable cultivars. Based on this study we recommend Agriseeds to do early generation testing at Gwebi and Harare, and then test in two sites per each natural region in order to reduce the time of releasing a cultivar due to the effects of GEI.

ACKNOWLEDGEMENTS

The authors express their thanks to Agriseeds for financing this study.

REFERENCES

Anderson SR, Lauer MJ, Schoper JB, Shibles RM (2004). Pollination timing effects on kernel set and silk receptivity in four maize hybrids. Crop. Sci. 44:464-473.

Banziger M, Setimela PS, Hodson D, Vivek B (2004). Breeding for improved drought tolerance in maize adapted to Southern Africa. New directions for a diverse planet: Proceedings of the 4th International Crop. Sci. Congress 26 September-1 October 2004. Brisbane, Australia, P. 10.

Bassetti P, Westgate ME (1993). Emergence, elongation and senescence of maize. Crop. Sci. 33:271-275.

Bernardo R (2002). Breeding for quantitative traits in plants. Stemma Press, Minnesota, P. 369.

Bolanos J, Edmeades GO (1993). Eight cycles of selection for drought tolerance in tropical maize. II. Responses in reproductive behavior. Field Crops Res. 31:253-268.

Caliskan ME, Erturk E, Sogut T, Boydak E, Arioglu H (2007). Genotype x environment interaction and stability analysis of sweet potato (Ipomea batatas) genotypes. New Zea. Crop. Hort. Sci. 35(1):87-99.

Edmeades GO, Bolanos J, Hernandez M, Bello S (1993). Causes of silk delay in a lowland tropical maize population. Crop. Sci. 33:1029-1035.

Fan X, Kang MS, Chen H, Zhang Y, Tan J, Xu C (2007). Yield stability of maize hybrids evaluated in multi- environmental trials in Yunan, China. Agron. J. 99:220-228.

Fehr WR (1987). Principles of cultivar development, Volume 1.Macmillan Publishing Company, New York. P. 536.

Finlay KW, Wilkinson GN (1963). The analysis of adaptation in a plant breeding programme. Aust. J. Agric. Res. 14:742-754.

Genstat (2010). Genstat Release 13.3 (PC/ Windows Vistal). VSN International Ltd.

Hallauer AR, Miranda (1988). Quantitative genetics in maize breeding. 2nd ed. Iowa State University Press, Ames.

Hassanpanah D (2009). Analysis of G X E Interaction by Using the Additive Main Effects and Multiplicative Interaction in Potato Cultivars. Int. J. Plant Breed. Gen. 1:1-7.

Kang VMS (2003). GGE biplot analysis. A graphical tool for plant breeders, geneticists, and agronomists. CRC Press, USA, P. 262.

Lee EA, Tollanaar M (2007). Physiological basis of successful breeding strategies for maize grain yield. Crop. Sci. 47:202-215.

Mohammadi R, Haghparast R (2010). Evaluation of promising rainfed wheat breding lines on farmers' fields in the west of Iran. Inter. J. Plant Breed. 5(1):30-60.

Muungani D, Setimela P, Dimairo M (2007). Analysis of multi-environment, mother-baby trial data using GGE biplots. Afr. Crop. Sci. Conf. Proc. 8:103-112.

Nyamapfene K (1991). Soils of Zimbabwe. Nehanda Publishers, Union Avenue, Harare. P. 179.

Rukuni M, Tawonezvi P, Eicher C, Munyuki-Hungwi M , Matondi P (2006). Zimbabwe's agricultural revolution revisited, University of Zimbabwe publications, Sable press private limited, Zimbabwe.

Setimela P, Chitalu Z, Jonazi J, Mambo A, Hodson D, Banziger M (2005). Environmental classification of maize-testing sites in the SADC region and its implication for collaborative maize breeding strategies in the subcontinent. Euphytica 145:123-132.

Tiawari DK, Panday P, Singh RK, Singh SP, Singh SB (2011). Genotype x environment interaction and stability analysis in elite clones of sugarcane (Saccharum officinarum L.). Int. J. Plant Breed. Gen. 5(1):93-98.

Wang G, Kang MS, Moreno O (1999). Genetic analysis of grain filling rate and duration in maize. Field crops Res. 61:211-222.

Yan W, Tinker NA (2006). Biplot analysis of multi-environmental trial data: Principles and applications. Can. J. Plant Sci. 86:623-645.

Yan W, Cornelius PL, Crossa J,Hunt LA (2001). Two types of GGE biplots for analyzing multi-environmental trial data. Crop Sci. 41: 656-663.

Yang R, Crossa J, Cornelius PL, Burgueno J (2009). Biplot analysis of genotype x environment interactions: Proceed with caution. Crop Sci. 49:1564-1576.

15

Generation mean analysis and heritability of drought resistance in common bean (*Phaseolus vulgaris* L.)

Abebe Hinkossa[1], Setegn Gebeyehu[2] and Habtamu Zeleke [3]

[1]Bule Hora University, P. O. Box 144 Bule Hora, Ethiopia.
[2]National Bean Research Program, Melkassa Agricultural Research Center, P. O. Box, 436, Adama, Ethiopia.
[3]Haramaya University, P. O. Box 76, Haramaya, Ethiopia.

Information on the availability of genetic variability and mode of gene action are critically important for choosing effective breeding methods that result in appreciable improvement in performance under drought stress. The objectives of this study were to estimate the gene action for drought resistance of quantitative traits and also to estimate the components of variance and heritability of drought resistance in common bean. Field experiment was carried out using six generations of two populations made of crosses between pairs of drought resistant and susceptible genotypes (Roba-1 × SER-16; Melka-Dima × SAB623). The treatments were laid in a split plot design with three replications, where watering regime was assigned to the main plot and generations to the sub-plot. Drought stress was initiated at flowering stage by withholding application of irrigation water. Scaling test and generation mean analysis brought out that individual crosses greatly differed for the gene action and on an overall basis all the types of gene action, additive, dominance and epistasis were important for drought resistance in common bean. Both additive and non-additive types of gene action were important in governing the inheritance of the traits considered. However, additive types of gene actions were important in the inheritance of number of pod per plant in Roba-1 × SER-16 and above ground biomass in Melka-Dima × SAB-623 under drought stress. Medium to high broad and narrow sense heritability were found for most of the traits under both watering regimes. Evidences have unfolded that chances to find stress tolerant breeding material in segregating populations of the two crosses were promising. The presence of significant amount of all types of gene action for the important traits imply that methods which can utilize all of them such as recurrent selection and multiple cross could be employed in breeding beans for drought environments.

Key words: Common bean, drought resistance, gene action, generation mean analysis, heritability, scale test.

INTRODUCTION

It is estimated that 60% of the bean crop is cultivated under the risk of either intermittent or terminal drought (Thung and Rao, 1999). Genotypic differences for drought resistance have been reported for common bean (Abebe et al., 1998). The choice of an efficient breeding program depends to a large extent on knowledge of the type of gene action involved in the expression of the character (Dabholkar, 1999). The efficiency of breeding program increases by careful choice of parents and populations capable of producing progeny with desirable trait combinations (Abreu et al., 2002; Cristina et al., 2002). Before embarking on any improvement program, genetic information regarding the inheritance of quantitative characters, particularly the nature and magnitude of gene action governing the inheritance of the character should be determined. The type of gene action controlling a trait is very important in decisions regarding breeding method, cultivar type (inbred, hybrid, population, etc.), and interpretation of data from quantitative genetic experiments (Lamkey and Lee, 1993).

To formulate an efficient breeding program for developing drought tolerant varieties, it is essential to understand the mode of inheritance, the magnitude of gene effects and their mode of action (Farshadfar et al., 2008). Due to their quantitative nature, drought related traits cannot be studied in simpler way. Specialized biometrical techniques are required to work out the type of genetic variability associated with the traits. These biometrical techniques are dependent on different mating designs such as diallel, line x tester; North Carolina design and generation mean analysis for the estimation of type of genetic variability. Among these mating designs, generation mean analysis has been the most powerful biometrical analysis since it gives additional information about the epistatic interactions. Generation mean analysis is an approach which provides information about nature and magnitude of gene actions involved and used to estimate the component variance which provides information about the predominant type of gene action for the important characters of crop species (Ganesh and Sakila, 1999). It is based on the mean of six generations that is, P_1, P_2, F1, F2, BC_1 and BC_2. Information derived from these analyses can be further utilized for the formulation of an effective breeding strategy. The presence or absence of epistasis can be detected by generation means analysis using the scaling test, which measures epistasis accurately whether it is complimentary or duplicate at the digenic level (Sharmila et al., 2007). Besides gene effects, breeders would also like to know how much of the variation in a crop is genetic and to what extent this variation is heritable, because efficiency of selection mainly depends on additive genetic variance, influence of the environment and interaction between genotype and environment.

Information about the genetic components of variation helps the breeder in the selection of desirable parents for crossing programs and also in deciding a suitable breeding procedure for the genetic improvement of various quantitative traits (Singh and Narayanan, 1993). Generation means analysis has been used in common bean to study the inheritance of other complex traits such as leafhopper insect resistance (Kornegay and Temple, 1986), rate of ethylene production (Sauter et al., 1990), bean pod morphology (Chung et al., 1991), aschochyta leaf blight tolerance (Hanson et al., 1993), leaf trichome density (Park et al., 1994), and most recently heat tolerance (Rainey and Griffiths, 2005) and climbing ability (Checa et al., 2006). This study was initiated to estimate the gene action for drought resistance of quantitative traits and to estimate the component of variance and heritability of post flowering drought resistance in common bean.

MATERIALS AND METHODS

Site description

The field experiment was carried out at Melkassa Agricultural Research Center (MARC) which is found in the Central Rift Valley of Ethiopia (Figure 1). The center is located at 8° 24' N latitude and 39° 21' E longitudes at an altitude of 1550 masl. The climate of the area is characterized as semi-arid with mean monthly maximum and minimum temperature of 33 and 10.8°C, respectively. The area is characterized by low and erratic rainfall with unimodal pattern of distribution. The soil is sandy clay loam (Cambisol). According to Laike et al. (2006), common bean crop coefficient during the mid-season (matches with the duration of stress imposed in the present study) was 1.01 with ET_c and ET_o values of 234.74 and 231.75 mm.

Experimental materials

The parents include two drought resistant (SER- 16 and SAB-623) and two susceptible (Roba-1 and Melka-Dima) common bean genotypes. Roba-1 is a small seeded commercial cultivar sensitive to drought stress. SER-16 is small red seeded advanced breeding line from CIAT with good degree of resistance to drought. Melka-Dima is another drought susceptible commercial cultivar with medium size seed. SAB6-23 is an advanced breeding line from CIAT with medium seed size and good level of tolerance to drought. Initially, two single crosses were made using the four parents: Roba-1 × SER-16 and Melka-Dima × SAB-623. The experimental material consisted of six generations for each cross. These were parents (P_1 and P_2), the first and second filial generation (F_1 and F_2) and back crosses ($BC_1 = P_1 × F_1$ and $BC_2 = P_2 × F_1$).

Development of plant material

The four parents were planted in the field under optimal growth conditions. Normal production package and crop husbandry techniques were followed to raise the crop. The different generations that were used in this study (F_1, F_2, BC_1, and BC_2) were developed through a stepwise crossing from February to September, 2010. Using the four parents, two F_1 cross combinations between tolerant and susceptible parents were made from February to June, 2010. The F_1 seeds were planted to generate F_2 population through selfing and the backcross populations (BC_1 and BC_2) were simultaneously generated through crossing F_1 of the crosses back to parent 1 (P_1) and parent 2 (P_2) from June, 2010 to September, 2010. List of parents and crosses produced are given in Table 1.

Experimental design and treatments

The six treatments (P_1, P_2, F_1, F_2, BC_1, and BC_1) for each population were planted in split plot design with watering regime as a main plot and generations were assigned to sub-plots. Overall, twelve generations ($2P_1$, $2P_2$, $2F_1$, $2 F_2$, $2BC_1$ and $2BC_2$) were grown under two contrasting watering regimes, non-stress (NS) and drought-stress (DS) and the treatments were replicated three times. Planting was done late in season (4th week of September, 2010) to expose the drought stressed treatments to terminal stress when the main season rain ceases. For the stress treatments, terminal drought was induced stressed by withholding application of irrigation at flowering stage.

Soil moisture measurement (centibars) was taken using water

Figure 1. Location of Melkassa Agricultural Research Center.

Table 1. List of parents and crosses produced.

Generation	Parents and crosses (Population 1)	Parents and crosses (Population 2)
P_1	SER-16	SAB-623
P_2	Roba-1	Melka-Dima
F_1	Roba-1 × SER-16	Melka-Dima × SAB-623
F_2	Roba-1 × SER-16	Melka-Dima × SAB-623
BC_1	SER-16 × (Roba-1 × SER-16)	SAB-623 × (Melka-Dima × SAB-623)
BC_2	Roba-1 × (Roba-1 × SER-16)	Melka-Dima × (Melka-Dima × SAB-623)

mark (IRROMETER CAMPANY, INC). The water marks were installed at six levels of depth (0 to 5, 5 to 10, 10 to 20, 20 to 40, 40 to 60, and 60 to 80 cm) at randomly selected sites in each block/replication. Reading was taken twice between each irrigation with water mark readings 0 to 10 centibars (soil moisture is at field capacity), 10 to 30 centibars (soil is adequately wet), 30 to 60 centibars (soil usually range for irrigation), above 100 centibars (soil is becoming dangerously dry for maximum production). Using this information on soil moisture status, the irrigation frequency was adjusted accordingly. For control (non-stressed) treatments, the

Table 2. Scaling test for morpho-physiological characteristics in Roba-1 × SER-16 (C1) and Melka-Dima × SAB-623 (C2) crosses grown under drought stress (DS) and non- stress (NS) conditions at Melkassa, Central Rift Valley of Ethiopia.

Scaling test		Roba-1 × SER-16			Melka-Dima × SAB-623		
Cross		A	B	C	A	B	C
PH	NS	4.00 ± 3.71	-7.47 ± 4.39	-12.33 ± 6.05*	-1.00 ± 2.25	-8.40 ± 2.39**	-13.40 ±3.86**
	DS	1.40 ± 2.75	-1.87 ± 2.75	-16.87 ± 4.22**	-4.10 ± 9.63	-0.70 ± 9.25	-11.33± 20.04
PL	NS	-0.71 ± 0.28*	-1.14 ± 0.34**	-4.19 ± 6.49	-1.02 ± 3.73	-1.14 ± 3.67	-2.98 ± 7.50
	DS	-1.07 ± 2.65	-1.48 ± 0.37**	-3.80 ± 6.69	-1.18 ± 3.96	-1.52 ± 4.15	-3.82 ± 8.10
gs	NS	-64.59 ± 41.18	-69.03 ± 45.03	-166.45 ± 97.61	-42.74 ± 9.60**	-14.52 ± 10.78	-225.5 ± 18.34**
	DS	-47.45 ± 35.79	-89.00 ± 40.82*	-236.65 ± 85.63**	-28.29 ± 10.08**	-16.86 ± 9.67	-91.01 ± 87.70
CT	NS	-1.35 ± 0. 47*	-1.38 ± 0.834	-0.06 ± 1.282	-2.49 ± 0.78**	1.07 ± 0.86	1.69 ± 1.34
	DS	-2.87 ± 0.88*	1.19 ± 0.94	6.86 ± 1.41**	2.49 ± 0.97*	0.91 ± 0.96	6.97 ± 7.64
QY	NS	-0.051 ± 0.079	-0.035 ± 0.081	-0.074 ± 0.172	-0.06 ± 0.07	-0.05 ± 0.02*	-0.26 ± 0.18
	DS	-0.04 ± 0.02*	-0.028 ± 0.02	-0.19 ± 0.03*	-0.05 ± 0.02*	-0.03 ± 0.02	-0.22 ± 0.17
LA	NS	-13.88 ± 14.18	-10.83 ± 3.47**	-58.30 ± 30.27	-16.94 ± 13.81	-15.83 ± 3.46**	-52.78 ± 33.35
	DS	-6.522 ± 13.91	-8.878 ± 3.60*	-48.289 ± 28.92	-13.06 ± 14.07	-16.6 ± 4.07**	-50.8 ± 33.05

PH, Plant height (cm); PL, pod length (cm); gs, stomatal conductance (mmol $m^{-2}s^{-1}$); CT, canopy temperature (°C); Qy, quantum yield; LA, leaf area (cm^2); NS, non- stress; DS, drought stress; * and ** significant at 5 and 1% level of probability, respectively.

plots were irrigated to restore the soil moisture status to field capacity until physiological maturity. Since the non-segregating generation represents the homozygous population while segregating generation represents heterozygous population, the number of plants used for the different generations was varied. Accordingly, there were two rows per plot for P_1, P_2, F_1, BC_1 and BC_2 generations and four rows per plot for F_2 generation. The row length was 2 m and the rows were kept at 0.6 m apart. Within row spacing (distance between plants) was 0.1 m. Fertilizers were applied at planting using the rate of 46 P_2O_5 kg ha^{-1} in the form of DAP and other crop management was carried out as recommended for the area.

Data collection

Pod length, plant height, number of pods per plant, seeds per pod and seed yield per plant were determined on ten and five randomly selected plants for F_2 and the other five generations, respectively. Leaf area (LA, cm^2) was measured using a non-destructive method developed as standard system for the evaluation of bean leaf area (Habtu, 1994). Seed yield per hectare was obtained by converting plot yield and adjusting seed moisture content at 10%. Above ground biomass (AGB, gm $plant^{-1}$) was considered as weight of above ground parts (stem + leaves +pod wall + seed) at harvest after drying for 48 h at 85°C randomly selected plants. Harvest index (HI) was determined as proportion of seed weight to the AGB at harvesting dry weight (stem + leaves +pod wall + seed) at harvest × 100.

Leaf chlorophyll content (SPAD value) was measured by using a non-destructive, hand-held chlorophyll meter (SPAD-502 chlorophyll meter, Minolta Camera Co., Ltd., Japan). Canopy temperature (CT; °C) was considered as the difference in temperature between the leaf canopy and the surrounding air temperature measured using an infrared thermometer (Telatemp model AG-42D, Telatemp CA, USA). Stomatal conductance (mmol $m^{-2}s^{-1}$) for water vapor was measured using a portable leaf porometer (leaf porometer, Decagon Devices INC). The photo system II quantum yield (quantum yield, QY) was measured by using a non-destructive, hand-held Qy meter (Fluorpen, FP100, Photo systems Instruments).

Data analysis

Data were subjected to the generation mean analysis to determine the type of genetic variation associated with each trait under the two watering regimes. Mather (1949) scaling tests were employed to detect the presence of epistasis. To obtain information on the nature of gene action governing the traits under study, all the six components of generation means were computed. In the presence of epistasis, generation mean analyses were carried out according to Hayman (1958). The significance of the different parameters was tested with the help of 't' values, which were calculated for each component by dividing the gene effect of respective components by their standard errors (SE). Data were subjected to the estimation of various components (environmental, genotypic, additive and dominance) variances as per Mather and Jinks (1971). Broad and narrow sense heritabilities were calculated following the method used by Mather (1949).

RESULTS

Estimates of gene action

Scaling test for Roba-1 × SER-16 cross has demonstrated that all morpho-physiological characters (except stomatal conductance and quantum yield under non-stress) were significant under both watering regimes (Table 2). In Melka-Dima × SAB-623 cross, all

Table 3. Gene effects for morpho-physiological characteristics in Roba-1 × SER-16 (C1) and Melka-Dima X SAB-623 (C2) crosses grown under drought stress (DS) and non-stress (NS) conditions at Melkassa, Central Rift Valley of Ethiopia.

Cross	Gene effect	PH NS	PH DS	PL NS	PL DS	gs NS	gs DS	CT NS	CT DS	Qy NS	Qy DS	LA NS	LA DS
C1	m	43.3 ± 1.12**	38.3 ± 0.77**	8.2 ± 0.09**	7.3 ± 0.10**	233.6 ± 3.25**	176.2 ± 2.9**	15.6 ± 0.23**	19.5 ± 0.25**	0.45 ± 0.005**	0.387 ± 0.005**	68.9±1.04**	61.4 ± 1.02**
	d	-1.1 ± 2.24	-1.8 ± 1.55	-0.2 ± 0.19	0.4 ± 0.20	-3.9 ± 6.51	30.1 ± 5.9**	-0.8 ± 0.46	-4.0 ± 0.51**	-0.042 ± 0.011*	0.008 ± 0.011	-3.6±2.09	6.2 ± 2.05**
	h	12.0 ± 6.64	20.6 ± 4.58**	2.5 ± 0.57**	2.2 ± 0.61**	72.4 ± 19.27**	143.1 ± 17.4**	-3.5 ± 1.36*	-9.8 ± 1.53*	0.042 ± 0.035	0.155 ± 0.033**	38.2±6.19**	38.2 ± 6.09**
	i	8.9 ± 6.34	16.4 ± 4.38*	2.3 ± 0.54**	1.3 ± 0.58*	32.8 ± 18.42	100.2 ± 16.7**	-2.7 ± 1.30*	-8.5.46**	-.011 ± 0.033	0.122 ± 0.031**	33.6±5.92**	32.9 ± 5.82**
	j	11.5 ± 5.01*	3.3 ± 3.46	0.4 ± 0.43	0.4 ± 0.46	4.4 ± 14.56	41.5 ± 13.2**	0.03 ± 1.03	-4.1 ± 1.16**	-0.015 ± 0.026	-0.009 ± 0.025**	-3.1 ± 4.68	2.4 ± 4.60
	l	-5.4 ± 10.76	-15.9 ± 7.43*	-0.5 ± 0.92	1.3 ± 0.99	100.8 ± 31.24	36.3 ± 28.3	5.4 ± 2.21*	10.2 ± 2.49**	0.096 ± 0.057	-0.056 ± 0.053	-8.9 ± 10.03	-17.5 ± 9.87
C2	m	37.6 ± 0.69**	34.5 ± 0.68**	10.0 ± 0.12**	9.4 ± 0.13**	230.0 ± 3.15**	211.3 ± 2.9**	16.6 ± 0.24**	20.7 ± 0.26**	0.389 ± 0.005**	0.379 ± 0.005**	77.2 ± 1.08**	74.4 ± 1.16
	d	-0.7±1.39	-2.6 ±1.34*	-0.04 ± 0.24	0.3 ± 0.26	-21.9 ± 6.31**	5.4 ± 5.8	-1.9 ± 0.48**	-0.13 ± 0.53	-0.008 ± 0.011	0.009 ± 0.011	-7.8 ± 2.17**	9.4 ± 2.32
	h	5.9±4.12	8.9 ± 4.04*	1.5 ± 0.72*	1.5 ± 0.78*	188.5 ± 18.68**	66. ± 17.2**	-4.3 ± 1.42**	-3.8 ±1.57*	0.190 ± 0.033**	0.191 ± 0.034**	24.2 ± 6.44**	35.1 ± 6.89
	i	4.0±3.94	6.5 ± 3.85	0.8 ± 0.69	1.12 ± 0.75	168.2 ± 17.86**	45.84 ± 16.5**	-3.1 ± 1.36*	-3.6 ±1.50*	0.148 ± 0.032**	0.137 ± 0.032**	20.0 ± 6.16**	21.1 ± 6.58
	j	7.4±3.11*	-3.4 ± 3.05	0.12 ± 0.54	0.34 ± 0.59	-28.2 ± 14.11*	-8.57 ±13.0	-3.6 ± 1.07**	1.6 ±1.19	-0.008 ± 0.025	-0.021 ± 0.025	-1.1 ± 4.87	3.6 ± 5.20
	l	5.4±6.69	-1.7 ± 6.54	1.35 ± 1.17	1.58 ± 1.27	-110.9 ± 30.28*	-0.68 ± 27.9	4.5 ± 2.31*	0.17 ± 2.55	-0.036 ± 0.054	-0.054 ± 0.055	12.8 ± 10.45	8.6 ± 11.17

PH, Plant height; PL, pod length; gs, stomatal conductance (mmol m^{-2} s^{-1}); LA, leaf area (cm^2); Qy, quantum yield; CT, canopy temperature (°c); NS, non-stress; DS, drought stress; * and **, significant at 5 and 1% probability level, respectively; C1, cross 1 (Roba-1 *SER-16); C2, cross 2 (Melka Dima * SAB-623); m, mean of the generation; d, additive effect; h, dominance effect; l, additive × additive effect; j, additive × dominance effect; l, dominance × dominance effect.

morpho-physiological characters (except pod length) were significant for at least one of the scale tests under both watering regimes. Estimates of genetic effects for the six parameter model indicated that mean effect (m) of each cross under the two watering regimes was significant (Table 3). Additive × dominance (j) was significant under non-stress in both crosses for plant height. Under drought stress dominance, additive × additive and dominance × dominance for Roba-1 × SER-16 cross with duplicate types of epistasis and additive and dominance for Melka-Dima × SAB-623 cross were significant (Table 3).

Dominance and additive × additive were significant for pod length in Roba-1 × SER-16 cross under both watering regimes. In the Melka-Dima × SAB-623 cross, only dominance was significant under both watering regimes. In the Roba-1 × SER-16 cross, dominance (h) under non-stress and additive, dominance, additive × additive and additive × dominance under drought stress were significant for stomatal conductance. Additive, dominance, additive × additive, additive × dominance and dominance × dominance with duplicate type of epistasis under non-stress and dominance, additive × additive under drought stress were significant in Melka-Dima × SAB-623 cross. For canopy temperature dominance, additive × additive and dominance × dominance with duplicate types of epistasis under non-stress and significant additive, additive, dominance, additive × additive, additive × dominance and dominance × dominance with duplicate types of epistasis under drought stress were observed in Roba-1 × SER-16 cross. In Melka-Dima × SAB-623 cross, additive, dominance, additive × additive, additive × dominance and dominance with duplicate types of epistasis under non-stress and dominance and additive × additive under drought stress were significant.

Scaling test for A, B, and C was done for yield and yield components in the two crosses for each watering regime (Table 4). At least one of the scales was significant in Roba-1 × SER-16 cross for seed per plant, seed yield, above ground biomass and harvest index under both growth conditions whereas pod per plant was non-significant under both watering regimes. In Melka-Dima × SAB-623 cross, all yield and yield components except above ground biomass were significant under the two watering regimes. Results of types of gene action estimated by generation mean analysis as genetic effects in six model parameter for yield and yield components are presented in Table 5. Results indicated that mean effect (m) of each cross under the two watering regimes was significant. Additive gene

Table 4. Scaling test for yield and yield components in two common bean crosses grown under drought stress (DS) and non-stress (NS) conditions at Melkassa, Central Rift Valley of Ethiopia.

Cross	WR	Roba-1 × SER-16			Melka-Dima × SAB-623		
		A	B	C	A	B	C
NPPP	NS	-9.2 ± 10.4	-4.6 ± 11.7	-10.6 ± 34.7	4.3 ± 1.8*	0.2 ± 2.1	-16.4 ± 3.4**
	DS	-7.3 ± 9.4	-7.7 ± 10.8	-14.8 ± 23.6	4.5 ± 2.0	-5.5 ± 1.9	-14.9 ± 5.5**
NSPPI	NS	-2.0 ± 24.1	-16.4 ± 6.7*	-58.1 ± 75.7	13.7 ± 5.9*	2.9 ± 6.3	-53.9 ± 9.2**
	DS	-1.1 ± 8.8	7.3 ± 9.1	-70.4 ± 14.1**	3.9 ± 7.2	-6.6 ± 5.6	-31.6 ± 8.1**
SYPP	NS	-11.2 ± 2.7**	-7.4 ± 2.9**	-31.7 ± 93. 6	7.9 ± 2.7**	-16.2 ± 2.3**	-32.2 ± 51.2
	DS	-6.3 ± 3.0*	-1.6 ± 2.2	-17.4 ± 42.0	-6.4 ± 2.9**	-1.9 ± 1.8	-26.0 ± 35.2
AGB	NS	-103.3 ± 15.5**	-67.5 ± 22.8**	-250.5 ± 177.4	10.3 ± 78.9	-184.4 ± 65.7*	-281.1 ± 187.1
	DS	-29.0 ± 12.7*	-29.2 ± 17.3	-225.6 ± 25.2	-7.2 ± 17.4	-10.2 ± 13.1	-6.9 ± 27.9
HI	NS	-6.2 ± 3.8	-6.4 ± 3.2*	-24.9 ± 32.3	-4.1 ± 3.6	-6.4 ± 3.4	-24.3 ± 5.8**
	DS	-5.2 ± 3.5	1.4 ± 3.3	-16.7 ± 5.4**	-2.4 ± 3.5	-0.5 ± 3.1	-19.2 ± 5.4**

NPPP, Number of pod per plant; NSPPI, number of seed per plant; SYPP, seed yield per plant; AGB, above ground biomass; HI, harvest index; WR, watering regime; NS, non-stress; DS, drought stress; * and **, significant at 5 and 1% level of probability, respectively.

action was significant for pod per plant under non-stress and drought stress in the Roba-1 × SER-16 cross. On the other hand, dominance, additive × additive and dominance × dominance with duplicate types of epistasis under non-stress and additive, dominance, additive × additive, additive × dominance and dominance × dominance with duplicate types of epistasis under drought stress were significant in Melka-Dima × SAB-623 cross. For seed per plant, additive × additive under non-stress and dominance and additive × additive under drought stress were significant in Roba-1 × SER-16 cross. In the Melka-Dima × SAB-623 cross, dominance, additive × additive and dominance × dominance with duplicate types of epistasis under non-stress and additive, dominance, additive × additive and additive × dominance under drought stress were significant for seed per plant. The additive, dominance and additive × additive under non-stress and dominance, additive × additive and dominance × dominance with duplicate types of epistasis under drought stress were significant for seed yield in Roba-1 × SER-16 cross. Similarly, additive, dominance, additive × additive, additive x dominance and dominance × dominance with duplicate types of epistasis under non-stress and additive, dominance and additive × additive under drought stress were found in Melka-Dima × SAB-623 cross. Under non-stress condition additive, dominance and additive × additive and under drought stress dominance and additive × additive were significant for above ground biomass in Roba-1 × SER-16 cross. Additive, dominance, additive × additive and additive × dominance were significant in Melka-Dima × SAB-623 cross under

non-stress but only additive was significant under drought stress. Significant epistatic additive × additive type of gene effects was detected for harvest index in Roba-1 × SER-16 cross under both watering regimes. On the other hand, dominance and additive × additive were significant for the same parameter in Melka-Dima × SAB-623 cross under both watering regimes.

Estimates of components of variance

Estimates of variance components varied considerably between crosses and watering regimes (Table 6). Environmental component variance was less than the additive and dominance component of variance in both crosses under both growth conditions for all morpho-physiological character considered. Additive variance was higher than dominance variance for stomatal conductance and quantum yield in both crosses under both watering regimes. The additive variance (VA) was larger than dominance variance (VD) for pod length under both watering regimes in Roba-1 × SER-16 cross. For leaf area, additive component of variance was higher than dominance component of variance in Melka-Dima × SAB-623 cross under both environmental condition and under non-stress condition in Roba-1 × SER-16 cross. Additive component of variance was less than dominance component of variance in Roba-1 × SER-16 cross for plant height and canopy temperature under the two watering regimes. Under drought stress, additive component of variance was less than dominance component of variance in Roba-1 × SER-16 cross for leaf

Table 5. Gene effects for yield and yield components in two common bean crosses grown under drought stress (DS) and non-stress (NS) conditions at Melkassa, Central Rift Valley of Ethiopia.

Cross	Gene effect	NPPP NS	NPPP DS	NSPPI NS	NSPPI DS	SYPP NS	SYPP DS	AGB NS	AGB DS	HI NS	HI DS
	m	29.9 ± 1.0**	24.1 ± 0.8**	123.1 ± 2.3**	97.3 ± 2.6**	26.9 ± 0.8**	24.0 ± 0.8**	236.3 ± 5.9**	178.9 ± 4.5**	56.7 ± 1.1**	52.6 ± 0.9**
	d	-3.9 ± 1.9*	3.9 ± 1.5*	-3.2 ± 4.6	7.4 ± 5.2	-4.0 ± 1.6*	0.5 ± 1.6	-43.8 ± 11.9**	11.2 ± 9.0	-1.7 ± 2.2	1.7 ± 1.9
Roba-1	h	-3.5 ± 6.1	4.6 ± 4.5	26.0 ± 3.7	81.9 ± 15.3**	12.6 ± 4.7**	16.4 ± 4.7**	72.7 ± 35.3	11.2 ± 9.0**	10.7 ± 6.4	15.3 ± 5.8
× SER-16	i	-3.2 ± 5.8	-0.1 ± 4.3	39.6 ± 13.1**	76.7 ± 14.6**	13.2 ± 4.5*	9.5 ± 4.5*	79.7 ± 33.7*	167.3 ± 25.4**	12.3 ± 6.1*	12.9 ± 5.6**
	j	-4.6 ± 4.6	0.4 ± 3.4	14.3 ± 10.4	-8.4 ± 11.5	-3.8 ± 3.6	-4.7 ± 3.6	-35.8 ± 26.6	0.2 ± 20.1	0.23 ± 4.8	-6.6 ± 4.4
	l	17.1 ± 9.8	15.1 ± 7.3	-21.2 ± 22.3	-82.8 ± 24.7	5.3 ± 7.7	-1.5 ± 7.6*	91.1 ± 57.2	-109.1 ± 43.1*	0.3 ± 10.3	-9.3 ± 9.4
	m	19.5 ± 0.6**	15.5 ± 0.6**	58.9 ± 1.7**	45.0 ± 1.9**	28.4 ± 0.7**	17.5 ± 0.7**	251.3 ± 5.9**	198.1 ± 4.8**	56.2 ± 1.0**	52.3 ± 0.9**
	d	0.9 ± 1.2	8.5 ± 1.2**	3.5 ± 3.4	12.9 ± 3.9**	10.7 ± 1.5*	4.2 ± 1.5**	80.4 ± 11.8**	56.7 ± 9.7**	-0.8 ± 2.1	1.0 ± 1.9
Melka-Dima	h	25.0 ± 3.4**	14.9 ± 3.5*	82.1 ± 10.1**	36.5 ± 11.6**	22.7 ± 4.5**	18.5 ± 4.4**	112.2 ± 35.0**	-5.0 ± 28.7	13.7 ± 6.1*	17.2 ± 5.7**
× SAB-623	i	20.9 ± 3.3**	14.0 ± 3.3*	70.5 ± 9.7**	28.9 ± 11.1**	23.9 ± 4.3**	17.7 ± 4.3**	106.9 ± 33.5**	-10.5 ± 27.4	13.8 ± 5.8**	16.3 ± 5.5**
	j	4.2 ± 2.6	10.0 ± 2.6**	10.8 ± 7.7	10.6 ± 8.8**	24.2 ± 3.4**	-4.4 ± 3.4	194.8 ± 26.4**	2.9 ± 21.7	2.3 ± 4.6	-1.9 ± 4.3
	l	-25.4 ± 5.5*	-13.1 ± 5.6*	-87.1 ± 16.4**	-26.2 ± 18.8	-15.7 ± 7.4*	-9.4 ± 7.3	67.5 ± 56.7	27.9 ± 46.6	-3.4 ± 9.9	-13.4 ± 9.3

NPPP, number of pod per plant; NSPPI, number of seed per plant; SYPP, (seed yield per plant); AGB, above ground biomass; HI, harvest index; NS, non-stress; DS, drought stress; * and **, significant at 5 and 1% probability level, respectively; C1, cross 1 (Roba-1 × SER-16); C2, cross 2 (Melka-Dima × SAB-623); m, mean of the generation, d, additive effect; h, dominance effect; I, additive × additive effect; j, additive × dominance effect; l, dominance × dominance effect.

area. In Melka-Dima × SAB-623 cross, dominance component of variance was higher than additive component of variance for canopy temperature under drought stress. The VA was larger than VD for pod per plant, seed per plant, seed yield per plant, seed yield (kg/ha), AGB and harvest index in both crosses under the two watering regimes. Environmental component of variance was less than the VA and VD for pod per plant, seed per plant, seed yield per plant, seed yield (kg/ha), AGB and harvest index for both crosses under the contrasting soil moisture regimes.

Heritability

Considerable differences were also found between the two crosses and the two soil moisture regimes for broad and narrow sense heritability of the growth and yield related traits (Table 7). Broad sense heritability ranged from 75.6% (leaf area) to 95.4% (for plant height) under non-stress and from 68.6% (leaf area) to 90.1% (for plant height) under drought stress in Roba-1 × SER-16 cross. In Melka-Dima × SAB-623 cross, it ranged from 83.6% (quantum yield) to 89.6% (canopy temperature) under non-stress and from 71.3% (leaf area) to 93.5% (plant height) under drought stress condition (Table 7). Narrow-sense heritabilities ranged from 36.1% (quantum yield) to 49.9% (stomatal conductance) under non-stress, and from 24.6% (leaf area) to 49.6% (stomatal conductance) under drought stress in the Roba-1 × SER-16 cross. In Melka-Dima

× SAB-623 cross, narrow sense heritability was low to moderate under both moisture regimes, ranging from 18.5% (pod length) to 78.9% (quantum yield) under non-stress and from 8.7% (canopy temperature) to 58.2% (quantum yield) under drought stress. Broad sense heritabilities were high in Roba-1 × SER-16 under both growth conditions for all yield and yield components, ranging from 62.7% (above ground biomass) to 88.6% (pod per plant) under non-stress and from 72.02% (pod per plant) to 92.4% (AGB) under drought stress condition. In contrast, broad sense heritabilities in Melka-Dima × SAB-623 cross were moderate to high under non-stress (36.1 to 93.04%) and low to high under drought stress (15.24 to 93.2%). In Roba-1 × SER-16 cross, narrow sense heritabilities ranged between 37.1%

Table 6. Components of variance for morpho-physiological, yield and yield related traits in two crosses of common bean grown under drought stress (DS) and non-stress (NS) conditions at Melkassa, Central Rift Valley of Ethiopia.

Cross	Traits	VE		VG		VA		VD	
		NS	DS	NS	DS	NS	DS	NS	DS
Roba-1 × SER-16	PH	3.1	3.0	63.2	27.6	11.3	7.9	51.9	19.6
	PL	0.5	0.9	1.4	0.9	1.6	0.9	0.3	0.1
	gs	95.6	93.7	460.9	334.1	277.7	212.5	183.2	121.6
	CT	0.3	0.4	2.6	3.1	1.1	1.1	1.4	1.9
	QY	0.0	0.0	0.001	0.001	0.001	0.003	0.001	0.003
	LA	12.8	14.7	39.9	32.2	22.8	11.5	17.1	20.6
	NPPP	8.1	7.7	62.6	19.9	41.6	12.7	12.7	21.0
	NSPPI	67.3	64.8	270.3	263.7	265.4	118.3	118.3	4.9
	SYPP	7.2	8.6	23.3	22.9	11.3	18.3	18.3	12.0
	AGB	639.7	87.3	1073.6	1060.3	1031.2	670.1	670.1	42.4
	HI	11.0	4.9	49.8	47.4	29.7	25.0	25.0	20.1
Melka-Dima × SAB-623	PH	3.6	1.6	22.3	22.9	14.7	6.1	7.6	16.9
	PL	1.3	1.8	1.7	1.6	0.6	0.4	1.2	1.3
	gs	78.7	60.3	491.5	397.5	437.1	243.9	54.4	153.6
	CT	0.3	0.4	2.8	3.1	1.5	0.3	1.3	2.8
	QY	0.0	0.0	0.002	0.001	0.001	0.001	0.0	0.0
	LA	10.1	18.0	55.9	44.7	49.5	31.8	6.4	12.9
	NPPP	8.2	8.7	4.6	2.7	2.8	1.7	1.8	1.1
	NSPPI	18.2	94.9	129.3	17.1	37.2	3.9	92.2	13.2
	SYPP	9.8	6.4	19.4	24.9	18.6	23.13	0.8	1.8
	AGB	149.8	185.4	2002.6	1174.9	1848.1	1120.3	154.6	54.7
	HI	5.1	3.6	55.2	49.3	34.3	27.6	20.9	21.7

PH, Plant height; PL, pod length; gs, stomatal conductance; CT, canopy temperature; Qy, quantum yield; LA, leaf area; NPPP, number of pod per plant; NSPPI, number of seed per plant; SYPP, seed yield per plant; AGB, above ground biomass; HI, harvest index; VE, environmental variance; VG, genotypic variance; VA, additive variance; VD, dominance variance.

(for seed yield per plant) to 78.6% (for seed per plant) under non-stress and between 36.01% (seed yield per plant) and 58.39% (AGB) under drought stress condition. In MelkaDima × SAB623 cross, it was moderate to high under non-stress (22.05% for pod per plant to 85.8% for AGB) and low to high under drought stress 3.5% (for seed per plant) to 82.4% (for AGB).

DISCUSSION

Analysis of variance depicted significant variation among generations and generation × watering regimes for most of the characters considered indicating the presence of genetic variability and possibility of selection for drought resistance. Information about the genetic components of variation helps the breeder in the selection of desirable parents for crossing programs and also in deciding a suitable breeding procedure for the genetic improvement of various quantitative traits (Singh and Narayanan, 1993). In both crosses, all morpho-physiological characters (except stomatal conductance and quantum yield under non-stress in Roba-1 × SER-16 cross and pod length in Melka-Dima × SAB-623 cross) were significant for at least one of the scaling tests under both watering regimes. The significance of any one of the scale reveals the presence of non-allelic interaction, revealing that the estimate of genetic parameters of the trait does not fit to the additive-dominance model. In both crosses, most of the yield related traits were significant under both watering regimes indicating absence of epistatic interaction effect. For plant height, pod length, stomatal conductance, canopy temperature, leaf area, pods per plant, seed per plant, seed yield (gm/plant), above ground biomass and harvest index, epistatic gene effect was present in one of the two crosses. Hence, additive genetic model was not sufficient to explain most of the genetic variation for the expression of these traits concurring the findings of Asrat and Kimani (2005). Under such conditions, epistatic effects have contributed to the inheritance of these traits in both crosses.

Because of the presence of epistasis, generation mean analyses were carried out according to Hayman (1958). The additive, dominance and epistatic types of gene interaction in each cross for different trait were found to be different from each other under different watering regimes. The generation mean analysis has brought out that individual crosses greatly differed for the gene action and on an overall basis all types of gene action, additive, dominance and epistasis were important. The results

Table 7. Broad and narrow sense heritabilities of morpho-physiological, yield and yield related traits in two crosses of common bean grown under drought stress (DS) and non-stress (NS) conditions at Melkassa, Central Rift Valley of Ethiopia.

Cross	Traits	h^2b		h^2n	
		NS	DS	NS	DS
	PH	95.4	90.1	17.1	25.9
	PL	72.1	50.8	58.4	45.2
	gs	82.8	78.1	49.9	49.7
	CT	89.8	88.6	39.3	31.7
	QY	80.1	84.2	36.1	29.7
Roba-1 × SER-16	LA	75.7	68.6	43.2	24.6
	NPPP	88.6	72.0	58.9	46.0
	NSPPI	80.1	80.3	78.6	36.0
	SYPP	76.4	72.7	37.1	58.4
	AGB	62.7	92.4	60.2	58.4
	HI	81.9	90.6	48.8	47.8
	PH	86.2	93.5	56.9	24.7
	PL	57.5	47.8	18.5	11.1
	gs	86.2	86.8	76.7	53.3
	CT	89. 6	88.9	47.7	8.7
	QY	83.6	76.3	78.9	58.0
Melka-Dima × SAB-623	LA	84.7	71.3	75.0	50.7
	NPPP	36.1	23.8	22.1	14.6
	NSPPI	87.6	15.2	25.2	3.5
	SYPP	66.5	79.7	63.9	73.9
	AGB	93.0	86.4	85.9	82.4
	HI	91.5	93.2	56.9	52.3

PH, Plant height; PL, pod length; gs, stomatal conductance; CT, canopy temperature; Qy, quantum yield; LA, leaf area; NPPP, number of pod per plant; NSPPI, number of seed per plant; SYPP, seed yield per plant; AGB, above ground biomass; HI, harvest index; h^2b, broad sense heritability; h^2n, narrow sense heritability.

indicated that mean effect (m) of each cross under the two watering regimes was significant for all characters implying the difference in these characters among the parents. High significance for the estimated values of mean effects (m) indicated that all the traits considered were quantitatively inherited under the contrasting soil moisture regimes. The significant additive × dominance (j) gene action found for plant height under non-stress condition in both crosses is in agreement with the reports of Vaid et al. (1985) and Melaku (1993) who found both additive and non-additive (dominance and epistasis) gene actions for the same trait. Dominance and additive × additive (in Roba-1 × SER-16 cross) and only dominance (in Melka-Dima × SAB-623 cross) gene actions for pod length under both watering regimes appear to slightly deviate from the results obtained by Arunga et al. (2010) and Carvalho et al. (1999) where epistatic effects were involved in addition to dominance in genetic control of pod length in snap bean.

The complex gene actions (involvement of both additive and non-additive) for stomatal conductance in both crosses and under the two soil moisture regimes are comparable with the expression of the trait in other crops

such as wheat (Rebetzke et al., 2003). Similarly, the gene effects for the other physiological characteristics, canopy temperature, were complex in both crosses grown under drought stress and non-stress conditions conferring the results obtained for leaf temperature as drought resistance trait in cowpea (Chozin et al., 2006). The involvement of several gene actions for quantum yield and leaf area in both crosses imply that additive and dominance as well as epistasis gene action were important for inheritance of the physiological characters under different watering regimes.

The significant values of additive (d) and absence of digenic non allelic interaction in Roba-1 × SER-16 for pod per plant revealed that selection for this trait for drought resistance would be useful to start from the early segregating generation under drought stress condition. Similarly, significant values of additive, and the non-allelic gene interaction for pod per plant in Melka-Dima × SAB-623 cross also showed less complexity in the inheritance of this trait for drought resistance. The dominance (h) and dominance x dominance (l) effects were in the opposite direction, suggesting that duplicate-type epistasis occurred in most cases and indicating predominantly dispersed alleles at the interacting loci. Dominance gene effects were found to be relatively more important, as indicated by the fact that the dominance (h) values were higher than the additive (d) values (Jinks and Jones, 1958). For seed per plant additive × additive under non-stress and dominance and additive x additive under drought stress were significant in. Presence of dominance (h) gene effect and additive × additive (i) components under drought stress for Roba-1 × SER-16 cross suggests that the selection for seed per plant would be delayed till dominance and epistatic components are reduced selfing for drought resistance. According to Kunkaew et al. (2007), seed yield per plant is controlled by genes with significance in additive, dominance, and epistatic effects in adzuki bean suggesting that an effective selection to improve this trait should be mild in earlier generations and intense in later generations. The presence of additive × additive effects for harvest index indicates the possibility of transgressive segregates in the later selfed generation besides being the only epistatic effect, at least theoretically, which could be effectively utilized in selection (Zimmerman et al., 1985). Similarly, Nigam et al. (2001) reported that in addition to additive and dominance effects, additive × additive types of epistasis which can be fixed in self pollinated crops was significant for harvest index.

Broad and narrow sense heritability estimates were found to be high in most of morpho-physiological characters for both crosses and under the two growth conditions. Similarly, Singh et al. (1994) reported the highest broad sense heritability for pod length and plant height. High heritability indicates that the environment least influenced these characters and selection based on mean would be successful in improving these traits. The low to high narrow sense heritability values under both watering regimes indicate medium chances of transmitting to the offspring the traits that determine bean productivity and improvement for these traits. Broad sense heritabilities were high in Roba-1 × SER-16 under both growth conditions and low to high under drought stress and moderate to high in Melka-Dima × SAB-623 cross for all yield and yield components. According to Escribano et al. (1994), heritabilities for seed yield and yield components varied from low to high in F2 and F3 generations of common bean. Farshadfar et al. (2008) also reported moderate narrow-sense heritability estimates for biological yield, harvest index, seed weight, and number of seed per plant in chick pea grown under drought stress. In both crosses, heritability in the broad sense as well as the narrow sense was slightly higher under non-stress compared with drought stress for all character (except seed yield), which is in agreement with the findings of Szilagyi (2003). According to Schneider et al. (1997), similar heritability in the broad and narrow sense found for yield and yield related traits between drought and non-stress conditions suggest that selection should be equally effective under different levels of stress.

Conclusion

The result of the present study generally showed that both additive and non-additive (dominance and epistatic) type of gene action were important in governing the inheritance of the characters studied. Since both additive and non-additive gene effects were of great importance in expression of the different traits, it is recommended that breeding methods, which make the best use of additive effects such as recurrent selection or diallel selection mating (DSM) and the pedigree method are applied to develop lines with resistance to drought in common bean. Presence of non-additive gene action for most of the yield related characters implies that conventional selection procedure may not be effective enough for improvement of yield for drought resistance. Therefore, postponement of selection in later generations or inter mating among the selected segregants followed by one or two generations of selfing could be suggested to break the undesirable linkage and allow the accumulation of favorable alleles for the improvement of desirable traits.

ACKNOWLEDGEMENTS

We thank the Ministry of Education of Ethiopia for the financial support. We are also grateful to Melkassa Agricultural Research Center for providing the germplasm and experimental facilities. The technical support of staff of the National Bean Research Program of Ethiopia is duly acknowledged.

REFERENCES

Abebe A, Brick MA, Kirkby R (1998). Comparison of selection induces to identify productive dry lines under diverse environmental conditions. Field Crops Res. 58:15-23.

Abreu de F, Ramalho MAP, Dos Santos J (2002). Prediction of seed-yield potential of common bean populations. Gen. Mol. Biol. 25(3):323-327.

Arunga EE, Van Rheenen HA, Owuoche JO (2010). Diallel analysis of Snap bean (Phaseolus vulgaris L.) varieties for important traits. Afr. J. Agri. Res. 5(15):1951-1957.

Asrat A, Kimani PM (2005). Estimation of genetic parameters for some quantitative traits in large seeded bean (Phaseolus vulgaris L.) lines by factorial analysis of generation means. Afr. Crop Sci. Conf. Proc. 6:85-89.

Carvalho AC, Leal PP, Rodrigues NR, Costa FA (1999). Capacidade de combinação paraoito caracteres agronômicos em cultivares rasteiras de feijão-de-vagem. Hortic. Bras. 17:102-105.

Checa O, Ceballos HN, Blair MW (2006). Generation Means Analysis of Climbing Ability in Common Bean (Phaseolus vulgaris L.). J. Hered. 97(5):456-465.

Chozin M, Garner JO, Watson CE (2006). Inheritance of traits associated with drought resistance in cowpea. Jurnal Ilmu Ilmu Pertenian Indonesia. 8(1):1-5.

Chung WJ, Baggett, JR, Rowe KE (1991). Inheritance of pod cross-section in beans (Phaseolus vulgaris L.). Euphytica 53:159-164.

Cristina de FM, dos Santos JB, de Sousa Nunes GH, Ramalho MAP

(2002). Choice of common bean parents based on combining ability estimates. Gen. Mol. Biol. 25(2):179-183.

Dabholkar AR (1999). Elements of Biometrical Genetics. Ashak Kumar Mitta, Concept Publishing Co. pp. 57-116.

Escribano MR, de AM, Amurrio JM (1994). Diversity in agronomical traits in common bean populations from Northwestern Spain. Euphytica 76:1-6.

Farshadfar E, Sabaghpour SH, Khaksar N (2008) Inheritance of drought tolerance in chickpea (Cicer arietinum L.) using joint scaling test. J. Appl. Sci. 8:3931-3937.

Ganesh SK, Sakila M (1999). Generation mean analysis in sesame (Sessamum indicum L.) crosses. Sesame and Safflower. News Lett. 14:8-14.

Hanson P, Pastor-Corrales M, Kornegay JL (1993). Heritability and sources of Ascochyta blight resistance in common bean. Plant Dis. 77:711-714.

Hayman BI (1958). The separation of epistatic from additive and dominance variation in generation mean. Heredity 12:371-390.

Jinks JL, Jones RM (1958). Estimation of the components of heterosis. Genetics 43:223-224.

Kornegay JL, Temple SR (1986). Inheritance and combining ability of leafhopper defense mechanisms in common bean. Crop Sci. 26:1153-1158.

Kunkaew W, Julsrigival S, Senthong CH, Karladee D (2007). Inheritance of seed yield in adzuki bean (Vigna angularis Willd). CMU. J. Nat. Sci. 6(2):341.

Laike S, Tilahun K, Hordofa T (2006). Crop coefficient of haricot bean at Melkassa, Central Rift Valley of Ethiopia. J. Agric. Rural Dev. Trop. Subtrop. 107(1):33-40.

Lamkey KR, Lee M (1993). Quantitative genetics, molecular markers, and plant improvement In: Imrie BC, Hacker, JB (ed.) Focused Plant Improvement. pp. 104-115.

Mather K (1949). Biometrical Genetics. In: The Study of Continuous Variation, Methun and Company Limited, London. P. 102.

Mather K, Jinks JL (1971). Biometrical Genetics: The Study of Continuous Variations. Chapman and Hall Ltd., London. P. 382.

Melaku A (1993). Heterosis and combining ability for yield and other quantitative characters in haricot bean (Phaseolus vulgaris L.). An MSc Thesis Presented to the School of Graduate Studies of Haramaya Univeristy. P. 103.

Nigam SN, Upadhyaya HD, Chandra S, Nageswarao RC, Wright GC (2001). Gene effect for specific leaf area and harvest index in three crosses of Groundnut. Ann. Appl. Bio. 139(30):1-306.

Park SJ, Timmins PR, Quiring DT, Jui PY (1994). Inheritance of leaf area and hooked trichome density of the first trifoliolate leaf in common bean (Phaseolus vulgaris L.). Can. J. Plant Sci. 74:235-240.

Rainey KM, Griffiths PD (2005). Inheritance of heat tolerance during reproductive development in snap bean (Phaseolus vulgaris L.). J. Am. Soc. Hortic. Sci. 30:700-706.

Rebetzke GJ, Condon AG, Richards RA, Farquhar GD (2003). Gene action for leaf conductance in three wheat crosses. Aust. J. Agri. Res. 54(4):381-387.

Sauter KJ, Davis DW, Li PH, Wallerstein IS (1990). Leaf ethylene evolution level following high-temperature stress in common bean. Hort. Sci. 25:1282-1284.

Schneider KA, Rosales-Serna R, Ibbara-Perez F, Cazares-Enriquez B, Acosta-Gallegos JA, Ramirez-Vallejo P, Wassimi N, Kelly JD (1997). Improving common bean performance under drought stress. Crop Sci. 37:43-50.

Sharmila V, Ganesh KS, Gunasekaran M (2007). Generation mean analysis for quantitative traits in sesame (Sesamum indicum L.) crosses. Gen. Mol. Biol. 30(1):80-84.

Singh DN, Nandi A, Tripathy P (1994). Genetic variability and character association in french bean (P. vulgaris). Indian J. Agric. Sci. 64:114-116.

Singh P, Narayanan SS (1993). Biometrical Techniques in Plant Breeding. 1st Edn. Kalayani Publishers NewDehli, India.

Szilagyi L (2003). Influence of drought on seed yield components in common bean. Bulg. J. Plant Physiol. pp. 320-330.

Thung M, Rao IM (1999). Integrated management of abiotic stresses. In S.P. Singh (ed.) Common bean improvement in the twenty-first century. Kluwer Academic Publishers, Dordrecht, the Netherlands. pp. 331-370.

Vaid K, Singh RM, Gupta VP (1985). Combining ability in dry beans (Phaseolus vulgaris L.). J. Crop Improv. 12(2):255-258.

Zimmerman MJ, Rosielle ODA, Foster KW, Waines JG (1985). Gene action for grain yield and harvest index of common bean grown as sole crop and intercrop with maize. Field Crop Res. 12:319-329.

16

Antioxidant enzyme and morphological characteristics of roots of three *Nicotiana Tabacum* L. genotype seedlings under chilling stress

C. Cui* and Q. Y. Zhou

College of Agronomy and Biotechnology, Southwest University, 400715 Chongqing, P. R. China.

In order to study the morphological and physiological ecological response of root of Flue-cured tobacco seedling after low temperature stress, this experiment was conducted with different stressful period which was 2, 4 and 6 days under 5 to 7°C (day) by the material named Yunyan87, Msk326 and Yunyan85, respectively. The results showed that comparing with favorable temperature (23 to 25°Cday), activities of superoxide dismutase (SOD) and peroxidase (POD) and ascorbate peroxidase (CAT) had different change trends with difference variety, in which three parameters increased for YY87, and except activities of SOD rise for YY85, other parameters decreased after low temperature stress for short-term (2 days) as for K326 and YY85. When low-temperature stress period prolonged, these parameters come to rise gradually. As for morphological parameters, average diagram, surface area and volume of root declined significant after low-temperature stress for 2 days for YY87, while these morphological parameters are similar to normal level if prolonging stress to 6 days. However all morphological parameters are opposite for K326 comparing with YY87. Surface area, volume of root declined to some extent for YY85. These conclusion indicate different morphological and physiological ecological response mechanisms for roots of Flue-cured tobacco seedling of different variety after chilling stress.

Key words: Chilling stress, flue-cured tobacco, morphology, physiology, roots.

INTRODUCTION

Low temperature is a major factor limiting the productivity and geographical distribution of chilling-sensitive plant species (Thomashow, 1998; Sui et al., 2007; Zhang et al., 2008). Chilling can also result in weak growth, and further reduce crop production. Chilling stress could impair membrane permeability by the transition of membrane lipids from a liquid–crystalline phase to a gel phase (Murata and Los, 1997; Parvaiz and Prasad, 2012), numerous investigations experiments have suggested that chilling tolerance is related to the composition and structure of plant membrane lipids (Örvar et al., 2000). It was similarly observed in maize (Hodges et al., 1997; Takáč, 2004), coffee (Queiroz et al., 1998) and rice (Huang and Guo,

2005; Guo et al., 2006; Morsy et al., 2007). Chilling stress result in oxidative stress (Nayyar et al., 2005; Wang et al., 2009; Parvaiz and Prasad, 2012), higher plants have developed several strategies to cope with oxidative stress (Foyer and Noctor, 2005; Zhou et al., 2012). Antioxidant enzymes have been found to play a vital role in improving chilling tolerance (Bolkhina et al., 2003; Li and Zhang 2012), especially, superoxide dismutase (SOD), peroxidase (POD) and catalase (CAT) may protect the cellular membranes against the deleterious effects of reactive oxygen species (ROS) (Bowler et al., 1992). Activities of oxygen-scavenging enzymes under chilling stress have been correlated with tolerance to the stress.

Chilling stress can induce the overproduction of reactive oxygen species (ROS) in plants, which then in turn negatively affects cellular structures and metabolism by oxidative stress (Wang et al., 2009; Parvaiz and Prasad, 2012). Although much information is available

---

*Corresponding author. E-mail: cuigreeny@163.com.

about the effects of sub- or supranormal temperatures on the physiological metabolism and functioning in the aerial parts (Hendrickson et al., 2004; Munro et al., 2004; Xu and Zhou, 2006; Dwyer et al., 2007), relatively little is known about the adaptation by plant roots to changes in substrate temperature (Xu and Huang, 2000; Rachmilevitch et al., 2006). Thus, metabolic damage caused by temperature stress could be alleviated to a great extent by keeping the root system at an optimal temperature since injury to the plants may be partly mediated by the disruption of root functions (Zhang et al., 2008). Roots are major organ of plants which is not only responsible for absorbing water, synthesizing, transmitting and storing, but also affect above-ground growth of plants. Roots are the primary site of perception and injury for several types of abiotic stress. In many circumstances, it is the stress-sensitivity of the root that limits the productivity of the entire plant (Steppuhn and Raney, 2005). In general, a highly structured root system is associated with vigorous plant development at the early stages (Richner et al., 1997). For example in cucumber, net photosynthetic rate (PN) and photochemical efficiency of photosystem 2 (PSII) were considerably decreased when the root temperature was lower than 15°C though the aerial temperature was kept optimal (Ahn et al., 1999). Therefore, roots are important organ for plants. Abiotic stresses, such as salinity (Cheng et al., 2009), P (Li et al., 2001), heavy metal ion (Chen et al., 2012) and so on, affect growth and development of roots. However, a few studies of root were found under abiotic stress.

Tobacco (Nicotiana tabacum L.) is one of the most important economic crops in China. Poor and erratic germination at suboptimal temperature is the most important hindrance in its sowing of early spring (Xu et al., 2011). The growth of tobacco seedlings would be inhibited when exposed to nonfreezing temperature below approximate 5°C (Gechev et al., 2003). Thus, understanding the mechanisms of plants tolerance to chilling stress during early growth stage is a crucial environmental research topic, however, no systematic studies had, until now, been conducted to characterize the adaptation of tobacco roots to chilling in term of their physiological metabolism. In this present study, to better understand the adaptability of roots of tobacco seedlings chilling stress at early spring, we evaluated the stress tolerance of tobacco cultivars under 5 to 7°C treatment by assessing seedling growth and antioxidant enzyme activity in roots.

MATERIALS AND METHODS

Plant material and experiment design

Three varieties of flue-cured tobacco, Msk326 (K326), Yunyan87 (YY87) and Yunyan85 (YY85), were used for experiment materials which were cultivated widely in Chinese tobacco region. After seeds of similar size were germinated for 14 days in a Petri dish containing 2 layer of filter paper and distilled water at the 25°C, Per 3 young seedlings were then planted in a plastic containers filled with commercial soil, and reared in a growth chamber, and supplementary lighting (12 h photoperiod), and irrigated water with 1/2 MS solution.

Tobacco seedlings with 5 to 6 true leaves were divided in several groups. One group remained at the initial temperature and illumination conditions. Others plant flue-cured seedling were put in a growth chamber with in low temperature 5 to 7°C for 2, 4 and 6 days for chilling stress treatments, and control grew in temperature at 23 to 25°C with other similar conditions. 90 replicate plants (30 plastic containers) in each treatment. At the end of each treatment, 36 randomly selected replicate plants (12 containers) of each treatment were examined for antioxidant enzyme activities (SOD, CAT, POD). The other seedlings were transferred to a growth chamber and grew at 23 to 25°C with other similar conditions for 10 days to measure roots characteristics. All treatments were done in four replicates.

Antioxidant enzymes extraction and assay

1.0 g fresh roots of flue-cured tobacco from each treatment were homogenized in a pestle and mortar with 0.05 M sodium phosphate buffer (pH 7.8) at the end of treated days. The homogenate was centrifuged at $10,000 \times g$ for 20 min, and the supernatant was used for analyzing SOD, POD, and CAT. The aforementioned steps were carried out at 4°C.

The SOD activity was detected according to the modified method of Zhang et al. (2005). The reaction mixture was made of 1.5 ml phosphate buffer (pH 7.8), 0.3 ml 130 mmol/L methionine, 0.3 ml 750 µmol/L nitroblue tetrazolium chloride (NBT), 0.3ml 100 µmol/L EDTA-Na$_2$, and 0.3 ml 20 µmol/L riboflavin. Appropriate quantity of enzyme extract was added to the reaction mixture. The reaction started by placing tubes below two 15 W fluorescent lamps for 15 min. Reaction stopped by keeping the tubes in dark for 10 min. Absorbance was recorded at 560 nm. One unit of SOD enzyme activity was defined as the quantity of SOD enzyme required to produce a 50% inhibition of reduction of NBT under the experimental conditions, and the specific enzyme activity was expressed as units per gram fresh weight (FW) of root.

The POD activity was examined according to the modified method of Zhang et al. (2005). The reaction mixture in a total volume of 6.9 ml 0.1 M of sodium phosphate buffer (pH5.5) containing 1 ml H$_2$O$_2$ (30 %), 2 ml deionized H$_2$O, and 1 ml 0.05 M guaiacol was prepared immediately before use. Then, 0.2 ml enzyme extract was added to reaction mixture. Increase in absorbance was measured at 470 nm at 1 min intervals up to 4 min using a UV-Vis spectrophotometer. Enzyme specific activity is defined as units (one peroxidase activity unit defined as absorbance at 470 nm changes per minute) per gram of fresh weight of roots.

The CAT activity was assayed according to the method of Zhang et al. (2005). CAT activity was determined at 25°C in 2.7 ml reaction mixture containing 2.25 ml 0.05 M sodium phosphate buffer (pH07.8), 1.5 ml deionized water, and 0.45 ml 0.1 M H$_2$O$_2$ prepared immediately before use, and then 0.3 ml enzyme extract was added. The CAT activity was measured by monitoring the decrease in absorbance at 240 nm as a consequence of H$_2$O$_2$ consumption. Activity was expressed as units (one catalase activity unit defined as absorbance at 240 nm changes per minute) per gram of fresh weight of roots.

Roots characteristics

After flue-cured seedlings recovery grew for 10 days at 23 to 25°C, the roots were washed out of the soil; EPSON V750 was used for

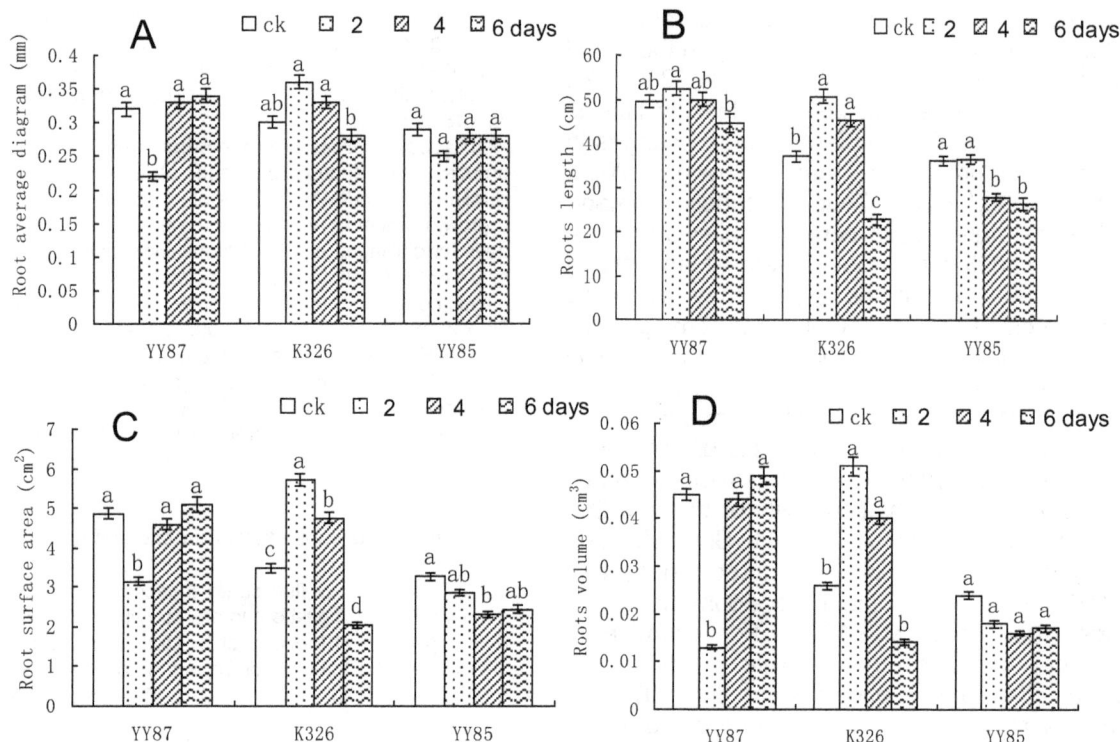

Figure 1. The effect of different cold periods on root morphological traits of tobacco seedling. A. Root average diagram; B. Root length; C. Root surface; D. Root volume.

image acquisition roots. Images were pre processed in Photoshop followed by digital images analysis in WinRHIZO (Regent Instrument Inc, Canada) to analyze roots characteristics, including root length, average diagram, surface area, volume and so on.

Statistical analysis

Analysis of variance (ANOVA) was used to detect the effects of chilling. Multiple comparisons were also performed to permit separation of effect means using the least significant difference test at significant level of P=0.05. All statistical analyses were done using the software statistical package (DPS) version 3.01.

RESULTS

Cold periods induced antioxidant enzymes systems of tobacco seedling

Antioxidant enzymes activities happened different changes according to varieties and cold stress periods (Figure 2). Superoxide dismutase (SOD) activities of YY87 and YY85 are beyond control seedling, furthermore, that is significant higher than control level for YY87 under 2 to 6 days chilling stress. SOD activities of K326 had a prominent decline under 2 to 4 days cold stress and increased significantly after that, which was the maximum under 6 day cold stress (Figure 2A). Change of CAT activities among three cultivars was

inconsistent (Figure 2B). CAT activities of YY87 increased significantly less than 2 to 6 days chilling stress comparing with control seedlings. But for K326 and YY85, CAT activities were below the control level. The effect of chilling on POD activities were presented in Figure 2C. POD activity of YY87 was beyond the control level, but it was not remarkable at 4 to 6 days chilling stress. POD activity of K326 had a significant decline under 2 to 4 days cold stress and increased significantly by 27.31% under 6 day cold stress. POD activity of YY85 significantly declined comparing with control.

Response of root morphological traits of tobacco seedling on cold periods

The effect of chilling on root growth of flue-cured seedling varied for different genotype and treatment time (Figure 1). Low temperature treatment induced difference at the aspect of root average diagram. The decrease of average diagrams were more pronounced at short-term (2 days) chilling for YY87 and YY85 while it was on the contrast for K326 (Figure 1A).

On the other hand, 5 to 7°C low temperature for 2 days helped for elongation of roots length of three varieties seedling, which were above that of control, especially for K326, roots length significantly increased by 36.82% (P<0.05) above that of control (Figure 1B). However,

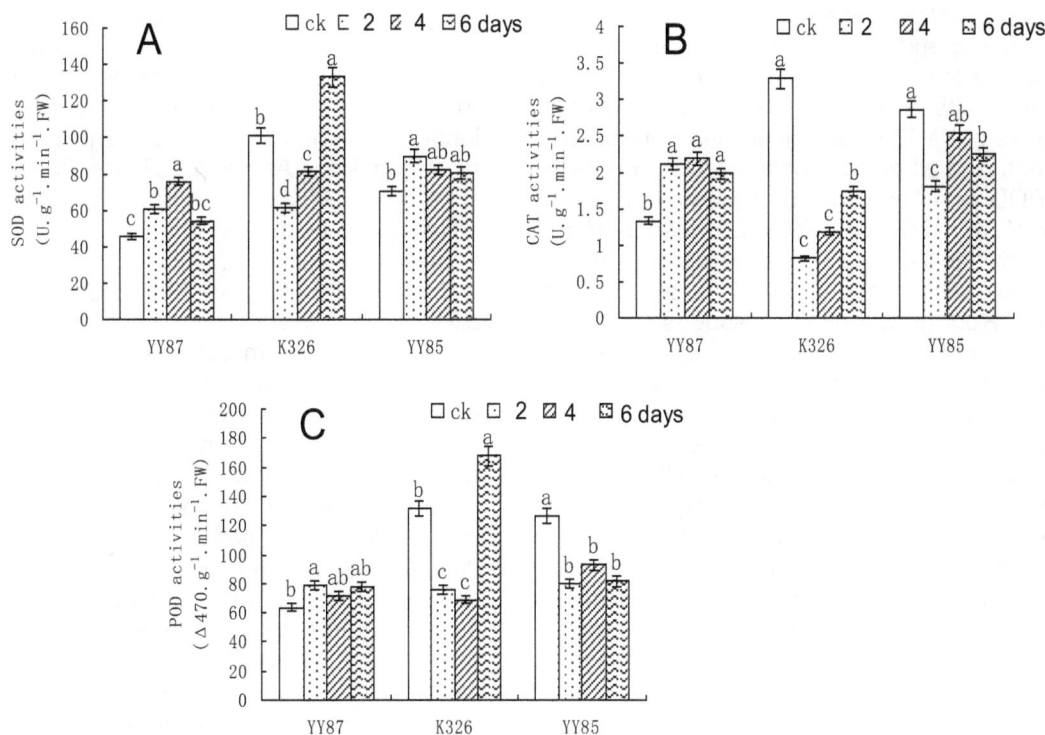

Figure 2. Effect of chilling on antioxidant enzyme activities of roots of tobacco seedling. A. SOD activities; B. CAT activities; C. POD activities.

when roots length for each of treatments decreased under chilling for 6 days, prominent influence were observed for K326 and YY85 (P<0.05). There were different results of roots surface area among varieties after chilling (Figure 1C). Short-term (2 to 4 days) chilling led to decline of roots surface area for YY87 and YY85, however, helped for increasing of surface area of roots for K326, which ascended by 64.08 and 36.49% comparing with control, respectively.

Roots volume at first (2 days) decreased significantly, and were similar to control seedlings at 4 and 6 days treatment for YY87; Progressive increased in roots volume due to 2 and 4 days chilling were apparent with for K326, which was enhanced by 96.15 and 53.84% of control samples, respectively, while roots volume declined the minimum after 6 days chilling; It was descendant but not significant for YY85 after 2, 4 and 6 days due to chilling (Figure 1D).

DISCUSSION

Antioxidant enzymes systems

Antioxidant enzymes are endogenous factors that protect cells from oxidative damage caused by ROS (Chiang et al., 2006). SOD catalyzes the dismutation of the O_2^- to molecular oxygen and H_2O_2, which in turn is metabolized to harmless water and oxygen by CAT (Chiang et al.,

2006). Although, roots that were rapidly increase in SOD activities for YY87 and YY85 significantly decreased for K326. Previous studies showed that the level of CAT decreased during chilling (EL-SAHT, 1998; Lee et al., 2004). Our result was not in completely accordance because CAT activities of YY87 significant enhancement under chilling stress. Peroxidase activity can be induced under a variety of stress conditions, for example, drought, chilling, salinity, γ-radiation, and toxic contamination (Qadir et al., 2004; Kim et al., 2005; D'Arcy-Lameta et al., 2006). POD is generally involved not only in scavenging of H_2O_2 but also in diverse plant physiological processes, such as plant growth, development, lignification, and suberization (Passardi et al., 2005). In a previous study, Wang et al. (2009) reported that POD activity in roots of alfalfa decreased when subjected to chilling stress, and both cultivars showed similar levels of POD activity. In this present study, POD activities showed different change trend among three tobacco cultivars. Abiotic stresses are limiting in crops production. Much of the injury to plants caused by stress is associated with oxidative damage at cellular level (Bowler et al., 1992). Activities of antioxidant enzymes under chilling stress have been correlated with tolerance to the stress. Tolerant plant species generally have a better capacity to protect themselves from chilling-induced oxidative stress, *via* the enhancement of antioxidant enzyme activity. Higher contents of defense enzymes were correlated with Higher chilling tolerance. Chilling-tolerant cultivars had higher antioxidation activities than the susceptible cultivar (Takáč, 2004; Huang and Guo, 2005). Activities of SOD, POD and CAT were inconsistent among the three tobacco varieties. After chilling stress, three antioxidant enzyme activities of YY87 were higher than control level to certain extent. And all of them decreased except raise of SOD and POD activities under 6 day's cold stress for K326. As for YY85, CAT and POD activities declined despite SOD activities increased under cold stress. In this study, these results suggest that YY87 is a variety of Chilling-tolerant. Root growth of YY87 suggests that it is functional to attenuate oxidative stress by increasing antioxidant enzyme activities. Antioxidant enzymes activities affected by genotypes of flue-cured and chilling periods, which depend on comprehensive ecophysiological adaptation because of difference of seedlings physiological situations, triggering of gene of antioxidant enzyme expression and cold resistance of varieties. In other words, it might be possible that triggering of different antioxidant enzyme protection mechanism is different among three flue-cured tobacco cultivars.

Root characteristics analysis

Soil temperature is below optimal temperature in which reduce water and nutrient uptake and limits growth of roots. As a consequence, growth rate of plant come to slow. Queiroz et al. (1998) reported that root tissue exposed to 10°C evolved significantly lower rates of metabolic heat compared with controls grown at 25°C, and the values were closely associated with the observed root growth inhibition. In cucumber, net photosynthesis and the photochemical efficiency of PSII are considerably decreased at root temperatures below 15°C, even if the aerial temperature remains optimal (Ahn et al., 1999). Plants possess a series of detoxification systems that break down highly toxic ROS *via* antioxidant enzymes, for example, superoxide dismutase (SOD), peroxidase (POD) and catalase (CAT), as well as by non-enzyme antioxidants, such as glutathione (GSH), ascorbate (AsA), á-tocopherol, and carotenoids, thereby limiting oxidative damage under stress conditions (Zhang et al., 2007). One of main aims of raise seedling is to cultivate healthy and strong seedling with developed roots, so that converted to heat and light energy for re-growth after transplant as soon as possible. In this present study, there were big differences at the aspect of roots growth among the three cultivars. Short-term (2 days) chilling prominent decreased average diagram, surface area and volume of root for YY87, inversely; it helped for enhancement to root average diagram, length, surface area and volume of K326. On the other hand, long-term chilling (6 days) increased root average diagram, surface area and volume of YY87, but it is on the converse for K326. This might be because root of K326 have stronger tolerance of short-term chilling than that for YY87 and

YY85, which result in great accumulation of root comparing with the control. The result also showed that long-term chilling (6 days) limited root growth for K326, which could not absorb enough water and nutrient at long-term chilling stress and maintain normal the process of metabolism. As for YY87, root was sensitive to the short-term (2 days) chilling, but it can gradually well-adjust at molecule and physiological level and adapt the unfavorable situations.

Our results support the fact that there are big differences at the aspects of antioxidant defense systems and roots growth among the three cultivars. YY87 present strong cold resistance, and had well-adaptation by increasing activities of three enzymes to keep roots prolonging for long-term chilling (6 days). The roots of K326 grew well for short-term chilling though decreasing activities of three enzymes. This might partly explain that the result from genetic composition and expression is difference in space-time to adapt chilling stress among cultivars. Therefore, improvement of chilling tolerance of crops requires a detailed knowledge about tolerance mechanisms in plants, which comprise a wide range of responses on molecular, cellular, whole plant levels, and include others the synthesis of compatible osmolytes and radical scavenging mechanisms, all of them need further research.

ACKNOWLEDGMENTS

This study was supported by the Science and Technique Foundation of CQ (CSTC, 2011jjA0326), "the fundamental research funds for the central university (XDJK2011C002)" and the "111" project (B12006).

REFERENCES

Ahn SJ, Im YJ, Chung GC, Cho BH, Suh SR (1999). Physiological responses of grafted-cucumber leaves and rootstock roots affected by low root temperature. Sci. Hort. 81(4):397-408.
Bolkhina O, Virolainen E, Fagerstedt K (2003). Antioxidants, oxidative damage and oxygen deprivation stress: a review. Ann. Bot. 91(2):179-194.
Bowler C, Van Montagu M, Inze D (1992). Superoxide dismutase and stress tolerance. Annu. Rev. Plant Physiol. Plant Mol. Biol. 43:83-116.
Chen YA, Chi WC, Huang TL, Lin CY, Nguyeh TTQ, Hsiung YC, Chia LC, Huang HJ(2012) . Mercury-induced biochemical and proteomic changes in rice roots. Plant Physiol. Bioch.55:23-32.
Cheng YW, Qi YC, Zhu Q, Chen X, Wang N, Zhao X, Chen HY, Cui XJ, Xu LL, Zhang W (2009). New changes in the plasma-membrane-associated proteome of rice roots under salt stress. Proteomics 9(11):3100–3114.
Chiang AN, Wu HL, Yeh HI, Chu CS, Lin HC, Lee WC (2006). Antioxidant effects of black rice extract through the induction of superoxide dismutase and catalase activities. Lipids 41(8):797-803.
D`Arcy-Lameta A, Ferrari-Iliou R, Contour-Ansel D, Pham-Thi A, Zuily-Fodil Y (2006). Isolation and characterization of four ascorbate peroxidase cDNAs responsive to water deficit in cowpea leaves. Ann. Bot. 97(1):133-140.
Dwyer SA, Ghannoum O, Nicotra A, von Caemmerer S (2007). High temperature acclimation of C-4 photosynthesis is linked to changes in photosynthetic biochemistry. Plant Cell Environ. 30(1):53-66.

EL-SAHT HM (1998). Responses to chilling stress in French bean seedling: antioxidant compounds. Biol. Plantarum 41(3):395-402.

Foyer CH, Noctor G (2005). Oxidant and antioxidant signaling in plants: a re-evaluation of the concept of oxidative stress in a physiological context. Plant Cell Environ. 28(8):1056-1071.

Gechev T, Willekens H, Van Montagu M, Inzé D, Van Camp W, Toneva V, Minkov I (2003). Different responses of tobacco antioxidant enzymes to light and chilling stress. J. Plant Physiol. 160(5):509-515.

Guo Z, Ou W, Lu S, Zhong Q (2006). Differential responses of antioxidative system to chilling and drought in four rice cultivars differing in sensitivity. Plant Physiol. Biochem. 44:828-836.

Hendrickson L, Ball MC, Wood JT, Chow WS, Furbank RT (2004). Low temperature effects on photosynthesis and growth of grapevine. Plant Cell Environ. 27(7): 795-809.

Hodges DM, Andrews CJ, Johnson DA, Hamilton RI (1997). Antioxidant enzyme responses to chilling stress in differentially sensitive inbred maize lines. J. Exp. Bot. 48(5):1105-1113.

Huang M, Guo Z (2005). Responses of antioxidative system to chilling stress in two rice cultivars differing in sensitivity. Biol Plant 49(1):81-84.

Kim JH, Chung BY, Kim JS, Wi SG (2005). Effects of in planta gamma-irradiation on growth, photosynthesis, and antioxidative capacity of red pepper (Capsicum annuum L.) plants. J. Plant Biol. 48(1):47-56.

Lee MA, Chun HS, Kim JW, Lee H, Lee DH, Lee CB (2004). Changes in antioxidant enzyme activities in detached leaves of cucumber exposed to chilling. J. Plant Biol. 47(2):117-123.

Li F, Li MY, Pan XH, Zhu AF (2001). Biochemical and physiological characteristics in seedlings roots of different rice cuhivars under low-phosphorus stress. Chinese. J. Rice Sci.18(1):48-52.

Li HY, Zhang WZ (2012) . Abscisic acid-induced chilling tolerance in maize seedlings is mediated by nitric oxide and associated with antioxidant system. Adv. Mater. Res. 378-379:423-427.

Morsy MR, Jouve L, Hausman JF, Hoffmann L, Stewart JM (2007). Alteration of oxidative and carbohydrate metabolism under abiotic stress in two rice (Oryza sativa L.) genotypes contrasting in chilling tolerance. J. Plant Physiol. 164(2):157-167.

Munro KD, Hodges DM, De Long JM, Forney CF, Kristie DN (2004). Low temperature effects on ubiquinone content, respiration rates and lipid peroxidation levels of etiolated seedlings of two differentially chilling-sensitive species. Physiol. Plantarum. 121(3):488-497

Murata N, Los DA (1997). Membrane fluidity and temperature perception. Plant Physiol. 115(3):875-879.

Nayyar H, Bains TS, Kumar S (2005). Chilling stressed chickpea seedlings: effect of cold acclimation, calcium and abscisic acid on cryoprotective solutes and oxidative damage. Environ. Exp. Bot. 54(3):275-285.

Örvar BL, Sangwan V, OmannF, Dhindsa RS (2000). Early steps in cold sensing by plant cells: the role of actin cytoskeleton and membrane fluidity. Plant J. 23(6):785-794.

Passardi F, Cosio C, Penel C, Dunand C (2005). Peroxidases have more functions than a Swiss army knife. Plant Cell Rep. 24(5):255-265.

Parvaiz A, Prasad MNV (2012). Environmental Adaptations and Stress Tolerance Of Plants In The Era Of Climate Change. In Zoldan et al. Understanding chilling tolerance traits using arabidopsis chilling-sensitive mutants. Springer. pp. 159-171.

Qadir S, Qureshi MI, Javed S, Abdin MZ (2004). Genotypic variation in phytoremediation potential of Brassica juncea cultivars exposed to Cd stress. Plant Sci. 167(5):1171-1181

Queiroz CGS, Alonso A, Mares-guia m, Magalahes AC (1998). Chilling-induced changes in membrane fluidity and antioxidant enzyme activities in coffea arobica L. Roots. Biol. Plantarum 41(3):403-413.

Rachmilevitch S, Lambers H, Huang BR (2006). Root respiratory characteristics associated with plant adaptation to high soil temperature for geothermal and turf-type Agrostis species. J. Exp. Bot. 57(3):623-631

Richner W, Kiel C, Stamp P (1997). Is seedling root morphology predictive of seasonal accumulation of shoot dry matter in maize. Crop Sci. 37(4):1237-1241.

Steppuhn H, Raney JP (2005). Emergence, height, and yield of canola and barley grown in saline root zones. Can. J. Plant Sci. 85(4):815-827.

Sui N, Li M, Zhao SJ, Li F, Liang H, Meng QW (2007). Overexpression of glycerol-3-phosphate acyltransferase gene improves chilling tolerance in tomato. Planta 226:1097-1108

Takáč T (2004). The relationship of antioxidant enzymes and some physiological parameters in maize during chilling. Plant Soil Environ. 50:27-32

Thomashow MF (1998) Role of cold-responsive genes in plant freezing tolerance. Plant Physiol. 118:1-7

Wang WB, Kim YH, Lee HS, Deng XP, Kwak SS (2009). Differential antioxidation activities in two alfalfa cultivars under chilling stress. Plant Biotechnol. Rep. 3:301-307

Xu Q, Huang B (2000). Growth and physiological responses of creeping bentgrass to changes in air and soil temperatures.Crop Sci. 40(5):1363-1368.

Xu S, Hu J, Li Y, Ma W, Zheng Y, Zhu S (2011). Chilling tolerance in Nicotiana tabacum induced by seed priming with putrescine. Plant Growth. Regul. 63:279-290.

Xu ZZ, Zhou GS (2006). Combined effects of water stress and high temperature on photosynthesis, nitrogen metabolism and lipid peroxidation of a perennial grass Leymus chinensis. Planta. 224(5):1080-1090.

Zhang HY, Jiang YN, He ZY, Ma M (2005). Cadmium accumulation and oxidative burst in garlic (Allium sativum). J. Plant Physiol. 162:977-984.

Zhang YP, Zhang YL, Zhou YH,Yu JQ (2007). Adaptation of Cucurbit Species to Changes in Substrate Temperature: Root Growth, Antioxidants, and Peroxidation. J. Plant Biol. 50:527-532.

Zhang YP, Qiao YX, Zhang YL, Zhou YH, Yu JQ (2008). Effects of root temperature on leaf gas exchange and xylem sap abscisic acid concentrations in six Cucurbitaceae species. Photosynthetica 46(3):356-362.

Zhou J, Wang J, Shi K, Xia XJ, Zhou YH, Yu JQ(2012). Hydrogen peroxide is involved in the cold acclimation-induced chilling tolerance of tomato plants. Plant Physiol. Biochem. 60:141–149

17

Water stress and temperature effects on germination and early seedling growth of *Digitaria eriantha*

Brevedan R. E.[1,2], Busso C. A.[1,2], Fioretti M. N.[1], Toribio M. B.[1], Baioni S. S.[1], Torres Y. A.[1,2], Fernández O. A.[1,2], Giorgetti H. D.[3], Bentivegna D.[2], Entío J.[4], Ithurrart L.[1,2], Montenegro O.[3], Mujica M. de las M.[4], Rodríguez G.[3] and Tucat G.[1,2]

[1]Department of Agronomía, Universidad Nacional del Sur (UNS), 8000 Bahía Blanca, Argentina.
[2]CERZOS (CONICET), 8000 Bahía Blanca, Argentina.
[3]Chacra Experimental de Patagones, Ministerio de Asuntos Agrarios, (8504) Carmen de Patagones, Argentina.
[4]Facultad de Ciencias Agrarias, Universidad Nacional de La Plata, (1900) La Plata, Argentina.

This study focused on the two major processes critical for plant establishment: Seed germination and seedling survival. We determined the effects of (1) water stress and temperature on the germination, and (2) water stress on early seedling growth of *Digitaria eriantha* cv 'Irene'. Seeds harvested in 2007 were used for temperature studies, and those coming from 2006 and 2007 for water stress studies. In 2009, viability decreased by 65.4% from 2006 to 2007. During the first twenty-four hours, germination was more than 50% at constant (30 or 35°C) than alternating (10/30 or 10/35°C) temperatures, although total germination was about 80% for all temperature treatments. Polyethylene glycol 8000 was used to impose water stress conditions. Germination percentages and coefficients of velocity decreased with decreasing water potentials. Early seedling growth was smaller at lower water potentials. *D. eriantha* cv 'Irene' appeared to germinate within a wide range of temperatures, but it varied greatly in germination response to water potentials. Results suggest that this species could be planted in late spring-early summer, when seedbed temperatures are increasing and soil moisture might still be adequate.

Key words: Perennial grasses, rangelands, seeds, forage, arid zones, Argentina.

INTRODUCTION

In the rangelands of central Argentina, warm-season, native perennial grass genotypes palatable to cattle are scarce (Busso et al., 2004). These genotypes are the major food source for these animals, which reject most of the other monocots and dicots that are part of the plant community. In rangelands at the south of Buenos Aires Province (e.g., the Chacra Experimental de Patagones, Patagones, Argentina (40°39'49,7"S, 62°53'6,4"W; 40 m.a.s.l.), within the Phytogeographical Province of the Monte: Cabrera 1976), *Pappophorum vaginatum* Buckey is almost the only native, warm-season, C_4, palatable perennial grass available for livestock grazing (Giorgetti et al., 1997). Need of increasing cattle forage production during the warm season in rangelands of central Argentina is thus critical. This could be achieved via perennial grass species introduction into the region (Anderson, 1980). Success in artificial range seeding requires knowledge of many parameters, including optimum temperature and moisture conditions for germination (Sabo et al., 1979).

Digitaria eriantha (Steudel) subsp. *eriantha* is a C$_4$, palatable perennial grass that was introduced in Argentina from South Africa in 1991 (Di Giambatista et al., 2010). Rimieri (1997) bred the cv. 'Irene' from the South African cv. 'Sudafricana'; 'Irene' showed a better adaptation than the subsp. *eriantha* to the soil and climatic conditions of the semiarid regions in Argentina. Besides South Africa (Du Toit, 2000), this species has spread worldwide [USA: Sanderson et al. (1999); Australia: Hacker et al., 1993); China and Europe: http://www.ehow.com/info_8407286_genus-species-crabgrass.html)]. It characterizes for its resistance to drought, healthiness, high warm-season forage production and adaptability to different soils (Dannhauser, 1988). Lavin and Johnsen (1977) determined that *D. eriantha* might attain a rating of excellent under warm and dry conditions at a site with 330 mm annual precipitation.

Periods of water stress and high temperature are common in rangelands of northwestern Patagonia during late spring and summer (Torres et al., 2011). It means that *D. eriantha* could be a good potential species to introduce in rangelands of central Argentina. Although some research has been conducted on *D. eriantha* in this country (Gargano et al., 2004; Pedranzani et al., 2005), no studies have yet addressed the importance of water potential and temperature on the most sensitive morphological developmental stages to water stress and high temperatures during summers: seed germination and early seedling growth (Brown, 1995).

Optimum temperatures will shorten the time to germination, and they might occur under momentarily favorable moisture conditions (Bonvissuto and Busso, 2007). Fulbright (1988) reported that the warm-season bunchgrass *Sorghastrum nutans* (L.) Nash appeared to germinate within a similar range of temperatures, but it varied in germination responses to water potentials. Several studies showed similar or greater germination percentages at constant than alternating temperatures in various range perennial grasses (Hylton and Bement, 1961; Young and Evans, 1982; Fernández et al., 1991). For example, Brown (1987) reported that alternating temperatures improved overall germination of the perennial bunchgrass *Aristida armata*, but there was little variation in the rate of germination with incubation under constant temperature. Ellern and Tadmor (1966) and Heydecker (1960) stated that speedy germination does not necessarily have to coincide with high germination percentages.

Plants are susceptible to water stress throughout their life-cycle, but are particularly vulnerable during seed germination, and seedling emergence and early growth (Brown, 1995; Almansouri et al., 2001). Seedling development of *S. nutans* cvs. 'Lometa', 'Cheyenne', 'Llano', 'Oto', and 'Tejas', for example, decreased linearly with decreasing water potential (Fulbright, 1988). Another critical factor in successful plant establishment is time to germination, especially in arid and semiarid environments where favorable conditions of water availability and temperature regimes may only occur over short time periods (Owens and Call, 1985). Germination may be part of the plant life cycle that requires the highest water potentials (Brown, 1995). Temperature also appears to determine the optimal and minimum water potentials for germination in various species (Brown, 1995). This is why studies on the effects of water stress and temperature on seed germination and early seedling growth are so important when thinking in introducing a rangeland species, like *D. eriantha*, in water stress-prone rangelands of northwestern Patagonia, Argentina. This information is actually lacking and essential for trying to increase warm-season forage production for livestock in rangelands of central Argentina. There is also a paucity of information describing the germination ecology of this perennial.

In arid and semiarid areas, the occurrence of optimum conditions for seed germination can vary within and among years (Romo and Eddleman, 1988). The greater the water stress conditions, the lower will be the total germination, and the time to germination (Brown, 1995). Some seeds will not germinate even under favorable conditions because they are dormant; the need for an after-ripening may help to avoid seed germination under unfavorable climate conditions (Brown, 1995).

Specific hypotheses were that (1) total germination percentage is similar under constant or alternating temperatures, (2) coefficients of velocity are higher at constant than alternating temperatures, and (3) seedling growth is higher at greater than lower water potentials. Our objectives were to (1) identify germination responses of *D. eriantha* to various temperatures and water potentials, and (2) determine the effect of water stress on its initial seedling growth. This information will assist those charged with developing strategies for reestablishing and managing degraded, central rangelands of northwestern Patagonia, Argentina.

MATERIALS AND METHODS

Study site

Characteristics of the study site (the Chacra Experimental de Patagones) where *D. eriantha* could potentially be introduced follow. Climate is temperate semiarid, with higher precipitations during the spring and fall seasons (Giorgetti et al., 2000). Long-term (1981-2010) annual rainfall is 416.7 mm Figure 1), with a mean annual temperature of 14.6°C, absolute minimum temperature of -7.6°C (August), absolute maximum temperature of 43°C (January), mean annual relative humidity of 60%, and mean annual wind speed of 13 km h^{-1}. Long-term (2006-2010) mean monthly diurnal and nocturnal temperatures, precipitations for various years, and long-term (1981-2010) mean monthly evapotranspiration are presented in Figure 1.

Soil is a typical Haplocalcid. Average pH is 7, and depth is not a constraint in the soil profile. The plant community is characterized by an open, shrubby stratum which includes different-quality,

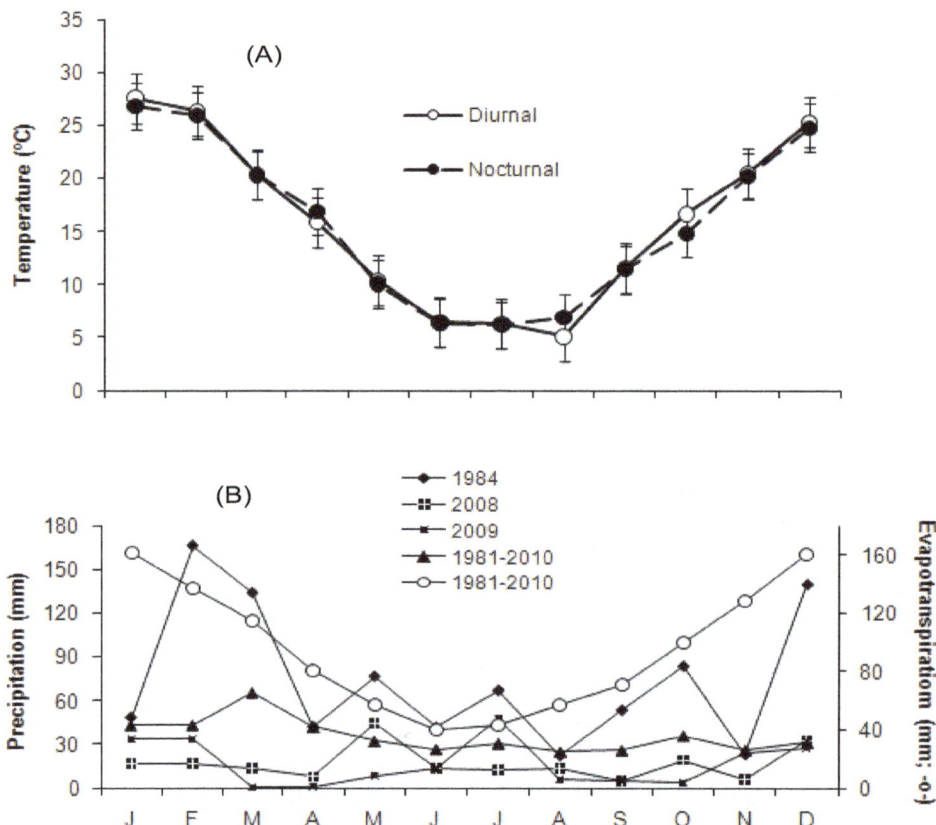

Figure 1. (A) Averages during 2006-2010 of mean monthly diurnal and nocturnal soil temperatures at 5 cm soil depth from January to December, and (B) monthly precipitation at various years (black symbols), long-term (1981-2010) mean monthly precipitation (black symbols), and long-term (1981-2010) mean monthly potential evapotranspiration (open symbols), in the Chacra Experimental de Patagones, Patagones, Argentina (40°39'49,7"S, 62°53'6,4"W; 40 m.a.s.l.). Mean annual standard errors for long-term precipitation and evapotranspiration were 29.0 and 38.9 mm, respectively.

herbaceous species for cattle production (Giorgetti et al., 1997). Dominance of a particular grass or shrubby species in the study region is partially dependent on grazing history and fire frequency and intensity (Distel and Bóo, 1996).

Storage time and conditions

Seeds of *D. eriantha* cv. 'Irene' used in this study were harvested in 2006 and 2007. Weight of 1000 seeds was 0.339 g in 2006 and 0.329 g in 2007. Seeds were kept in a seed storage room at 20°C until the experiment was conducted in 2009. At this later time, seed viability was tested using the 2,3,5-triphenyl tetrazolium chloride test (TTC; Peters, 2000). Five sets of 50 seeds each were immersed in water in Petri dishes during 24 h. Seeds were then incubated in 20 ml-Petri dishes containing a 0.1% solution of TTC at 35°C under darkness, and embryos were thereafter examined to establish their viability.

Temperature effects

A bar of stainless steel (0.95 m length, 0.20 m wide, 0.20 m height) was used to create the thermal gradient. It was achieved submerging one extreme of the bar in cool water (3°C) and the other extreme in hot water (36°C). This extreme was maintained hot

using an electrical resistance. This generated a temperature gradient such as various constant temperatures were obtained throughout the width every 3-increments along the bar length. Thermocouples, and an infrared thermometer, were used to determine these constant temperature conditions. An absorbent paper was placed on the bar that remained saturated with water during the experiment. The stainless steel bar had only one side (at the top) that could be opened throughout its length; this side was a piece of transparent acrylic (5 mm thick) that was in immediate contact with a piece of rubber (5 mm thick; between the acrylic and the stainless steel bar) that prevented water losses via evaporation from the germination chamber. Water vapor loss from germination containers has been shown to change the water potential of PEG solutions: Berkat and Briske (1982). Temperature was measured using a Delta T multi-channel data-logger. Seed germination was tested at different temperatures using seeds from the 2007 harvest. Seeds were placed on top paper on the thermal gradient with a temperature range from 13.9 to 36.9°C. Cumulative germination was recorded by counting germinated seeds every 12 h for six days. A seed was considered germinated once its radicle reached a length of at least 2 mm (Emmerich and Hardegree, 1990).

In another test, seeds were placed on top filter paper in plastic boxes (53 mm length × 59 mm wide × 15 mm height). Distilled water was added, and boxes were placed in a germination cabinet that was set to four temperature regimes: continuous (30 or 35°C), and alternating [30/10°C (14 h light-10 h darkness) or 35/10°C

Table 1. Germination percentage and coefficient of velocity (CV) of *D. eriantha* cv. 'Irene' as function of different temperature treatments.

Temperature (°C)	Time (hours)					
	24	48	72	96	Total	CV
30	48.7 ± 2.1^b	24.7 ± 1.7^a	7.0 ± 1.2^a	0.3 ± 0.3^a	80.7 ± 1.7^a	67.2 ± 1.3^c
30-10	11.7 ± 2.8^a	50.3 ± 3.9^c	18.0 ± 1.5^b	0.0 ± 0.0^a	80.0 ± 3.4^a	47.2 ± 1.0^a
35	54.0 ± 1.6^b	24.3 ± 0.6^a	7.7 ± 1.0^a	0.3 ± 0.3^a	86.3 ± 3.0^a	68.2 ± 0.9^c
35-10	23.7 ± 3.1^b	40.0 ± 3.0^b	16.7 ± 1.1^b	0.0 ± 0.0^a	80.3 ± 3.6^a	52.3 ± 1.0^b

Seeds of *D. eriantha* come from the 2007 harvest. Means within a column with the same letter are not significantly different according to the LSD test (p=0.05). Values are the mean ±1 EE of n=6.

(14 to 10 h)] temperatures. Photosynthetically active radiation within the cabinet was 25 to 45 µmol/m^2/s. The study was a completely randomized design, with four temperature treatments. There were six replications (boxes) of 50 seed each. Observations of germinating seeds were carried out every twenty-four hours during four days. After each counting time, boxes from all treatments were randomly distributed within the germination cabinet.

Water potential effects

The germination studies under different water potentials were carried out using seeds from the 2006 and 2007 harvests. Solutions were prepared by adding polyethylene glycol 8000 formerly PEG 6000 (Michel, 1983) to distilled water, such as seeds were in direct contact with the PEG 8000 solution, without any support material, in petri dishes. PEG solution-seed contact does not reduce seed germination (Emmerich and Hardegree, 1990). Osmotic potential was determined by a Wescor 5500 osmometer, after calibration with standard KCl solutions. Water potential treatments were 0 (controls), -0.4, -0.6, -0.8, -1.0, -1.2, -1.5 and -2.0 MPa, although no germination was recorded at the lastest two water potentials; six replications (petri dishes) of 50 seeds each were used for each water stress level. Each Petri dish was sealed with Parafilm to prevent water vapor loss, and placed within a Forma Scientific 3770 germination chamber set at 35±1°C. At each counting time, the petri dishes from all water level treatments were randomly distributed on the chamber shelves. Both the cumulative germination and germination coefficient of velocity were recorded by counting and removing germinated seeds every three hours up to the sixth day when the experiment was ended.

The coefficient of velocity (CV) was calculated as:

CV = 100 [∑Ni/ ∑NiTi]

Where N is the number of seeds which germinated on day i, and T is the number of days from sowing (Scott et al., 1984). One advantage of using the coefficient of velocity as a measure of the vigor index is the simplicity of the formula; CV increases are often obtained as more seeds germinate and with shorter germination time (Scott et al., 1984). Root and shoot lengths of three randomly selected seedlings were measured per Petri dish on each water potential treatment. An average of these measurements was considered to be one replication.

Statistical analysis

A completely randomized experimental design with six replications was utilized. Temperature data were analyzed using one-way ANOVA. Percentage germination data were transformed to arc-sin of the square root to comply with the normality and homoscedasticity assumptions of variance. Within any given study time from imbibition, two-way ANOVA (year x water potential) were used to analyze germination and coefficients of velocity data. When the interaction was significant, each year or water potential was analyzed separately. Growth of young seedlings was analyzed using three-way ANOVA (year x plant part x water potential). Untransformed values are presented in Tables. When F test were significant (p<0.05), treatments were compared using LSD. Data were analyzed using the statistical software INFOSTAT version 2009 (Di Rienzo et al., 2009).

RESULTS

Storage time

Seeds harvested in 2006 showed 33% viability, while it was 93% for those harvested in 2007.

Germination percentage

Temperature

After one day of study initiation, the lowest (p<0.05) germination percentage was reached at the 30-10°C alternating temperature (Table 1). At the same time, percentage germination was about 50% at 30 and 35°C, although percentages at these temperatures were not different (p>0.05) than that at the alternating temperature of 35-10°C (Table 1). As a result, germination percentages were greater (p<0.05) at the alternating than at the constant temperatures after two and three days from initiation of the study (Table 1). Despite these differences among temperature treatments, total germination was similar (p>0.05; aprox. 80%) in all of them after four days of initiating the study (Table 1).

Water potential

Changes in germination percentages with time

Eighteen out of one-hundred and twelve comparisons of all germination percentages at all study water potentials

Table 2. Percentage germination of *D. eriantha* seeds in 2006 and 2007 (on an hourly basis from study initiation, and Total, cumulative germination) as a function of water potential. Measurements were conducted at 35±1°C.

Year 2006

Water potential (MPa)	Hours								Total
	15	18	24	36	48	60	72	84	
0.0	2.5 ± 1.1 [a,bc]	4.3 ± 0.8 [a,bc]	6.8 ± 0.9 [a,a]	5.8 ± 1.2 [a,c]	2.0 ± 0.4 [a,abcd]	1.7 ± 0.6 [a,a]	0.3 ± 0.2 [a,a]	0.3 ± 0.3 [a,a]	26.0 ± 0.9 [a,b]
-0.2	4.3 ± 1.5 [a,c]	5.7 ± 1.7 [a,c]	3.7 ± 0.8 [a,b]	6.0 ± 1.2 [a,c]	1.3 ± 0.7 [a,abc]	1.3 ± 0.7 [a,a]	0.0 ± 0.0 [a,a]	0.0 ± 0.0 [a,a]	23.7 ± 1.7 [a,b]
-0.4	1.3 ± 0.7 [a,abc]	3.7 ± 1.0 [a,abc]	6.3 ± 1.0 [a,b]	3.7 ± 1.0 [a,bc]	1.0 ± 0.4 [a,ab]	2.0 ± 0.9 [a,a]	0.3 ± 0.3 [a,a]	0.0 ± 0.0 [a,a]	19.0 ± 1.8 [a,b]
-0.6	1.0 ± 0.7 [a,ab]	7.3 ± 1.2 [a,ab]	7.7 ± 2.5 [a,b]	6.7 ± 1.7 [a,c]	0.0 ± 0.0 [a,a]	1.0 ± 0.7 [a,a]	0.0 ± 0.0 [a,a]	0.3 ± 0.3 [a,a]	25.7 ± 3.4 [a,b]
-0.8	0.0 ± 0.0 [a,a]	0.0 ± 0.0 [a,a]	3.7 ± 1.2 [a,b]	5.7 ± 1.6 [a,c]	2.3 ± 0.8 [a,bcd]	1.7 ± 0.6 [a,a]	1.0 ± 0.7 [a,a]	0.7 ± 0.4 [a,a]	20.3 ± 2.7 [a,b]
-1.0	0.0 ± 0.0 [a,a]	0.0 ± 0.0 [a,a]	0.3 ± 0.3 [a,a]	1.7 ± 1.0 [a,ab]	4.0 ± 1.2 [a,d]	1.3 ± 0.8 [a,a]	1.0 ± 0.4 [a,a]	1.7 ± 0.6 [a,a]	12.3 ± 2.8 [a,a]
-1.2	0.0 ± 0.0 [a,a]	0.0 ± 0.0 [a,a]	0.0 ± 0.0 [a,a]	0.0 ± 0.0 [a,a]	3.7 ± 1.2 [a,cd]	0.0 ± 0.0 [a,a]	1.7 ± 1.1 [a,a]	4.0 ± 1.5 [a,a]	10.3 ± 1.7 [a,a]

Year 2007

Water potential (MPa)	Hours								Total
	15	18	24	36	48	60	72	84	
0.0	8.7 ± 1.1 [b,d]	10.7 ± 1.5 [b,d]	17.7 ± 2.2 [b,c]	17.7 ± 1.2 [b,c]	3.0 ± 1.0 [a,bc]	5.7 ± 1.4 [b,c]	0.0 ± 0.0 [a,a]	0.0 ± 0.0 [a,a]	74.0 ± 2.6 [b,e]
-0.2	7.7 ± 1.2 [a,d]	10.3 ± 1.9 [a,d]	16.0 ± 2.2 [b,c]	10.3 ± 2.4 [a,b]	0.7 ± 0.4 [a,a]	2.7 ± 1.1 [b,b]	0.3 ± 0.3 [a,ab]	0.0 ± 0.0 [a,a]	52.3 ± 4.1 [b,cd]
-0.4	3.0 ± 0.7 [a,c]	10.0 ± 2.3 [a,c]	26.7 ± 2.5 [b,d]	13.3 ± 0.9 [b,bc]	2.0 ± 0.9 [a,ab]	3.7 ± 1.3 [b,b]	0.3 ± 1.0 [a,ab]	0.3 ± 0.3 [a,a]	66.3 ± 6.0 [b,de]
-0.6	1.0 ± 0.4 [a,b]	9.7 ± 1.9 [a,b]	18.0 ± 3.0 [b,c]	17.3 ± 0.4 [b,c]	0.7 ± 0.4 [a,a]	5.0 ± 1.2 [b,b]	1.7 ± 0.7 [a,abc]	1.7 ± 0.3 [a,a]	60.7 ± 2.4 [b,de]
-0.8	0.0 ± 0.0 [a,a]	0.0 ± 0.0 [a,a]	5.3 ± 1.4 [a,b]	8.3 ± 1.1 [a,b]	5.7 ± 1.1 [b,c]	6.0 ± 1.9 [b,b]	2.0 ± 1.3 [a,bc]	2.0 ± 1.0 [a,a]	39.7 ± 7.9 [a,bc]
-1.0	0.0 ± 0.0 [a,a]	0.0 ± 0.0 [a,a]	0.0 ± 0.7 [a,a]	1.7 ± 0.7 [a,a]	3.3 ± 0.7 [a,bc]	6.0 ± 1.5 [b,b]	7.7 ± 1.3 [b,d]	7.7 ± 1.4 [a,a]	27.7 ± 3.5 [b,b]
-1.2	0.0 ± 0.0 [a,a]	0.0 ± 0.0 [a,a]	0.0 ± 0.0 [a,a]	0.0 ± 0.7 [a,a]	1.0 ± 0.7 [a,a]	1.7 ± 0.8 [b,b]	4.0 ± 1.6 [a,c]	4.0 ± 1.4 [a,a]	12.0 ± 1.9 [a,a]

Each value is the mean ± EE of n=6. Different letters to the left of the comma indicate significant differences between years, and those to the right of the comma indicate significant differences among water potential treatments.

and times from imbibition in both study years were greater ($p<0.05$) in 2007 than in 2006 (Table 2). Except at -0.8 and -1.2 MPa, total germination was greater ($p<0.05$) in 2007 than in 2006 (Table 2). In 2006, germination percentages were greater ($p<0.05$) at 0 and -0.2 MPa than at -0.4 MPa and lower water potentials after 15 and 18 h from study initiation (Table 2). At 24 and 36 h after initiation of the study in 2006, germination percentages were similar ($p>0.05$) between 0 and -0.8 MPa, but these germination percentages were greater ($p<0.05$) than those at -1.0 and -1.2 MPa (Table 2). Differences in germination

percentages among water stress treatments diluted with increasing time from the initiation of the study, and they were similar ($p>0.05$) between 60 and 84 h from study initiation (Table 2). During 2007, differences in germination percentages among water potential treatments were more marked than in 2006 (Table 2). Until 24 h from study initiation, germination percentages were similar ($p>0.05$) at 0 than at -0.2 MPa, and decreased as water potentials decreased (Table 2). Twelve hours later, germination percentages were greater at 0 than at -0.2 MPa, and as from the beginning of the study, water potentials between

0 and -0.6 MPa were greater ($p<0.05$) than those at or below -0.8 MPa.

Cumulative germination

Cumulative (total) germination percentages were similar ($p>0.05$) among 0 and -0.8 MPa in 2006 (Table 2). However, these percentages decreased ($p<0.05$) at lower water potentials in that year (Table 2). During 2007, cumulative germination was greater ($p<0.05$) at 0 than -0.2 MPa, but values were similar ($p>0.05$) between -0.2 and -

Table 3. Coefficients velocity of germination in the various water potential treatments in 2006 and 2007.

Water potential (MPa)	Year	Coefficient of velocity
0	2006	$70.4 \pm 10.3^{a,c}$
	2007	$62.3 \pm 1.7^{a,c}$
-0.2	2006	$68.3 \pm 3.7^{a,c}$
	2007	$73.9 \pm 5.2^{a,c}$
-0.4	2006	$67.2 \pm 5.4^{a,c}$
	2007	$68.1 \pm 2.1^{a,c}$
-0.6	2006	$70.8 \pm 4.1^{a,c}$
	2007	$60.8 \pm 2.4^{a,c}$
-0.8	2006	$48.2 \pm 2.2^{a,b}$
	2007	$46.6 \pm 2.0^{a,b}$
-1.0	2006	$40.9 \pm 3.1^{a,a}$
	2007	$35.8 \pm 1.0^{a,a}$
-1.2	2006	$34.0 \pm 1.7^{a,a}$
	2007	$31.1 \pm 1.9^{a,a}$

Letters to the left of the comma indicate significant differences between years, and those to the right of the comma indicate significant differences among water potential treatments. Each value is the mean ± 1EE of n=6.

0.6 MPa (Table 2). At water potentials lower than -0.6 MPa, germination percentages decreased (p<0.05) as water potentials also decreased (Table 2).

Patterns in changes of germination percentages

Similar patterns of changes in germination Percentages with time were obtained in both study years (Table 2). Germination percentages followed a similar pattern with time from the initiation of the study in 2006 and 2007 at water potentials from 0 to -0.6 MPa: they increased until after approximately 60 h from incubation before leveling off. At lower water potentials (from -0.8 to -1.2 MPa), germination started to increase later until approximately 84 h from study initiation compared to the other water potential treatments (Table 2). Between -0.8 to -1.2 MPa, time to germination from study initiation delayed as medium water potential decreased (Table 2).

Coefficients of velocity of germination

No differences (p<0.05) were found between study years in the coefficients of velocity in any of the water potential treatments (Table 3). In both study years, coefficients of velocity were greater (p<0.05) from 0 to -0.6 MPa than at lower water potentials. The lowest (p<0.05) coefficients of velocity were found at -1.0 and -1.2 MPa (Table 3). At

-0.8 MPa, coefficients of velocity were greater (p<0.05) than at lower, but smaller (p<0.05) than at higher, water potentials (Table 3).

Early seedling growth

Early seedling growth was similar (p>0.05) between years (2006: 6.33 mm, n=72; 2007: 5.40 mm, n=72), and plant parts (shoots: 5.51, n=72; roots: 6.22 mm, n=72). It was greater (p<0.05) at 0 than at -0.6 MPa and lower water potentials (Table 4). Also, growth of early seedlings was greater (p<0.05) at -0.4 than at -0.8 MPa and lower water potentials (Table 4).

DISCUSSION

Storage time

Maintenance of seed quality in storage from the time of production until the seed is planted is imperative to assure its planting value. There was a marked decrease in seed viability with storage time in *D. eriantha*. This might have been partially the result of the storage conditions. The best alternative to avoid risks associated with storing seeds is to avoid storing seeds. For example, the grass seed industries in Oregon ship the seeds within a few months after harvesting. Another example include

Table 4. Early growth of *D. eriantha* seedlings (shoot + root) at various water potentials.

Water potential (MPa)	Early seedling growth (mm)
0	15.09[a]
-0.4	10.35[ab]
-0.6	6.48[bd]
-0.8	1.41[cd]
-1.0	1.29[cd]
-1.2	0.56[cd]

Different letters indicate significant differences (p<0.05) among water potentials. Each value is the mean of n=24.

Bolivia where the wheat seed harvested in the Highlands in April is being planted in May in the Lowlands, or Colombia where rice can be produced twice a year, which decreases the storage period (http://seedlab.oregonstate.edu/book/export/html/123).

These strategies are becoming highly desirable not only because they reduce storage, but especially because they make possible to market the seeds and meet financial obligations. However, there are times when seed growers and dealers carry over seed lots from one year to the next due to weak market, to insure an adequate supply the following year, and/or because the production system does not provide choices. Under such circumstances, the question is how to manage the seeds to maintain a high viability. If dry weather prevails during grass seed maturation and harvest, it should be allowed to harvest seeds not only with low moisture but also with high initial viability (Copeland and McDonald, 1995). This should be followed up by placing the seeds in cool and dry warehouses to lower the risks in storage (Copeland and McDonald, 1995). Storage of *D. eriantha* seeds at 20°C most likely caused some loss of seed viability after two years from harvesting. Losses of viability because of storage conditions contribute to explain the lower germination percentages in 2006 than in 2007.

Temperature effects

Total germination percentages were similar at constant than at alternating temperatures, in agreement with the first hypothesis. Similar to our results Hsu et al. (1985) found that chilled 'IG-2C-F1' *S. nutans* seeds germinated best at constant temperatures of 20 to 35°C. Fifteen-month-old seeds of the perennial, warm-season *Eragrostis curvula* (Schrad.) Nees cvs. 'Don Juan', 'Don Pablo' and 'Don Arturo' showed similar germination percentages when temperatures were (a) constant at 20 or 30°C under darkness, (b) alternate at 4/20, 10/20, 20/30 or 20/35°C under darkness, or (c) alternate as described in (b) but with a 9h-photoperiod at the maximum temperature (Fernàndez et al., 1991). These authors also reported that the highest germination

percentage in *E. curvula* cv. 'Morpa' occurred at constant temperatures of 30°C, while *E. curvula* cvs. 'Don Eduardo' and 'Tanganyika' were favored by either constant or alternating temperatures higher than 20°C. Young and Evans (1982) found that maximum germination percentages could be obtained under constant temperatures for most of the perennial grasses they studied. Hylton and Bement (1961) also found greatest germination of *Festuca octoflora* Walt. under constant 20°C than under various alternating temperature regimes. Terenti (2004) showed that the best germination (80%) in *D. eriantha* occurred at 30 and 35°C. We showed that high germination percentages occurred in this species at these high, constant temperatures in our study a day after study initiation. When temperatures were outside these values, a significant decrease in germination resulted. From these determinations it is conceivable that large annual differences in *D. eriantha* density could potentially be obtained to the prevailing temperature during moisture availability.

It is important to highlight that cumulative germination percentages found with the thermal gradient were above than 80% between 20.4 and 33.8°C, but these percentages decreased at lower or higher temperatures. There was no germination beyond 36.9°C. These results imply that seeding of *D. eriantha* could be made during late-spring, early-summer (December, January; Figure 1) at the study site, when soil temperatures are at or beyond 20°C, and precipitations keep the soil moist (see long-term precipitation data in Figure 1).

Overall coefficients of velocity supported that, over the short-term, more seeds germinated at constant than alternating temperatures, in agreement with the second hypothesis. These results suggest that 30 to 35°C could be within the optimum temperature for germination in this species. Terenti (2004) also reported the highest total germination for *D. eriantha* at these temperatures, when he only compared constant, but not constant versus alternating temperatures. We need to keep in mind, however, that the mean temperature was lower at the fluctuating temperature than under the constant temperature regimes. Therefore, differences between fluctuating and constant may be due to different mean temperatures rather than to temperatures being constant or fluctuating *per se*. This reasoning, that could help to explain differences between constant versus alternating temperatures, might extend to other germination studies, that have attributed either to constant or alternating temperatures their results on the germination response to temperature in various species.

Our results differ, however, with most studies on perennial grass species, where germination was stimulated most by alternating than constant temperatures. When *Elymus cinereus* cv. Magnar (Scribn. and Merr.) was exposed to different temperature profiles from 0 to 40°C, it showed greater germination percentages after exposure to alternating temperatures between 15 and 25°C (Evans and Young, 1983). A similar response was obtained with *Achnatherum robustum* (Vasey) Barkworth which optimum germination occurred with 20°C during 8 h, and 15°C during 16 h (Young et al., 2003). *Leymus cinereus* showed the greatest mean germination percentages at alternating temperatures of 15/25°C, and then at 10/20, 20/30 and 5/15°C in decreasing order of magnitude (Meyer et al., 1995).

Results from this study are important since more than 50% germination might occur at constant temperatures within a few hours after seeding, taking advantage of small rainfall events, which are common at the study site (Paez et al., 2005). These authors reported that 61% of the rainfall events were lower than 5 mm after an analysis of 18 years (1983-2000) of rainfall records at the Chacra Experimental de Patagones. After four days of study, however, germination was similar at all temperature treatments. These results suggest that temperature does not appear to place a major restriction on the germination of *D. eriantha*; seeds can be expected to germinate over a wide thermal gradient.

Water potential effects

Moisture availability imposed severe limitations on seed germination of *D. eriantha*, which has similar germination requirements that many mesophytic crops (Levitt, 1980; Bonvisutto and Busso, 2007): the lower the water potential, the lower the germination percentage and the velocity of germination in this species. The lower coefficients of velocity at lower water potentials are an indication of greater germination times (Scott et al., 1984); in fact, seeds started to germinate later at lower than higher water potentials (Table 2). Plants possessing seeds with exacting requirements for germination can establish more successfully than those with few restrictions (Hegarty, 1978). However, in an environment with changing moisture conditions the opportunities for germination may be reduced for seeds with specific moisture requirements. If moisture stress is low, seeds of *D. eriantha* can germinate over a wide range of temperatures; however, the more severe the water stress, the greater the reduction in germination percentage. This response presumably reflects an adaptive strategy because *D. eriantha* is generally restricted to habitats with moister conditions than those of the Phytogeographical Province of the Monte (Cano, 1988). Such an adaptation protects against germination under conditions of transient or low soil moisture, limiting most germination to periods with protracted conditions of high soil moisture.

Exposure to descending water stress reduced germination. Relatively warm soil temperatures and water stress are usually simultaneous events in the Chacra Experimental de Patagones, within the Phytogeographical Province of the Monte. Providing seed mortality does not occur, no germination induced by the combination of warm temperatures and water stress may act to preserve

a portion of the seedbank for germination at a later date. This lack of germination under unsuitable conditions may also serve to block germination that would otherwise predispose seedlings to temperature and moisture conditions that most likely would not be conducive to their growth and survival.

Seedling growth

Despite increasing temperatures provide conditions favorable for growth of seedlings (Brown, 1995), water potentials lower than -0.4 MPa greatly reduced seedling growth. These results, in agreement with the third hyphotesis, are in accordance with those reported by Brown (1995) in several perennial grasses. Our data showed that *D. eriantha* has the ability to germinate over a broad range of temperatures, but severe restrictions are imposed by reduced moisture availability. Seeding of this species in rangelands of central Argentina (e.g., the Chacra Experimental de Patagones) will most likely fail under water stress conditions (e.g., years 2008 and 2009 in Figure 1). Results suggest that this species should be planted in late-spring, early-summer, when seedbed temperatures are increasing and soil moisture might still be adequate (e.g., year 1984 in Figure 1).

REFERENCES

Almansouri M, Kinet JM, Lutts S (2001). Effect of salt and osmotic stresses on germination in durum wheat (*Triticum durum* Desf.). Plant Soil 231:243-254.

Anderson DL (1980). La recuperación y mejoramiento de los pastizales naturales. Ecol. Arg. 4:9-11.

Berkat O, Briske DD (1982). Water potential evaluation of three germination substrates utilizing polyethylene glycol 20,000. Agron. J. 74:518-521.

Bonvisutto G, Busso CA (2007). Germination of grasses and shrubs under various water stress and temperature conditions. Phyton. Int. J. Exp. Bot. 76:119-131.

Brown RF (1987). Germination of Aristida armata under constant and alternating temperatures and its analysis with the cumulative welbull distribution as a model. Aust. J. Bot. 35:581-591.

Brown RW (1995). The water relations of range plants: Adaptations to water deficit. In: Bedunah DJ, Sosebee RE (eds.) Wildland Plants: Physiological Ecology and Developmental Morphology. Denver: Soc. Range Manage. pp. 635-710.

Busso CA, Giorgetti HD, Montenegro OA, Rodríguez GD (2004). Perennial grass species richness and diversity on Argentine rangelands recovering from disturbance. Phyton. Int. J. Exp. Bot. 53:9-27.

Cabrera AL (1976). Regiones fitogeográficas Argentinas. In: Ferreira Sobral EF (ed.) Enciclopedia Argentina de Agricultura y Jardinería. Buenos Aires: ACME, Argentina. pp. 1-85.

Cano E (1988). Pastizales naturales de La Pampa. Descripción de las especies más importantes. Convenio AACREA-Provincia de La Pampa, Buenos Aires, Argentina.

Copeland L, McDonald M (1995). Principles of Seed Science and Technology. 3rd ed. Chapman & Hall.

Dannhauser CS (1988). A review on forage in the central grassveld with special reference to Digitaria eriantha. Tydskrif van die Weidingsvereniging van Suidelike Afrika, 5:193-196.

Di Giambatista G, Garbero M, Ruiz M, Giulietti M, Pedranzani H (2010). Germinación de Trichloris crinita y Digitaria eriantha en condiciones de estrés abiótico. Pastos y forrajes 33:1-7.

Di Rienzo JA, Casanoves F, Balzarini MG, Gonzalez L, Tablada M, Robledo CW (2009). INFOSTAT versión 2009. Grupo INFOSTAT, FCA, Universidad Nacional de Córdoba, Argentina.

Distel RA, Bóo RM (1996). Vegetation states and transitions in temperate semiarid rangelands of Argentina. In: West EN (ed) Proceedings V[th] International Rangeland Congress: Rangelands in a Sustainable Biosphere, Soc. Range Manage., Salt Lake City, USA, pp. 117-118.

Du Toit PCV (2000). Estimating grazing index values for plants from arid regions. J. Range Manage. 53:529-536.

Ellern SJ, Tadmor NH (1966). Germination of range plant seeds at fixed temperatures. J. Range Manage. 19:341-345.

Emmerich WE, Hardegree SP (1990). Polyethylene glycol solution contact effects on seed germination. Agron. J. 82:1103-1107.

Evans RA, Young JA (1983). Magnar' basin wildrye-germination in relation to temperature. J. Range Manage. 36:395-398.

Fernàndez OA, Brevedan RE, Gargano AO (1991). El pasto lloròn. Su biologìa y manejo. CERZOS-Universidad Nacional del Sur, Buenos Aires.

Fulbright TE (1988). Effects of temperature, water potential, and sodium chloride on indiangrass germination. J. Range Manage. 41:207-210.

Gargano AO, Adúriz MA, Busso CA (2004). Nitrogen Fertilization and Row Spacing Effects on Digitaria eriantha. J. Range Manage. 57:482-489.

Giorgetti HD, Manuel Z, Montenegro OA, Rodríguez GD, Busso CA (2000). Phenology of some herbaceous and woody species in central, semiarid Argentina. Phyton, Int. J. Exp. Bot. 69:91-108.

Giorgetti HD, Montenegro OA, Rodríguez GD, Busso CA, Montani T, Burgos MA, Flemmer AC, Toribio MB, Horvitz SS (1997). The comparative influence of past management and rainfall on range herbaceous standing crop in east-central Argentina: 14 years of observations. J. Arid Environ. 36:623-637.

Hacker JB, Wilson GPM, Ramírez L (1993). Breeding and evaluation of Digitaria eriantha for improved spring yield and seed production. Euphytica 68:193-204.

Hegarty TW (1978). The physiology of seed hydration and dehydration, and the relation between water stress and the control of germination: a review. Plant Cell Environ. 1:101-119.

Heydecker W (1960). Can we measure seedling vigour?. Proc. Int. Seed Test. Assn. 25:498-512.

Hsu FH, Nelson CJ, Matches AG (1985). Temperature effects on germination of warm-season forage grasses. Crop Sci. 25:215-220.

http://seedlab.oregonstate.edu/book/export/html/123. Maintaining grass seed viability in storage. A brief review of management principles with emphasis on grass seeds stored in Oregon. Access date: 1/12/2011.

http://www.ehow.com/info_8407286_genus-species-crabgrass.html. Genus and species of crabgrass. Access date: 19 July 2011.

Hylton LO, Bement RE (1961). Effects of environment on germination and occurrence of sixweeks fescue. J. Range Manage. 14:257-261.

Lavin F, Johnsen TNJr (1977). Species adapted for planting Arizona Pinyon-Juniper Woodland. J. Range. Manage. 30:410-415.

Levitt J (1980). Responses of plants to environmental stresses. Academic Press, New York.

Meyer SE, Beckstead J, Allen PS, Pullman H (1995). Germination ecophysiology of Leymus cinereus (Poaceae). Int. J. Plant Sci. 156:206-215.

Michel BE (1983). Evaluation of the water potentials of solutions of polyethylene glycol 8000 both in the absence and presence of other solutes. Plant Physiol. 72:66-70.

Owens DW, Call CA (1985). Germination characteristics of Helianthus maximilianai Schard. and Simsia calva (Engelm, & Gray). J. Range Manage. 38:336-339.

Paez A, Busso CA, Montenegro OA, Rodríguez GD, Giorgetti HD (2005). Seed weight variation and its effects on germination in Stipa species. Phyton. Int. J. Exp. Bot. 74:1-14.

Pedranzani HE, Ortiz OM, Garbero M, Terenti OA (2005). Efecto de baja temperatura sobre distintos parámetros de producción en Digitaria eriantha cv. mejorada INTA. Phyton, Int. J. Exp. Bot. 54:121-126.

Peters J (2000). Tetrazolium testing handbook. Contribution No. 29 to the handbook on seed testing. Association of Official Seed Analysts, Lincoln, NE. P. 176.

Rimieri P (1997). Creación de cultivares mejorados e identificables de Poa ligularis y Digitaria eriantha. Informe técnico de Proyectos, Área Producción Animal. INTA EEA San Luis, Argentina. P. 183.

Romo JT, Eddleman LE (1988). Germination of green and gray rubber rabbitbrush and their establishment on coal mined land. J. Range Manage. 41:491-495.

Sabo DG, Johnson GV, Martin WC, Aldon EF (1979). Germination of 19 species of arid land plants. USDA Forest Serv. Res. Pap. RM-210. USDA Forest Serv., Rocky Mnt. Forest Range Exp. Sta., Fort Collins, CO.

Sanderson MA, Voigt P, Jones RM (1999). Yield and quality of warm-season grasses in central Texas. J. Range Manage. 52:145-150.

Scott SJ, Jones RA, Williams WA (1984). Review of data analysis methods for seed germination. Crop Sci. 24:1192-1199.

Terenti OA (2004). Evolución del crecimiento y la calidad de semilla en Digitaria eriantha. Pastos y Forrajes 27:21-24.

Torres YA, Busso CA, Montenegro OA, Ithurrart L, Giorgetti H, Rodríguez G., Bentivegna D, Brevedan R, Fernández O, Mujica MM, Baioni S, Entío J, Fioretti M, Tucat G. (2011). Defoliation effects on the arbuscular mycorrhizas of ten perennial grass genotypes in arid Patagonia, Argentina. Appl. Soil Ecol. 49:208-214.

Young JA, Clements CD, Jones TA (2003). Germination of seeds of robust needlegrass. J. Range Manage. 56:247-250.

Young, JA, Evans RA (1982). Temperature profiles for germination of cool season range grasses. USDA, ARS, Agr. Res. Results. AAR-W-27, Oakland, CA.

18

The genetic variability using sequencing of the ribosomal internal transcribed spacer (ITS) region in cultivars of the cowpea [*Vigna unguiculata* L. (Walp).]

Ana Dolores S. de Freitas[1], Maria Luiza R. B. da Silva[2], Adália C. E. S. Mergulhão[2] and Maria do Carmo C. P. de Lyra[2]

[1]Soil Microbiology Laboratory, Agronomy Department, University Federal Rural of Pernambuco (UFRPE). Rua Dom Manoel de Medeiros, s/n, Dois Irmãos 52171-900 – Recife. PE – Brazil.
[2]Genomics Laboratory, Agronomical Institute of Pernambuco (IPA). Av. Gal San Martin 1371 Bonji 50761-000 – Recife. PE – Brazil.

Cowpea is a predominant crop in the small farms of the Brazilian semi-arid region, where several varieties of cowpea with high genetic variability are planted. Due to their genetic diversity, good adaptation to marginal environments and growth in low-input systems, these varieties, often called "landraces," are of great interest for use in breeding and biodiversity conservation programs. The present study describes twelve varieties and three cultivars of cowpea, chosen on the basis of their high frequency of planting by farmers in the Paraíba and Pernambuco States. Total DNA was extracted from the plant seeds. In order to observe the variability of the studied material, four Intersimple sequence repeat (ISSR) markers, the amplified ribosomal DNA restriction analysis (ARDRA) technique using various endonucleases (*Alu*I, *Hinf*I, *Hpa*II, *Rsa*I and *Nru*I), and amplification and sequencing of the ribosomal internal transcribed spacer (ITS1) and ITS2 regions were employed. The results show that two assays, the fingerprinting and the sequencing of the ribosomal ITS1 and ITS2 regions, were sufficient to detect the variability of the cowpea germplasm used by small farmers in the Brazilian semi-arid region. Although the varieties often received the same designation, high diversity may have occurred within a variety, according to the origin of the cowpea seed. This work represents efforts to guide preservation of cowpea biodiversity in semi-arid regions.

Key words: *Vigna unguiculata* L. (Walp.), coat seeds, intersimple sequence repeat (ISSR), amplified ribosomal DNA restriction analysis (ARDRA), diversity.

INTRODUCTION

Cowpea [*Vigna unguiculata* (L.) Walp.] is an important grain legume cultivated in all tropical and subtropical regions of the world, as well as in South-East Europe and in the United States. In small Brazilian semi-arid farms, cowpea is one of the predominant crops. In general, local varieties of cowpea are preferentially cultivated, as the government-supplied cultivars do not satisfactorily meet the needs of the producers and consumers due to either type of maturation and plant cycle or to the color, shape and size of the grains, despite the high productive potential and resistance to virus diseases of the cultivars developed by government agencies. Often, varieties with quite distinct phenotypic characteristics (such as color and size of seed or growth habit) have the same name in distinct micro-regions.

Local varieties are important in several regions and countries where cowpea is cultivated, particularly in Africa (Adjei-Nsiah et al., 2008; Ghalmi et al., 2010). Due to their genetic diversity, good adaptation to marginal farming environments and ability to grow in systems of low inputs, these varieties, often called "landraces", are of great interest for use in breeding programs and fordevelopment of new cropping systems.

Accordingly, various efforts to understand the genetic diversity of the preferred cultivars and to preserve this diversity have been undertaken by researchers from different countries (Ghalmi et al., 2010). These studies show that a complex interaction of factors is responsible for the observed diversity patterns in cowpea.

Molecular markers can be used as tools for studying the genetic diversity between individuals within a population and between populations, or between related species, and for determining the identity of a cultivar variety (Arnao et al., 2008). The Intersimple sequence repeat (ISSR) analysis has many advantages in assessing genetic diversity (Gupta et al., 1994) because it is based on the amplification of microsatellite sequences between adjacent regions of DNA via polymerase chain reaction (PCR). ISSR technology uses highly polymorphic targets, it is reproducible, does not require prior knowledge of the genome and is relatively inexpensive (Vieira et al., 2009).

Recently, analyses of the ribosomal DNA (rDNA) spacers and comparative studies of the nucleotide sequences of the rDNA genes using the Amplified ribosomal DNA restriction analysis (ARDRA) technique, have been applied to the study of the phylogeny and taxonomy of plants, fungi and bacteria (Yadav et al., 2011). While the regions of ribosomal genes are highly conserved within species, regions of the internal transcribed spacer (ITS) evolve faster; therefore, evolution may vary interspecifically on the sequence of bases and in length (Gerbi, 1985), being frequently used for taxonomy of species and genera (Menezes et al., 2010).

To date, no studies have been published that have used the ISSR markers, the ARDRA technique and sequencing of the 18S rDNA to examine cowpea cultivars in use by small family farmers in the Brazilian semi-arid region states. The objective of this study was to use molecular markers (ISSR and ARDRA) and sequencing the ITS region of rDNA to estimate the genetic diversity and investigate the genetic relationships between local varieties of cultivated cowpea in the States of Pernambuco and Paraíba.

MATERIALS AND METHODS

Thirteen local cowpea varieties were used, including four from Paraiba (Corujinha-PB, Sedinha-PB, Canapú-PB and Azul-PB) and eight from Pernambuco (Sempre Verde-PE, Sedinha-PE, Maratuã-

PE, Canapú-PE, São Sebastião-PE, Costela de Vaca Branca-PE, Corujinha-PE and Costela de Vaca Marrom-PE). In addition to these, the cultivars that were recommended for use were the BRS Pujante, the IPA206 and the IPA207. Some of the varieties in use have the same denomination in two states, but it has been observed that they are sufficiently divergent in color, format and size, suggesting possible genetic differences; for this reason, these varieties were included in the study. For easy identification of the studied materials, each variety was given the initials of its state of origin after the variety name. Total DNA was extracted from seeds. The Invisorb Spin Plant Mini Kit from Invitek was used and followed the manufacturer's suggestions with slight modifications in the preparation of samples. As the seed coat has high levels of compounds such as tannic acid, phenol and cyanide which causes interference in the amplification of DNA were removed from seeds were separated from the pistil. After this procedure, liquid nitrogen was added at a rate of 60 mg to the vegetable material to obtain a fine powder, and DNA extraction was performed. The quantification was performed in an agarose gel containing 0.8% agarose in 0.5 × Tris/Borate/EDTA (TBE) buffer using the bromothymol blue buffer and the SybrGold (Invitrogen) stain. Electrophoresis was run on 100 V, and the product was viewed under ultraviolet (UV) light using a Gel Doc L-Pix image-Loccus software system.

For genotypic characterization of cowpea varieties and cultivars by amplification of the ITS1 and ITS 2 regions, two primers were used: ITS1 (5'-TTC CGT AGG TGA ACC TGC GG-3') and ITS2 (5'-TCC TCC GCT TAT TGA TAT GC- 3') (White et al., 1990). For amplification by PCR, a 15 µl final volume was used under the following conditions: 2 µl of DNA (20 to 40 ng), 0.4 mM of each primer, 10 × Taq polymerase buffer, 10% DMSO, 1 U Taq polymerase, 200 mM of mix dNTP's, and 2.5 mM $MgCl_2$. The amplification cycles used included an initial denaturation (5 min at 94°C) and further 30 cycles of: 1 min at 94°C, 1 min at 55°C, 2 min at 72°C and one last extension step of 5 min at 72°C. The amplified fragments were separated by agarose gel electrophoresis on a gel containing 0.8% agarose in 0.5 × TBE buffer, in bromothymol blue buffer and stained with SybrGold (Invitrogen). The resultant fragments were viewed under UV light and photographed under UV light using a Gel Doc L-Pix image-Loccus software system. After purification with 7.5M ammonium acetate, the PCR products were treated separately with endonucleases: AluI, HinfI, HpaII, NruI and RsaI. Digestion was performed in a final volume of 20 µl (9 µl of water, 8 µl of amplified purified product, 2 µL of the specified buffer for each enzyme and 1 µl of endocuclease). The enzyme digestion temperature varied depending on the enzyme used, and all reactions were incubated overnight. Then, the restriction fragments were separated by agarose gel electrophoresis at 80 V for 3 h on a gel containing 2.5% agarose in 0.5 × TBE buffer. The running buffer was supplemented with bromothymol blue, and SybrGold was used for staining. Gels were then viewed under ultraviolet light and photographed in under UV light using a Gel Doc L-Pix image-Loccus software system. The PCR products of the ITS1 and ITS2 regions were purified after the reaction in a final volume of 100 µl.

This volume was subdivided: one part was used for the endonuclease digestions, and the other part was used for sequencing. The purification consisted of adding 8 µl of 7.5 M ammonium acetate and 208 µl of 100% ethanol to samples in 1.5 ml microtubes and centrifuging those at 13,000 rpm for 20 min at room temperature. The supernatant was removed and 150 µl of ice cold 70% ethanol was then added to before centrifugation for 5 min at 13,000 rpm. The microtube was poured onto a paper towel to allow the pellet to dry overnight. DNA-containing pellets were then re-suspended in 30 µl of ultrapure water and kept at -20°C until sequencing was performed. Total DNAs were used at a dilution of 1:500. The ISSR analyses used the following primers: UBC01-5´-AC^8-3´, UBC808 - 5´- AG^8C- 3´, UBC809 - 5´- AG^8G - 3´; UBC810 - 5´-GA^8T-3´. For amplification by PCR, a final volume of 15 µl was used and contained 20 to 40 ng of genomic DNA (1 µl), 0.4 µM

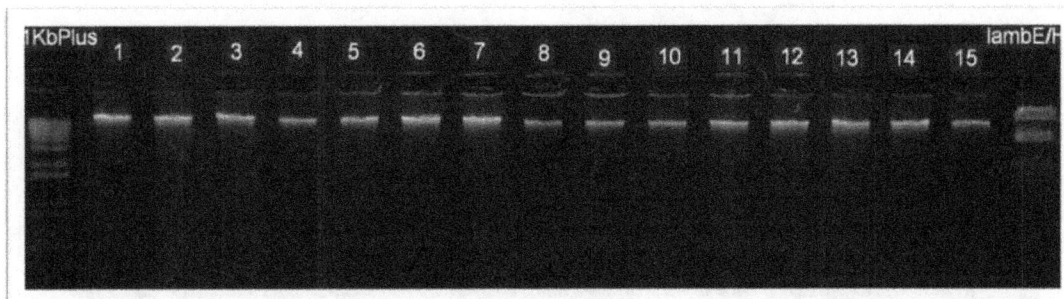

Figure 1. Extraction of genomic DNA from cowpea seeds. Lane 1. Corujinha-PB (CRJPB), Lane 2. Sedinha-PB (SDHPB), Lane 3. IPA207, Lane 4. Canapú-PB (CNPPB), Lane 5. Azul-PB, Lane 6. Sempre Verde-PE (SVPE), Lane 7. Sedinha-PE (SDHPE), Lane 8. Maratauã-PE (MRTPE), Lane 9. Canapú-PE (CNPPE), Lane 10. São Sebastião-PE (SSPE), Lane 11. BRS Pujante-PE (BRSPPE), Lane 12. Costela de Vaca Branca-PE (CVBPE), Lane 13. IPA206, Lane 14. Corujinha-PE (CRJPE), Lane 15. Costela de Vaca Marrom-PE (CVMPE). DNA ladder 1 Kb Plus (Invitrogen) and Lambda E/H.

primer, 10% Taq polymerase buffer (Invitrogen), 10% DMSO, 1U Taq polymerase enzyme (Invitrogen), 200 mM of each dNTP, and 2.5 mM of MgCl$_2$. The amplifications were conducted in a PCT-100 (MJ-Research-Peltier) thermal cycler. The amplification cycles used included an initial denaturation (5 min at 94°C) and further 30 cycles of 1 min at 94°C, 1 min at 55°C, 2 min at 72°C and a final extension step of 5 min at 72°C. To eliminate the possibility of contamination, a negative control was always used. The amplified products were separated by agarose gel electrophoresis on a gel containing 1.2% agarose in 0.5 × TBE buffer and were then stained with SybrGold and viewed under UV light. The repeated bands were punctuated and denoted as either present (1) or absent (0) in ISSR markers and ARDRA, and each feature was treated independently. For each primer, the number of different frequency bands and the bands were designated as polymorphics. A genetic similarity calculation was measured using the Simple Matching (SM) coefficient where N_{ij} is the number of bands present in both genotypes i and j, N_i is the number of bands present in genotype i, and N_j is the number of bands present in genotype j. The genetic similarity was converted to genetic dissimilarity. The dissimilarity matrix produce was used to generate a cophenetic matrix (Figure 7), the adjustment between the dissimilarity matrix and the dendrogram being estimated from the cophenetic correlation coefficient (r). The clustering analysis was performed by calculating the similarity of SM coefficient, and the results generated an array of genetic distance.

This array was visualized as dendrograms constructed using the NTSYSpc program (Numerical Taxonomy and Downloads Analysis System) version 2.1 (Rohlf, 1998) and by applying the SM coefficient of similarity. For the analysis of the genetic distance between variants, Unweighted pair group method with arithmetic mean (UPGMA) clustering was used with the parameters of the Sequential agglomerative hierarchical nested cluster analysis (SAHN) program, and the construction of a phylogenetic tree was completed using the TREE plot (Sokal and Sneath, 1963) program. For sequencing reactions, the oligonucleotides ITS1 and ITS4 were used. Sequencing was performed on the MegaBace 1000 DNA sequence (Amersham Biosciences) employing the DNA sequencing NTBIO platform of EMBRAPA Genetic Resources and Biotechnology - CENARGEN. The electrophoresis parameters used for sequencing were: 40 s injection of sample under 1 Kv voltage and 5 Kv running voltage for 240 min. For construction of the phylogenetic tree, the nucleotide sequences obtained from the different isolates were subjected to alignment by the program BioEdit (Hall, 1999). Nucleotide sequences were aligned with the ClustalW program, and phylogenetic analysis was conducted using

neighbor-joining (with 1000 bootstrap replications) and pairwise deletion of nucleotides according to the Tamura-Nei model made by MEGA program, version 4 (Tamura et al., 2004).

RESULTS

Extracted DNA from all seeds resulted in a high-quality material as can be viewed in Figure 1. The four primers used for ISSR showed to be polymorphic and the few tracks of amplified products that were observed are shown in Figure 2. These primers are characterized by their ability to generate profiles with strong banding; the amplified products range from 6 to 11 bands with polymorphic DNA bands ranging from 300 to 2000 oligonucleotides in length. The dendrogram of similarity (Figure 3) of the four ISSR markers showed the formation of three ISSRs clusters (Clusters 1, 2, and 3), with genetic distances of 67, 66 and 58%, respectively.

The first cluster contains two subclusters: 1A and 1B (Figure 3). In subcluster 1A, Corujinha-PB and Canapú-PB varieties showed 81% similarity to one another, while the Azul-PB showed 73% similarity in relation to the other two varieties. Along another branch, Maratauã-PE and Canapú-PB varieties showed 88% similarity, the greatest similarity presented using the SM coefficient, while São Sebastião-PE showed a less than 77% similarity to the other two varieties. In subcluster 1B, the highest level of similarity (85%) was observed between Corujinha-PE and Costela de Vaca Marrom-PE. These varieties showed 81% similarity with respect to the IPA206 and IPA207 cultivars used in this work as control *V. unguiculata* species, and 73% in relation to the Costela de Vaca Branca-PE variety. In this cluster, the two varieties, Costela de Vaca Branca and Costela de Vaca Marrom, both from Pernambuco, showed distinct genotypes.

In Cluster 2, Sedinha-PB variety and the cultivar IPA207 showed 82% similarity, and Costela de Vaca Branca and Costela de Vaca Marrom varieties showed

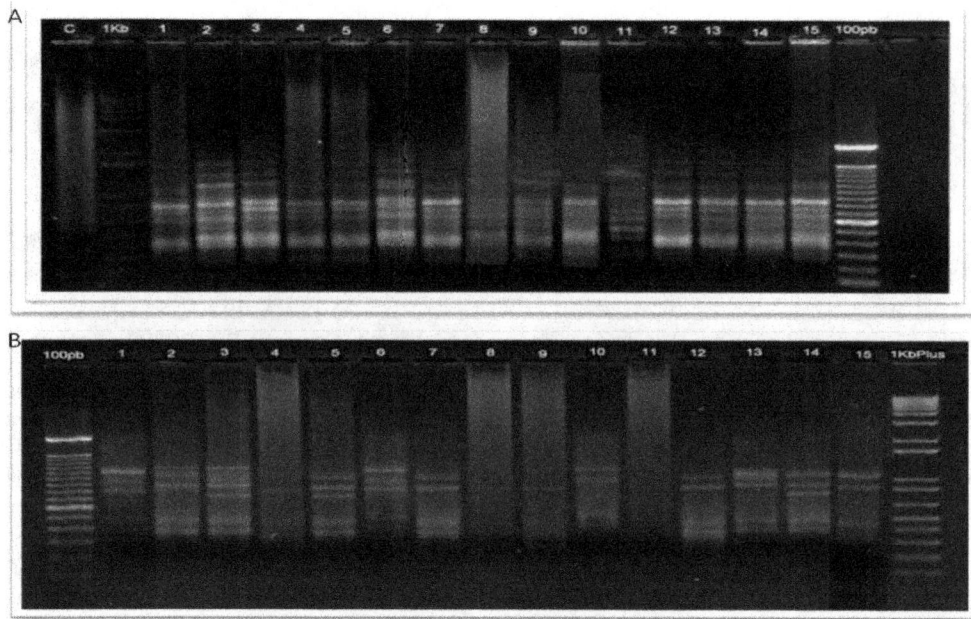

Figure 2. ISSR patterns of cowpea varieties and cultivars generated by primers: UBC 808 (A) and UBC 810 (B). Lane 1. Corujinha-PB (CRJPB), Lane 2. Sedinha-PB (SDHPB), Lane 3. IPA207, Lane 4. Canapú-PB (CNPPB), Lane 5. Azul-PB, Lane 6. Sempre Verde-PE (SVPE), Lane 7. Sedinha-PE (SDHPE), Lane 8. Maratauã-PE (MRTPE), Lane 9. Canapú-PE (CNPPE), Lane 10. São Sebastião-PE (SSPE), Lane 11. BRS Pujante-PE (BRSPPE), Lane 12. Costela de Vaca Branca-PE (CVBPE), Lane 13. IPA206, Lane 14. Corujinha-PE (CRJPE), Lane 15. Costela de Vaca Marrom-PE (CVMPE). DNA Ladder 1 Kb Plus and 100 pb (Invitrogen).

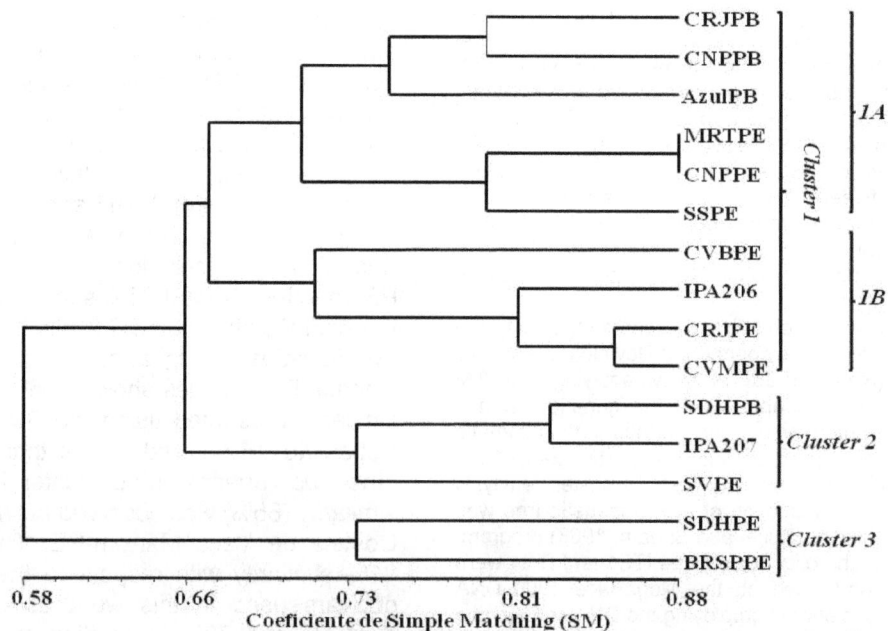

Figura 3. UPGMA dendrogram showing the relationship among cowpea varieties and cultivars using ISSR Markers (UBC 01, UBC 808 , UBC 809; UBC 810. 1. Corujinha-PB (CRJPB), 2. Sedinha-PB (SDHPB), 3. IPA207, 4. Canapú-PB (CNPPB), 5. Azul-PB, 6. Sempre Verde-PE (SVPE), 7. Sedinha-PE (SDHPE), 8. Maratauã-PE (MRTPE), 9. Canapú-PE (CNPPE), 10. São Sebastião-PE (SSPE), 11. BRS Pujante-PE (BRSPPE), 12. Costela de Vaca Branca-PE (CVBPE), 13. IPA206, 14. Corujinha-PE (CRJPE), 15. Costela de Vaca Marrom-PE (CVMPE).

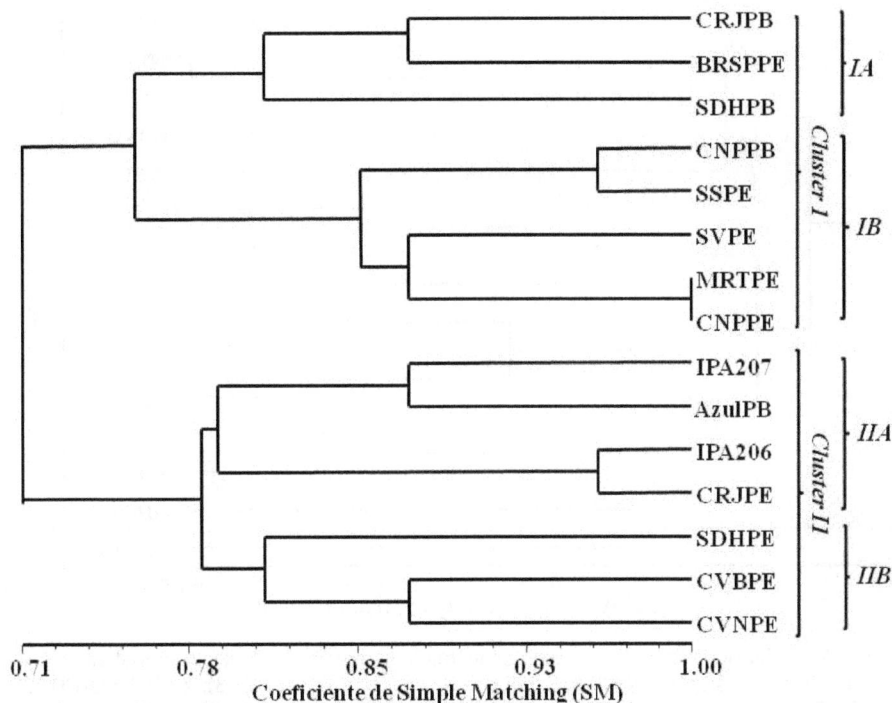

Figure 4. UPGMA dendrogram showing the relationship among cowpea varieties and cultivars using ARDRA techniques with six 06 different endonucleases (*Alu*l, *Hin*fl, *Hpa*ll, *Nru*l e *Rsa*l. 1. Corujinha-PB (CRJPB), 2. Sedinha-PB (SDHPB), 3. IPA207, 4. Canapú-PB (CNPPB), 5. Azul-PB, 6. Sempre Verde-PE (SVPE), 7. Sedinha-PE (SDHPE), 8. Maratauã-PE (MRTPE), 9. Canapú-PE (CNPPE), 10. São Sebastião-PE (SSPE), 11. BRS Pujante-PE (BRSPPE), 12. Costela de Vaca Branca-PE (CVBPE), 13. IPA206, 14. Corujinha-PE (CRJPE), 15. Costela de Vaca Marrom-PE (CVMPE).

73% similarity in relation to the other two. The third cluster is formed only by Sedinha-PE and BRS Pujante-PE varieties, and these shares 73% similarity (Figure 3).

In this paper, we show that cowpea varieties with the same title present different genetic traits. For example, in the case of varieties denominated Canapú (Canapú-PB and Canapú-PE), which, even positioned in the same Cluster 1, actually show a genetic distance of 73% (Figure 3).Using the ARDRA technique, a dendrogram with 2 clusters, each with 2 subclusters, was generated (Figure 4). Varieties with the same name, but with different origins (Pernambuco and Paraíba) such as Sedinha, Canapú and Corujinha, showed a large genetic distance. Sedinha and Corujinha from different states were grouped into different subclusters (IA and IIB, and IA and IIA respectively). However, varieties denominated as Canapú, but originated from two different states, showed less genetic distance and both clustered into subcluster IB. Once again, Maratauã-PE and Canapú-PB varieties showed maximal similarity. Canapú-PB and São Sebastião-PE, as well as Corujinha- PE and IPA206 varieties, showed high similarities to one another (greater than 95%). The other varieties, and cultivar IPA206, have higher genetic variability when analyzed by ARDRA techniques. These finding have not been previously reported in the literature using these techniques, for cowpeas.

Although the dendrogram formed by the compilation of the two techniques (Figure 5) also comprises three clusters, also it revealed an increase of the variability for those varieties and cultivars with genetic distances between 60 and 85% and generated three new subcluters (2A, 2B and 2C). Once again, both Canapu (PE and PB) varieties were grouped in the same cluster and Sedinha-PE and BRS Punjante-PE varieties formed the out group.

Data analysis using the dominant program NTSYS-pc showed that the method is to explore and to visualize similarities or dissimilarities of data was constructed for principal coordinate analysis of dominant data (Figure 7), and observed associations between fifteen cowpea cultivars obtained by analysis coordinate primary similarity coefficients of SM calculated from the 720 bands generated by two combinations of five endonucleases in ARDRA analyzes and four combinations of primers ISSR where cultivars CRJPB, SDHPB and IPA207 showed a distance of 1.0.

The sequencing of the ITS1 and ITS2 regions of the cowpea varieties and cultivars resulted in a phylogenetic tree (Figure 6), with two main Clusters (1 and 2). Along

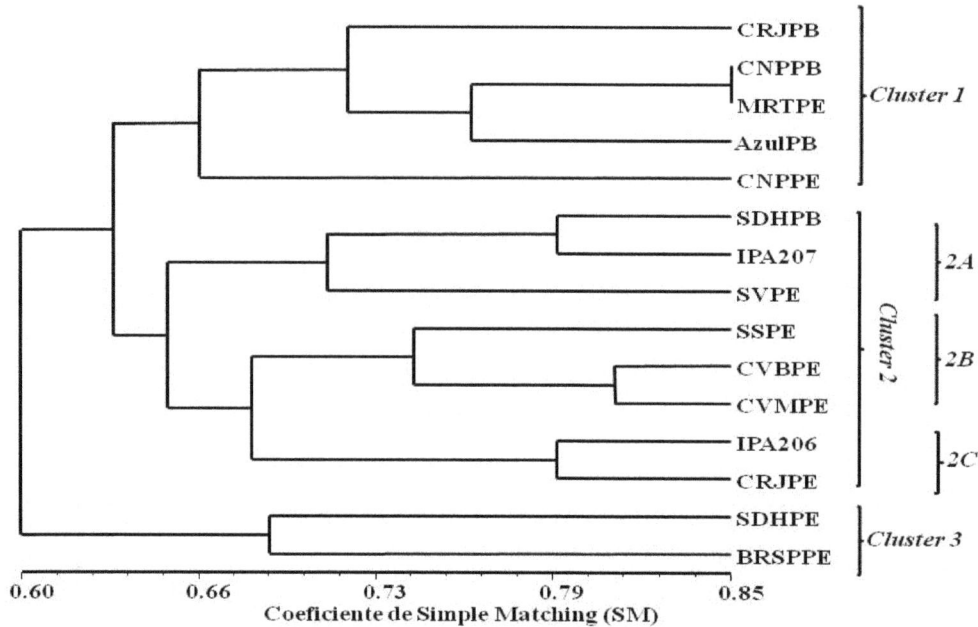

Figure 5. UPGMA dendrogram showing the relationship among cowpea varieties and cultivars using ISSR markers and ARDRA techniques. 1. Corujinha-PB (CRJPB), 2. Sedinha-PB (SDHPB), 3. IPA207, 4. Canapú-PB (CNPPB), 5. Azul-PB, 6. Sempre Verde-PE (SVPE), 7. Sedinha-PE (SDHPE), 8. Maratauã-PE (MRTPE), 9. Canapú-PE (CNPPE), 10. São Sebastião-PE (SSPE), 11. BRS Pujante-PE (BRSPPE), 12. Costela de Vaca Branca-PE (CVBPE), 13. IPA206, 14. Corujinha-PE (CRJPE), 15. Costela de Vaca Marrom-PE (CVMPE).

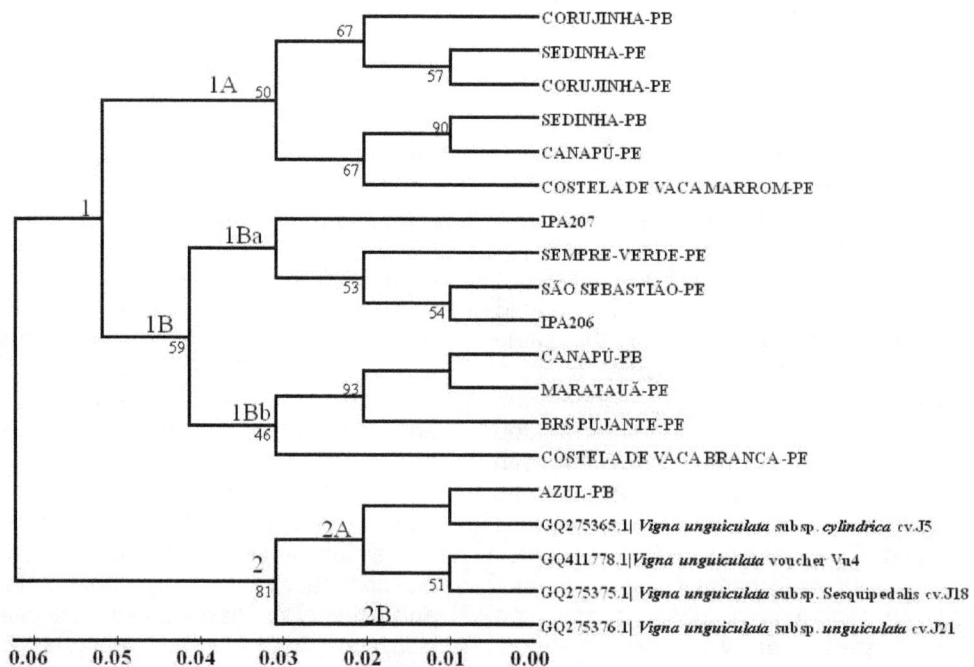

Figura 6. Phylogenetic tree of cowpea varieties and cultivars based in the sequencing of the ITS1 and ITS2 regions using BioEdit and ClustalW programs and Tamura-Nei model computed by the MEGA v.4 program. 1. Corujinha-PB (CRJPB), 2. Sedinha-PB (SDHPB), 3. IPA207, 4. Canapú-PB (CNPPB), 5. Azul-PB, 6. Sempre Verde-PE (SVPE), 7. Sedinha-PE (SDHPE), 8. Maratauã-PE (MRTPE), 9. Canapú-PE (CNPPE), 10. São Sebastião-PE (SSPE), 11. BRS Pujante-PE (BRSPPE), 12. Costela de Vaca Branca-PE (CVBPE), 13. IPA206, 14. Corujinha-PE (CRJPE), 15. Costela de Vaca Marrom-PE (CVMPE).

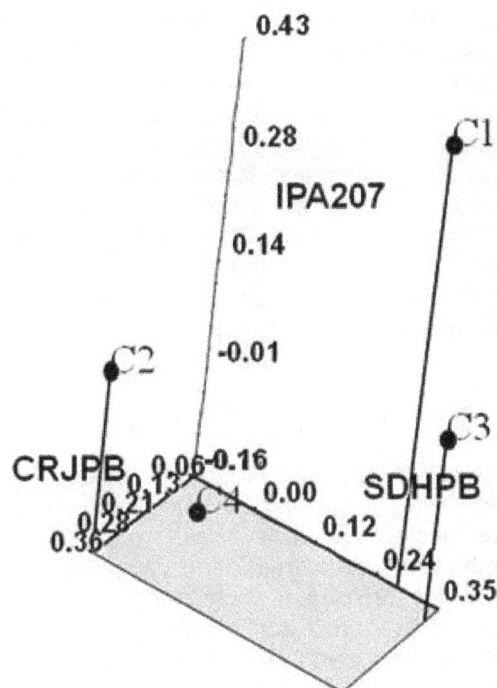

Graphic Matrix Cophenetics

Matrix Cophenetics

Cultivars	OTUi	OTUj	Dist
CRJPB	1,0000		
SDHPB	3,1864	1,0000	
IPA207	3,1864	6,2962	1,0000
CNPPB	3,6137	3,1864	3,1864
AZULPB	3,6137	3,1864	3,1864
SVPE	3,1864	4,9936	4,9936
SDHPE	3,0052	3,0052	3,0052
MRTPE	3,6137	3,1864	3,1864
CNPPE	3,0251	3,0251	3,0251
SSPE	3,1864	4,0989	4,0989
BRSPPE	1,3894	1,3894	1,3894
CVBPE	3,1864	4,0989	4,0989
IPA206	3,1864	4,0989	4,0989
CRJPE	3,1864	4,0989	4,0989
CVMPE	3,1964	4,0989	4,0989

Figure 7. Matrix cophenetics obtained of UltrametricDis and Graphics Mod3D plot among cowpea varieties and cultivars using ISSR markers and ARDRA techniques. 1. Corujinha-PB (CRJPB), 2. Sedinha-PB (SDHPB), 3. IPA207, 4. Canapú-PB (CNPPB), 5. Azul-PB, 6. Sempre Verde-PE (SVPE), 7. Sedinha-PE (SDHPE), 8. Maratauã-PE (MRTPE), 9. Canapú-PE (CNPPE), 10. São Sebastião-PE (SSPE), 11. BRS Pujante-PE (BRSPPE), 12. Costela de Vaca Branca-PE (CVBPE), 13. IPA206, 14. Corujinha-PE (CRJPE), 15. Costela de Vaca Marrom-PE (CVMPE).

with the studied varieties, four sequences of the *V. unguiculata* species, reiterated from GenBank at National Center for Biotechnology Information (NCBI) were also used. With the exception of the Azul-PE variety, all varieties and both studied cultivars are grouped in Cluster 1, composed by subclusters 1A (with Corujinha-PB, Sedinha-PE, Corujinha-PE, Sedinha-PB, Canapú-PE and Costela de Vaca Marrom-PE) and 1B (which consists of all other cultivars and varieties). The IPA206 cultivar showed 70% similarity to the IPA207 cultivar, and 80 and 90% similarity with Sempre Verde-PE and São Sebastião-PE varieties, respectively (Figure 6). In branch 1B, behaved as monophyletic crude to cultivar BRS Pujante-PE (80%) and the variety Costela de Vaca Branca-PE (70%). The Costela de Vaca Marrom-PE variety was highly genetically different than Costela de Vaca branca, and these cultivars appeared in different groups (1A), confirming that their genetic differences are related to their phenotypic differences. The Canapú-PB and Maratauã-PE varieties showed 90% similarity to each other.

The four subspecies of *V. unguiculata* (*cylindrica*, *voucher*, *sesquipedalis* and *unguiculata*), which were accessed through GenBank, formed the Cluster 2, along with the Azul-PB variety. *V. unguiculata unguiculata* cvJ21 demonstrated the largest genetic distance in relation to the other subspecies (70%). The subspecies *V. unguiculata* voucher Vu4 and the subspecies *sesquipedalis* cvJ18, demonstrated a similarity of 90% among themselves and 80% similarity to both the subspecies *cylindrica* cvJ5 and to Azul-PB variety.

DISCUSSION

The discriminatory potential of the ISSR markers depends on the variety and frequency of microsatellites, which change with the species; this factor explains why simple sequence repeats are the target of the present study. The sequences of nucleotide repeats were anchored to allow the analysis of multiple loci on a single reaction, that is, multiplexed (Ajibade et al., 2000). The reproducibility of fragments generated by ISSR markers exceeds those of analyses of arbitrary sequences of primers (RAPD). However, Xavier et al. (2005) observed in studies of genetic variability with 45 cowpea allows to discriminate the accessions from different countries, and groups of genotypic Brazilian landraces were grouped into a single group, suggesting a limitation of the genetic basis and there can be a tendency to group themselves according to their origins.

The study of genetic relationships of several species of the genus *Vigna* using the ISSR markers (Ajibade et al., 2000), showed that cultivated varieties of cowpea cluster closely with the subspecies *V. unguiculata*, and that, although clearly separated, these species were also very close to *V. triphylla* and *V. reticulata*. On a sub-generic level, these authors observed that the clustering of taxa differs from the currently accepted classification. Similar data studied with microsatellites genomic shows a powerful tool to evaluate local varieties not yet explored, what comes tbc the data found in this work (Badiane et al., 2012).

There is a growing concern about conservation of the genetic characteristics of cowpea varieties used by the small-scale farmers in the Brazilian semi-arid. In this sense, this work intends to contribute in this field. According to Yang et al. (1996), Culley and Wolfe (2001) and Arnao et al. (2008), the markers used in this work have great potential and population - level polymorphic the polyphasic studies could bring greater clarification of the genetic diversity of these varieties (Zhanou et al., 2008).

A very important factor in the genetic diversity of cowpea according to Nagalakshmi et al. (2010), which is the main contribution to a good rating in the genotypes and the environment that influences a particular character. The character that less influences these characteristics may reveal that it was the least affected in evolution. The results obtained from the sequencing of ITS1 and ITS2 regions suggest that the *V. unguiculata* studied varieties do not have genetic purity. While studying the cowpea lineages of the Sahara regions (likely the center of origin for the species) Pasquet (2000) observed a low variation within and between accessions of cowpea, probably due to the extreme isolation in the Oasis.

Asare et al. (2010) studied the genetic diversity in cowpea in Ghana using SSR markers observed results similar to those found in our study that there is a need for future studies that will conserve and gestionar the cowpea germplasm in this country having as starting point the selection of parental lines for breeding program. Studied data from phenotypic traits in cowpea through analysis of diversity using hybridization protocol and observed that the improvement of culture reveals the potential for reproduction and genetic improvement program in Nigeria (Adewale et al., 2011). In the Brazilian semi-arid, cowpea is grown from seeds obtained from the same field or purchased at fairs, without regard to the genetic purity of the material. The fingerprinting markers

and the ITS1 and ITS2 regions of the rDNA sequencing were sufficient to detect the variability of the cowpea germplasm used by small farmers of the Brazilian semi-arid region. Although cowpea varieties often receive the same denomination, a particular variety can contain high diversity, according to the source of the seed. A clear understanding of genetic variation and differences between populations may be helpful for the conservation of cowpea, and therefore efforts to preserve this biodiversity in the Brazilian semi-arid can be a great contribution to this endeavor.

Conclusions

The use of ISSR markers revealed greater genetic variability in the varieties and cultivars investigated and suggested that varieties with the same titles may have different genetic traits. The compilation of the ISSR and ARDRA techniques increased the variability among the studied varieties and cultivars, whose genetic distances varied between 60 and 85%, indicating the need for additional molecular markers. The sequencing of the ITS1 and ITS2 regions from cowpea varieties and cultivars showed that these do not have genetic purity in the Brazilian semi-arid region because the cowpea is grown from seeds obtained from the same cultivated field or acquired in trade fairs.

ACKNOWLEDGMENTS

This work was partially supported by PNPD/CAPES/FINEP (470911/2009-3). The author would like to thank the Nuclear Energy Department of UFPE for sending the cowpea seeds from the State of Paraíba and the University Federal Rural of Pernambuco for sending the material from the State of Pernambuco.

REFERENCES

Adewale BD, Adeigbe OO, Aremu, CO (2011). Genetic distance and diversity among some cowpea (*Vigna unguiculata* L. Walp) genotypes. Int. J. Res. Plant Sci. 1(2):9-14.
Adjei-Nsiah S, Kuyper TW, Leeuwis C, Abekoe MK, Cobbinah J, Sakyi-Dawson O, Giller KE (2008). Farmers' agronomic and social evaluation of productivity, yield and N2-fixation in different cowpea varieties and their subsequent residual N effects on a succeeding maize crop. Nutr. Cycl. Agroecosys. 80(3):199–209.
Ajibade SR, Weeden NF, Chite, SM (2000). Inter simple sequence repeat analysis of genetic relationships in the genus *Vigna*. Euphytica 111(1):47-55.
Arnao E, Jayaro Y, Hinrichsen P, Ramis C, Marín R C, Pérez-Almeida I (2008). Marcadores AFLP en la evaluación de la diversidad genética de variedades y líneas élites de arroz en Venezuela. Interciência. 33(5):359-364.
Asare AT, Gowda BS, Galyuon IKA, Aboagye LL, Takrama JF, Timko MP (2010). Assessment of the genetic diversity in cowpea (*Vigna unguiculata* L. Walp.) germplasm from Ghana using simple sequence repeat markers. Plant Gen. Res. 8(2):142-150.
Badiane FA, Gowda BS, Cissé N, Diouf D, Sadio O, Timko MP (2012).

The genetic variability using sequencing of the ribosomal internal transcribed spacer (ITS) region in cultivars...

159

Genetic relationship of cowpea (*Vigna unguiculata*) varieties from Senegal based on SSR markers. Genet. Mol. Res. 8:11(1):292-304.

Culley TM, Wolfe AD (2001). Population genetic structure of the cleistogamous plant species *Viola pubescens* Aiton (Violaceae), as indicated by allozyme and ISSR molecular markers. Heredity 86(1):545-556.

Gerbi SA (1985). Evolution of ribosomal DNA. In: Mcintyre RE (ed) Molecular evolutionary genetics New York: Plenum, Chapter 7, pp. 419-517.

Ghalmi N, Malice M, Jacquemin JM, Ounane SM, Mekliche L, Baudoin JP (2010). Morphological and molecular diversity within Algerian cowpea (*Vigna unguiculata* (L.) Walp.) landraces. Genet. Resour. Crop. Ev. 57(3):371-386.

Gupta M, Chyi Y-S, Romero-Severson J, Owen JL (1994). Amplification of DNA markers from evolutionarily diverse genomes using single primers of simple-sequence repeats. Theor. Appl. Genet. 89(7-8):998-1006.

Hall TA (1999). BioEdit: a user-friendly biological sequence alignment editor and analysis program for Windows 95/98/NT. Nucleic Acids Sym. Ser. 41:95-98.

Menezes JP, Lupatini M, Antoniolli ZI, Blume E; Junges E, Manzoni CG (2010). Genetic variability in rDNA ITS region of *Trichoderma* spp. (biocontrole agent) and *Fusarium oxysporum* f. *sp. chrysanthemi* isolates. Ciênc. Agrotec. 34(1):132-139.

Nagalakshmi RM, Kumari U, Boranayaka MB (2010). Assessment of genetic diversity in cowpea (*Vigna unguiculata*). Electron. J. Plant Breed. 1(4):453-461.

Pasquet RS (2000). Allozyme diversity of cultivated cowpea *Vigna* in assessing genetic variation among cowpea (*Vigna unguiculata unguiculata* (L.) Walp). Theor. Appl. Genet. 101(2):211-219.

Rohlf FJ (1998). NTSYSpc Numerical Taxonomy and Multivariate Analysis System Version 2.0 User Guide. Department of Ecology and Evolution State University of New York Stony Brook. P. 31.

Sokal RR, Sneath PHA (1963). Principles of numerical taxonomy. Freeman: San Francisco. P. 359.

Tamura K, Nei M, Kumar S (2004). Prospects for inferring very large phylogenies by using the Neighbor-Joining method. PNAS USA. 101(30):11030-11035.

Vieira ESN, Schuster I, Silva RB, Oliveira MAR (2009) Variabilidade genética em cultivares de soja determinada com marcadores microssatélites em gel de agarose. Pesq. Agropec. Bras. 44(11):1460-1466.

White TJ, Bruns T, Lee S, Taylor JW (1990). Amplification and direct sequencing of fungal ribosomal RNA genes for phylogenetics In: Innis MA, Gelfand DH, Sninsky JJ, White TJ.PCR Protocols: A Guide to Methods and Applications, (eds) Academic Press, Inc., New York. pp. 315-322.

Xavier GR, Martins LMV, Rumjanek NG, Freire Filho FR (2005). Cowpea genetic variability analyzed by RAPD markers. Pesq. Agropec. Bras. 40(4):353-359

Yadav S, Kaushik R, Saxena AK, Arora DK (2011). Diversity and phylogeny of plant growth-promoting bacilli from moderately acidic soil. J. Basic Microbiol. 51(1):98-106.

Yang W, Oliveira AC, Godwin I, Schertz K, Bennetzen JL (1996) Comparison of DNA marker technologies in characterizing plant genome diversity: variability in Chinese sorghums. Crop Sci. 36(6):1669-1676.

Zhanou A, Kossou DK, Ahanchede A, J., Zoundjihekpon J, Agbicodo E, Struik PC (2008). Genetic variability of cultivated cowpea in Benin assessed by random amplified polymorphic DNA. Afr. J. Biotechnol. 7(24):4407-4444.

19

Photosynthesis, ion accumulation, antioxidants activities and yield responses of different cotton genotypes to mixed salt stress

Guowei Zhang[1], Lei Zhang[2], Binglin Chen[3] and Zhiguo Zhou

Key Laboratory of Crop Physiology and Ecology in Southern China of Ministry of Agriculture, Nanjing Agricultural University, Nanjing, Jiangsu Province, 210095, P. R. China.

This study analyzes the effects of soil salinity on photosynthetic character, osmoregulation, content of pigment, $K^+/Na+$ ratio, lipoxygenase and antioxidants activities in functional leaves during the flowering and boll-forming stages of two cotton cultivars, namely, CCRI-44 (salt-tolerant) and Sumian12 (salt-sensitive), grown under different soil salinity conditions. In the control plants, non-significant differences were found in gas exchange, saturation irradiance (SI) and carotenoid (Car) content between the two cultivars. However, it showed higher K^+/Na^+ ratio, antioxidant enzyme activities, soluble sugar and protein contents, and lower chlorophyll (Chl) content and yield in CCRI-44. Salinity stresses remarkably increased soluble sugar and protein contents, lipoxygenase and the antioxidant activities, but decreased K^+/Na^+ ratio, Chl and Car contents, SI, photosynthetic capacities and yield, the extent being considerably larger in Sumian12 than CCRI-44. Although the soluble sugar, protein contents and the antioxidant activities of Sumian12 elevated more evidently under salt stresses, those variables never reached the levels of CCRI-44. Thus, CCRI-44 could maintain higher seed cotton yield than Sumian12 by sustaining higher osmoregulation and antioxidative abilities, which led to higher photosynthetic capacity. Hence, the salt-tolerant cotton cultivars could harmonize the relationship between CO_2 assimilation (source) and the seed cotton yield (sink) under the experimental conditions.

Key words: Cotton (*Gossypium hirsutum* L.), mixed salt stress, salt tolerance, photosynthesis, ion accumulation, antioxidant enzyme activity, seed cotton yield.

INTRODUCTION

Salinity is considered one of the major limiting factors for plant growth and agricultural productivity (Munns, 2005). Currently this stress is becoming even more prevalent as the intensity of land use increases in the world (Meloni et al., 2003). It reduces the growth of crops, at least partially, by leading to specific ion toxicity and enhances the generation of reactive oxygen species (ROS), which resulting in a decrease of photosynthetic capacity. The K^+/Na^+ ratio is important for the adjustment of cell osmoregulation, stomatal function, activation of enzymes, protein synthesis, oxidants metabolism, and photosynthesis

(Glenn et al., 1999). Normally, the K^+/Na^+ ratio tends to decrease under salinity stress as a result of either excessive Na^+ accumulation in plant tissue or enhanced K^+ leakage from the cell by activating K^+ efflux channels (Cuin and Shabala, 2007).

Although the adverse symptoms caused by salinity could be partially alleviated by implementing schemes, such as plastic film mulching (Dong et al., 2009), KNO_3 supply (Zheng et al., 2008), or foliar application of coronatine (Xie et al., 2008), the most promising strategy to overcome the problems of salty soil, is the use of salt-

Table 1. Selected physical and chemical properties of the soil used. Means ± SE (n = 3).

Characteristic	Value
Bulk density (g cm^{-3})	1.25 ± 0.09
Field water capacity (%)	28.6 ± 0.19
pH (H$_2$O)	7.3 ± 0.04
Organic matter (OM, g kg^{-1})	15.3 ± 0.14
Total nitrogen (N, g kg^{-1})	0.96 ± 0.08
Available phosphorus (Olsen-P, mg kg^{-1})	32.43 ± 0.51
Available potassium (K, mg kg^{-1})	202 ± 4.83
ECe (dS m^{-1})	1.25 ± 0.02

tolerant species. Thus, investigating the metabolism characteristics of salinity tolerant genotypes will be extremely important both for understanding the mechanism of salinity tolerance and selecting salt-tolerant cultivars.

Cotton is one of the most important economic crops in China. In the Yangtze River Valley, one of the largest cotton-growing areas in China, high temperature and evapotranspiration in summer and autumn, inadequate water management as well as saltwater intrusion close to the coastal area has contributed to an increase in soil salinity. Although cotton is classified as a salt-tolerant crop, its tolerance of salinity is not only limited, but also varied according to genotypes (Ashraf and Ahmad, 2000). Several past studies have been done analyzing physiological variations at seedling stage under either hydroculture or sand culture by NaCl treatment (Yang et al., 2010; Zhang et al., 2011). However, less information is available about the responses of different cotton genotypes grown in saline field soil (the major solutes comprising dissolved mineral salt are the cations Na$^+$, Ca^{2+}, Mg^{2+}, and K$^+$ and the anions Cl$^-$, SO$_4^{2-}$, HCO$_3^-$, CO3^{2-}) at flowering and boll-forming stage which is the key yield determinant period of cotton. The physiological responses caused by mixed salt stress differ from only NaCl stress since there is a significant interaction among ions (Ashraf and Ahmad, 2000; Chen et al., 2010). Thus, a careful study on the responses of cotton to mixed salt stress at this stage is urgently needed.

In order to understand why salt-tolerant cotton could relieve the saline adverse effects and obtain higher grain yield than the salt-sensitive one in saline field, two cotton genotypes (CCRI-44 (salt-tolerant) and Sumian12 (salt-sensitive)) were treated with various mixed salts (NaCl, MgCl$_2$, Na$_2$SO$_4$, MgSO$_4$, CaCl$_2$, Na$_2$CO$_3$, and NaHCO$_3$ at even molar ratio) levels (1.25, 5.80, 9.61, 13.23, and 14.65 dS m^{-1}). Photosynthetic capacities, K$^+$/Na$^+$ ratios, soluble sugars, and proteins, and antioxidant activities of both cultivars were measured at 70 days after transplanting (when all the cotton plants were at flowering and boll-forming stage). The objectives were to compare the differential responses of two genotypes to mixed salt stress at flowering and boll-forming stage, and try to

demonstrate a determinant of growth and yield determination under mixed salt conditions.

MATERIALS AND METHODS

Plant materials and growth conditions

Pot experiments were conducted in the summer of 2009 in a greenhouse at the Pailou experimental station of the Nanjing Agricultural University, located at Nanjing (32°02'N and 118°50'E), Jiangsu Province of China. The minimum and maximum air temperature was 19 and 34°C, respectively. The relative humidity ranged from 40 to 68%. Cotton (*Gossypium hirsutum* L.) cultivars planted were CCRI-44 (salt-tolerant) and Sumian12 (salt-sensitive) (Zhang et al., 2011), which are grown widely in the Yangtze River Valley in China. Cotton seeds were planted on 25 April 2009.

When the seedlings had three true leaves, individual healthy, uniform plants were transplanted into plastic pots, 50 cm high and 33 cm diameter, filled with 30 kg air dry soil. The yellow brown soil collected from the 0 to 30 cm topsoil layer from the experiment station was passed through a 2 mm sieve and packed in the pots. Selected physical and chemical properties of the soil are presented in Table 1. For each cotton cultivar, there were five salinity treatments. Seven kinds of salts (sodium carbonate, sodium bicarbonate, sodium chloride, calcium chloride, magnesium chloride, magnesium sulfate and sodium sulfate) were mixed into natural dried, sieved, selected soils at an even molar ratio before the experiment, forming soils with five levels of salinity (ECe, electrical conductivity of 1:5 soil/water extract), (CK) with 1.25 dS m^{-1} soil salinity, (S1) with 5.80 dS m^{-1} soil salinity, (S2) with 9.61 dS m^{-1} soil salinity, (S3) with 13.23 dS m^{-1} soil salinity, and (S4) with 14.65 dS m^{-1} soil salinity, respectively. The resulting ion compositions in the treated soil were similar to those observed in the local Coastal saline soil. Before filling the pots, 4.5 g N, 0.36 g P$_2$O$_5$, and 0.9 g K$_2$O per pot was applied into the soil. Another 4.5 g N per pot was top-dressed at the early flowering stage (35 days after transplanting) in every treatment.

The experiment was arranged in a completely random design, each treatment had 20 replications and one pot with single plant represented one replication.

Gas exchange and irradiance response

Simultaneous measurements of gas exchange and irradiance response were taken on functional leaves (4th leaf from top) at flowering and boll-forming stage (70 days after transplanting), with a portable open-flow gas exchange system LI-6400 (LI-COR Biosciences, Lincoln, USA). Photosynthetic rate (P_N), stomatal conductance (Gs) were measured at PAR of 1500 µmol m^{-2} s^{-1} of internal light source, 65±5% RH, 32±2°C leaf temperature and 380 µmol mol^{-1} CO_2 concentration.

The irradiance response curve was recorded automatically in the same leaf by means of operation program. During irradiance response measurements, CO_2 concentration was maintained at 380 µmol mol^{-1}. Photosynthesis versus PAR of 1 800, 1 600, 1 400, 1 200, 1 000, 800, 600, 400, 200, and 0 µmol m^{-2} s^{-1} was measured. Each PAR step lasted 3 min. The data obtained for each leaf were analyzed with the program photosynthetic assistant (Version 1.1, Dundee Scientific, Dundee, UK) to obtain saturation irradiance (SI).

Pigment analyses

After determined the gas exchange and irradiance response, six leaves (4th leaf from top) were harvested. Half of the samples were immediately frozen in liquid N_2 and stored at -70℃ for enzyme activity analysis; others were dried in an oven at 80°C to a constant weight. Frozen leaf samples (0.1 g) were extracted with pure acetone in the dark for 48 h at 4°C in order to guarantee the complete extraction of pigment from leaf. The concentration of Chlorophyll (Chl) and carotenoid (Car) was determined spectrophotometrically according to Shabala et al. (1998).

Determination of soluble sugar content

Soluble sugars were determined based on the method of phenolsulfuric acid (Dubois et al., 1956). Frozen leaf samples (0.3 g) were homogenized with deionized water, extract was filtered and the extract treated with 5% phenol and 98% sulfuric acid, mixture remained for 1h and then absorbance at 485 nm was determined by spectrophotometer (UV-2401, Shimadzu Corporation, Japan). Content of soluble sugar was determined by using glucose as standard.

Determination of Lipoxygenase activity

Lipoxygenase (LOX) was assayed spectrophotometrically at 234 nm according to You et al. (2009). Frozen leaf samples (0.2 g) were crushed into fine power in a mortar in an ice-bath. 5.0 ml of 0.1 mol L^{-1} phosphate buffer (including 0.5 mmol L^{-1}PMSF, 0.6 mmol L^{-1} EDTA, pH 7.0) was used as an extraction buffer. The homogenate was centrifuged at 15,000 × g for 15 min at 4°C, then the supernatants were used to measure, LOX activity was analyzed in 2.8 ml of 0.1 mol L^{-1} phosphate buffer containing 0.1 ml of 100 mmol L^{-1} sodium linoleate. The increase in absorbance at 234 nm was recorded after adding 0.2 ml enzyme extract.

Determination of antioxidant enzymes activity and soluble proteins and malondialdehyde (MDA) Content

Frozen leaf samples (0.3 g) were crushed into fine powder in a mortar in an ice-bath. To each sample, 5 ml of 0.05 mol l^{-1} phosphate buffer (pH 7.8) with 1% Poly venyl pyrrolidone (PVP) was used as an extraction buffer. The homogenate was centrifuged at 15000 g for 15 min at 4°C. The supernatant was used to measure protein and malondialdehyde (MDA) and antioxidant enzyme activities.

Protein content was determined according to Bradford (1976) with bovine serum albumin as the standard. Superoxide dismutase (SOD) activity was determined according to the method of Foster and Hess (1980): One unit of SOD activity was defined as the amount of enzyme required to cause 50% inhibition of nitro blue tetrazolium (NBT) reduction, measured with a spectrophotometer (UV-2401, Shimadzu Corporation, Japan) at 560 nm. Catalase (CAT) activity was determined by potassium permanganate titration (Giannopolitis and Ries, 1977). The action mixture contained 2.9 ml of 50 mM phosphate buffer (pH 7.0), 1.0 ml of 10 mM H_2O_2, and 100 ml of enzyme extract in tubes. Peroxidase (POD) activity was analysed in 2.9 ml of 0.05 mol l^{-1} phosphate buffer containing 1.0 ml of 0.05 mol l^{-1} guaiacol and 1.0 ml of 2% H_2O_2 (Tan et al., 2008). The increase in absorbance at 470 nm was recorded after adding 2.0 ml of 20% chloroacetic acid. Content of MDA content was made according to the method of Du and Bramlage (1992). Determination of K$^+$, Na$^+$ and Cl$^-$ contents.

The content of Na$^+$ and K$^+$ were determined in the extract obtained after digestion with HNO_3:$HClO_4$ (10:1, v/v) of dry leaf powder (0.2 g) by atomic absorption spectrophotometer (Liu et al., 2010). Cl$^-$ content was determined after treated with 10 ml of deionized water at 100°C for 60 min by ion chromatography (DX-300, Sunnyvale, CA, USA) (Liu et al., 2010). Growth parameters and yield components.

The measurements of plant height, leaf area and leaf water content were performed soon after the measurements of gas exchange and irradiance response. Leaf area per plant was measured with a portable area meter (Li 3000-A, Li-COR Inc., NE, USA). From fresh weight (FW) and dry weight (DW) measurements, leaf water contents ((FW-DW)/FW×100%) were obtained. Number of boll was counted at boll-opening stage. Boll weight and seed cotton yield were obtained after harvest. Statistical analysis.

Each determination was carried out with three replicates. Statistical treatment of the data was performed by One-way ANOVA method. Differences between means were established using a Duncan test ($p<0.05$). For these analyses SPSS 11.0 software (SPSS, Chicago, IL, USA) was used.

RESULTS

Gas exchange and SI Non-significant differences ($p<0.05$) were observed in net photosynthetic rate (P_N) between CCRI-44 and Sumian12 in CK plants (Figure 1). Mixed salt stress remarkably reduced the P_N of both cultivars, with the largest reductions occurring under S4 treatment. Such adverse effects were more serious in Sumian12 than in CCRI-44. In CCRI-44, P_N was reduced by 3.4, 13.6, 34.2 and 43.9% relative to CK when plants were under S1, S2, S3 and S4 treatments, respectively, whereas in Sumian12, P_N decreased by 5.4, 23.1, 48.6 and 50.1%, respectively. The responding trends of stomatal conductance (g_S) to mixed salt stress were consistent with P_N. The reductions of g_S were always smaller in Sumian12 than in CCRI-44 relative to CK. P_N and g_S in CCRI-44 were consistently higher than in Sumian12 under all treatments. Distinct PAR-P_N curves were established after analyzing the data set measured in different treatments (Figure 2). For CCRI-44, two PAR-P_N curves monitored under S1and S2 treatments were closed to CK, with SI of about 1500 µmol m^{-2} s^{-1}. However, the slope of PAR-P_N curve was much lower under S3 and S4 treatments, with SI of about 1300 and 1200 µmol m^{-2} s^{-1}. The extent of variation was larger in Sumian12 than CCRI-44. Two PAR-P_N curves of Sumian12 measured under CK and S1 treatment were

Table 2. Contents of soluble sugars and proteins (g kg^{-1}), K$^+$, and Na$^+$ (g kg^{-1}) and K$^+$/Na$^+$ ratio in functional leaves of salt-tolerant CCR-44 and salt-sensitive Sumian12 under salt treatments at flowering and boll-forming stage. Means±SE (n = 3).

Cultivar	Treatments	Soluble sugars	Soluble proteins	Na$^+$	K$^+$	Cl$^-$	K$^+$/Na$^+$
CCRI-44	CK	87.83 ± 2.51^c	8.90 ± 0.46^c	5.36 ± 1.03^d	16.46 ± 0.98^a	18.81 ± 0.93^e	3.14 ± 0.51^a
	S1	89.76 ± 3.26^c	9.10 ± 0.65^c	7.47 ± 0.71^d	16.11 ± 0.79^{ab}	22.15 ± 1.12^d	2.17 ± 0.31^b
	S2	94.10 ± 4.03^c	11.12 ± 0.48^b	13.16 ± 1.75^c	14.78 ± 0.95^b	29.22 ± 1.08^c	1.13 ± 0.21^c
	S3	113.90 ± 3.33^b	13.28 ± 0.52^a	15.48 ± 1.21^b	12.75 ± 0.91^c	37.46 ± 1.50^b	0.82 ± 0.09^{cd}
	S4	120.78 ± 4.23^a	12.57 ± 0.30^a	18.74 ± 1.02^a	9.51 ± 0.56^d	42.23 ± 2.16^a	0.51 ± 0.11^d
Sumian12	CK	74.37 ± 1.85^e	7.99 ± 0.21^e	5.89 ± 0.56^d	15.55 ± 0.70^a	18.12 ± 1.03^e	2.75 ± 0.20^a
	S1	79.49 ± 2.37^d	8.69 ± 0.14^d	10.07 ± 1.46^c	14.97 ± 0.51^a	24.04 ± 0.74^d	1.57 ± 0.15^b
	S2	85.78 ± 1.89^c	9.96 ± 0.33^c	14.88 ± 1.27^b	12.40 ± 0.46^b	33.84 ± 2.04^c	0.84 ± 0.07^c
	S3	102.75 ± 3.52^b	11.39 ± 0.38^b	16.39 ± 0.79^b	10.47 ± 0.86^c	42.05 ± 1.62^b	0.65 ± 0.11^c
	S4	115.56 ± 3.41^a	12.06 ± 0.22^a	21.15 ± 1.55^a	8.36 ± 0.69^d	46.74 ± 2.34^a	0.39 ± 0.09^d

Different letters in the same column indicate significant difference ($p<0.05$).

close, with SI of about 1300 µmol m^{-2} s^{-1}. The SI of Sumian12 was about 1300, 1100 and 1000 µmol m^{-2} s^{-1} in S2, S3 and S4 treatment, respectively.

Contents of pigment

Chlorophyll (Chl) content of CCRI-44 was lower (by 5 %) than that of Sumian12 under CK (Figure 3). Mixed salt stress caused decreases in Chl contents of both cultivars. However, the reductions in CCRI-44 were lower than those in Sumian12. In fact, the Chl contents of CCRI-44 were 7.2, 7.6, 13.2 and 8.6% higher than Sumian12 under S1, S2, S3 and S4 treatments, respectively. There was no significant difference in Car content between the two varieties under CK condition. Car content in Sumian12 kept on decreasing with increasing salinity levels while CCRI-44 maintained a steady content under S1 and S2 treatments compared with CK, however, a considerable decrement was observed in S3 and S4 treatments.

Contents of soluble sugars, and proteins

As important osmoregulatory compounds, soluble sugars and proteins accumulated significantly in both cultivars under salinity stresses (Table 2). Soluble sugar content of CCRI-44 was considerably higher (by 18.2%) than that of Sumian12 in CK plants. Non-significant changes were noted in CCRI-44 under S1 and S2 treatments compared with CK, however, considerable increment were observed in S3 and S4 treatments. Although the increments in Sumian12 were more significant considerable than those in CCRI-44 under mixed salt stress, they never reached the high levels of CCRI-44. The variation tendency of the soluble protein contents was similar to the trends of soluble sugar contents.

Contents of K$^+$, Na$^+$ and Cl$^-$

Even in CK plants, the Na$^+$ content was remarkably higher and K$^+$ was considerably lower, resulting in a higher K$^+$/Na$^+$ ratio in CCRI-44 than Sumian12 (Table 2). Mixed salt stress decreased K$^+$ content, but increased Na$^+$ content, with reduced K$^+$/Na$^+$ ratio in both cultivars. However, the variable extent of Na$^+$ and K$^+$ content was smaller in CCRI-44 than that in Sumian12, especially in CK and S1 treatments. With further increase of salinity level, the K$^+$ content and the K$^+$/Na$^+$ ratio decreased significantly in both cultivars, but they were still higher in CCRI-44 than Sumian12. There was no significant difference in Cl$^-$ content between the two varieties under CK condition. Mixed salt stress resulted in a higher accumulation of Cl$^-$ in CCRI-44 than Sumian12.

Antioxidant enzyme and lipoxygenase (LOX) activities, MDA content

Figure 4 shows that, the activity of SOD was significantly higher (by 23.1%) in CCRI-44 than that in Sumian12 even in CK plants. Mixed salt stress remarkably elevated the activities of those antioxidant enzymes in both cultivars. Although the SOD activities of Sumian12 increased considerably before the salinity level approached S4 (14.65 dS m^{-1}), they were always lower than those of CCRI-44.

POD activity in CCRI-44 was lower (by 10.8%) than Sumian12 in CK. POD activity increased with increasing salinity levels only in CCRI-44, whereas in Sumian12, it remained constant under all salinity levels. In fact, POD activity was higher in CCRI-44 than that in Sumian12 under all salinity levels (Figure 4).

Non-significant differences ($p<0.05$) were observed in CAT activity between CCRI-44 and Sumian12 in CK plants. A continuous increase in CAT activity was associated with increased salinity in CCRI-44. However,

Table 3. Plant height (cm), leaf area (m^2 plant^{-1}), leaf water content(%) and yield components, *i.e.* number of boll per plant, boll weight (g) and seed cotton yield (g plant^{-1}) in salt-tolerant CCRI-44 and salt-sensitive Sumian12 under control conditions (CK) or increased salinity levels (S1, S2, S3 and S4). Means ± SE (*n* = 3).

Cultivar	Treatments	Plant height	Leaf area	water content	Number of bolls	Boll weight	Seed cotton yield
CCRI-44	CK	111.79.83 ± 4.01[a]	0.45 ± 0.02[a]	83.81 ± 0.43[a]	19.40 ± 1.81[a]	4.41 ± 0.27[a]	86.74 ± 5.05[a]
	S1	106.27 ± 3.03[a]	0.43 ± 0.02[a]	83.54 ± 0.28[a]	18.20 ± 1.10[a]	4.30 ± 0.24[ab]	80.21 ± 4.73[a]
	S2	83.74 ± 2.03[b]	0.36 ± 0.02[b]	82.04± 0.44[b]	15.40 ± 1.51[b]	3.92 ± 0.18[bc]	63.20 ± 4.03[b]
	S3	63.88 ± 3.50[c]	0.28 ± 0.01[c]	81.06 ± 0.26[c]	10.80 ± 1.30[c]	3.72 ± 0.24[c]	41.30 ± 3.53[c]
	S4	51.71 ± 2.52[d]	0.22 ± 0.01[d]	78.74 ± 0.24[d]	8.60 ± 1.14[d]	3.20 ± 0.17[d]	26.73 ± 3.80[d]
Sumian12	CK	117.18.37± 2.02[a]	0.47 ± 0.03[a]	83.14 ± 0.38[a]	19.80 ± 2.04[a]	4.56 ± 0.20[a]	90.11 ± 7.17[a]
	S1	98.24 ± 3.52[b]	0.40 ± 0.02[b]	82.83 ± 0.50[a]	17.60 ± 1.14[b]	4.17 ± 0.30[b]	73.80 ± 5.68[b]
	S2	75.07 ± 2.60[c]	0.29 ± 0.02[c]	81.52 ± 0.39[b]	12.20 ± 1.48[c]	3.63 ± 0.21[c]	41.16 ± 4.68[c]
	S3	54.75 ± 2.54[d]	0.24 ± 0.02[d]	80.25 ± 0.39[c]	8.80 ± 1.10[d]	3.34 ± 0.31[cd]	28.58 ± 2.62[d]
	S4	48.46 ± 2.50[e]	0.20 ± 0.01[e]	78.57 ± 0.33[d]	7.40 ± 1.14[d]	3.02 ± 0.09[d]	21.96 ± 2.21[d]

Different letters in the same column indicate significant difference (*p*<0.05).

for Sumian12, CAT activity increased under S1, S2 and S3 treatments followed by a decline under S4 treatment. Anyway, CAT activity was higher in CCRI-44 than that in Sumian12 under all salinity levels (Figure 4).

LOX activity and MDA content increased gradually with increasing salinity levels in both cultivars (Figure 5). When salinity level was lower than S1, they did not show significant difference between CCRI-44 and Sumian12, while higher than S1, LOX activity and MDA content in CCRI-44 were markedly higher than that in Sumian12.

Growth and yield component

Plant height and leaf area were lower, but leaf water content was higher in CCRI-44 than Sumian12 under CK treatment (Table 3). Mixed salt stress caused decreases in the above mentioned parameters of both cultivars. However, the extents of reduction were larger in Sumian12 than those in CCRI-44. Number of bolls, boll weight and seed cotton yield of CCRI-44 were lower than those of Sumian12 under CK treatment. However, salt-induced decreases in yield components of CCRI-44 were lower than those of Sumian12, especially at S2 and S3 treatment. There was no significant change in seed cotton yield of CCRI-44 under S1 treatment compared with CK, whereas it decreased significantly (being 19.2% lower than that in CK) in Sumian12. Nevertheless, the seed cotton yield of both cultivars decreased significantly under higher salinity. Although the yield of CCRI-44 was lower by 4.4%) than that of Suamin12 in CK, it was higher than that of Suamin12 under each salt treatment.

DISCUSSION

Previous studies have shown that, CCRI-44 could maintain higher net photosynthetic rate and dry mass than Sumian12 after stressed by 150 mmol L^{-1} NaCl at seeding stage (Zhang et al., 2011). According to Kao et al. (2006) and Moradi and Ismail (2007), the relatively higher salt-tolerant species would have less reduced in P_N, our results proved that salinity induced less reduction in P_N, leaf area and seed cotton yield (Figure 1 and Table 3) conferred CCRI-44 higher salt-tolerance at flowering and boll-forming stage. Significant positive correlations existed between P_N and g_S (Ma et al., 2006). Mixed salt stress drastically reduced g_S, this might be attributed to the lower leaf water potential and a reduction in leaf water content, which resulted in loss of turgor, which leads to reduced photosynthetic rate. However, the salt-tolerant cotton cultivar CCRI-44 performed higher ability in maintaining leaf water content (Figure 3). Thus, it could maintain higher g_S and P_N under salinity.

The value of SI reflects the efficiency of photon energy utilization (Evans et al., 1993). SI of the salt-sensitive cotton cultivar Sumian12 decreased more significantly than that of salt-tolerant cotton cultivar CCRI-44 under salinity stress (Figure 2). Several studies suggested that photon energy efficiency mostly determines the ability of photosystem II, which plays a key role in response of photosynthesis to salinity stress (Xu et al., 1995; Didenko and Suslick., 2002). The salt-tolerant cotton cultivar CCRI-44 had higher ability in catching photon energy which might be attributed to lower salt-induced degradation of CP43 or by higher salt-enhanced synthesis of D1 protein than those in Sumian12 (Sairam et al., 2002).

The decrease in Chl and Car content under mixed salt stress in both cotton cultivars were consistent with those results obtained from results by NaCl treatment at seeding stage (Meloni et al., 2003; Yang et al., 2010; Zhang et al., 2011), which could be attributed to increased activity of the chlorophyll-degrading enzyme

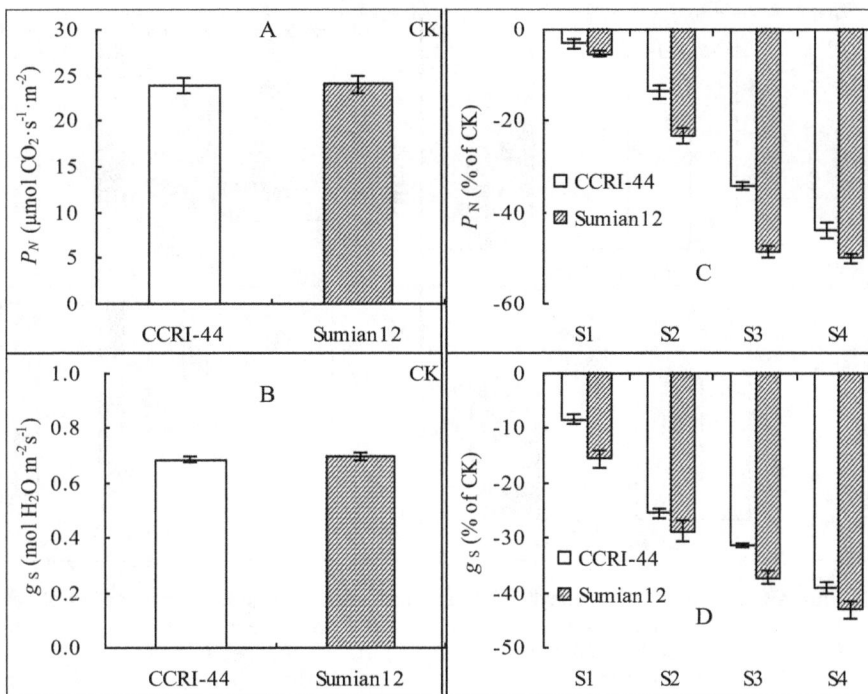

Figure 1. Net photosynthetic rate (P_N) and stomatal conductance (g_S) in functional leaves of salt-tolerant CCRI-44 and salt-sensitive Sumian12 at flowering and boll-forming stage. (A, B) Plants grown under control conditions (CK), (C, D) changes of contents (%) under S1, S2, S3 and S4 in relation to CK, Vertical bars indicate SE (n = 3).

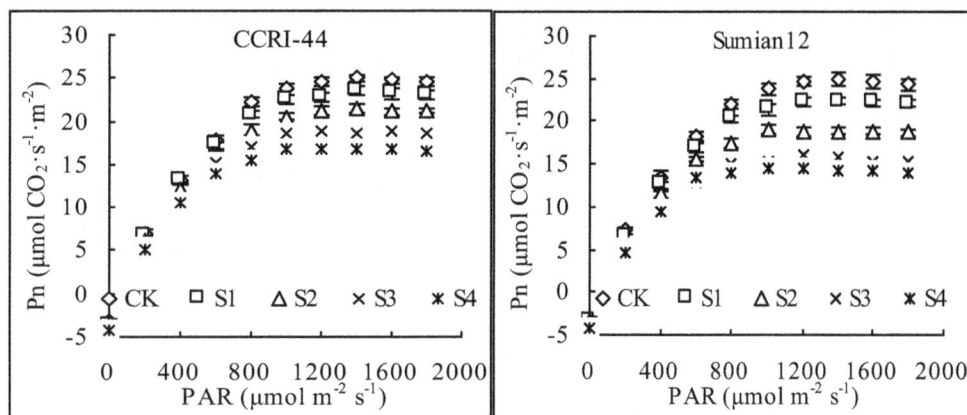

Figure 2. Functional relationship between net photosynthetic rate (P_N) and photosynthetically active radiation (PAR) in functional leaves of salt-tolerant CCRI-44 and salt-sensitive Sumian12 at flowering and boll-forming stage. Error bars show S.E., n=3.

such as chlorophyllase and Mg-chelatase, and toxic ion accumulation in leaves (Garcia-Sanchez et al., 2002). Parida and Das (2005) reported Chl content as one of the parameters of salt tolerance in crop plants. Car is responsible for quenching of singlet oxygen (Koyro, 2006), the decrease in Car under salinity stress leads to degradation of β-carotene and formation of zeaxanthins, which are apparently involved in protection against photoinhibition (Sultana et al., 1999). The salt-tolerant cotton cultivar CCRI-44 retained higher Chl and Car content than salt-sensitive cotton cultivar Sumian12 under salinity stress. Hence, their comparative levels in Chl and Car may determine its relative tolerance.

Salinity stress can impair plant growth by specific ion

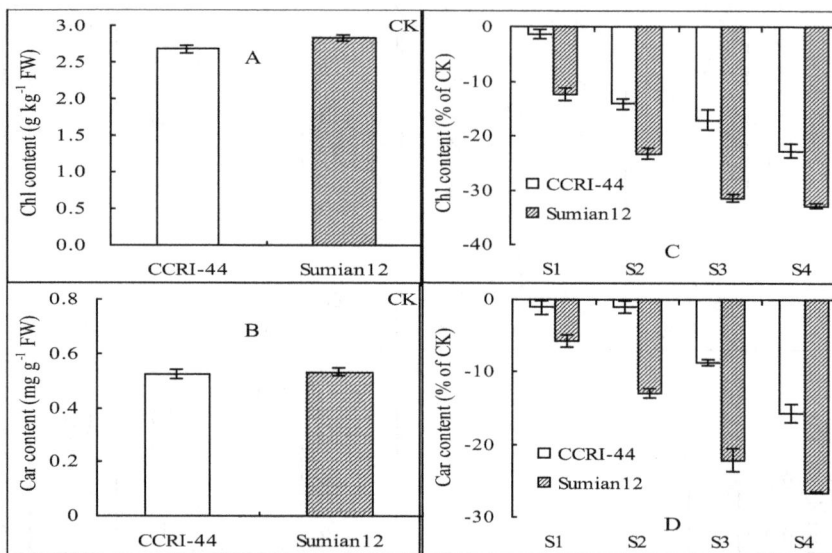

Figure 3. Content of Chlorophyll (Chl) and carotenoid (Car) in functional leaves of salt-tolerant CCRI-44 and salt-sensitive Sumian12 at flowering and boll-forming stage. (A, B) Plants grown under control conditions (CK), (*C, D*) changes of contents (%) under S1, S2, S3 and S4 in relation to CK, Vertical bars indicate SE (*n* = 3).

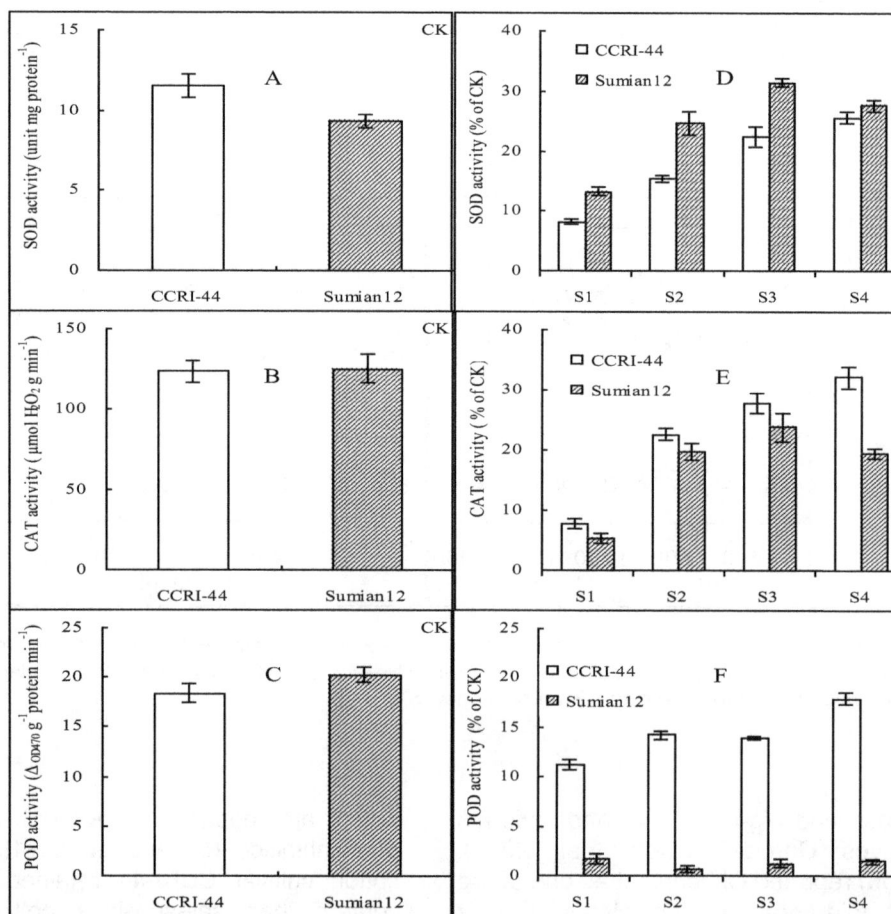

Figure 4. Superoxide dismutase (SOD), peroxidase (POD), and catalase (CAT) activities in functional leaves of salt-tolerant CCRI-44 and salt-sensitive Sumian12 at flowering and boll-forming stage. (A-C) Plants grown under control conditions (CK), (*D-F*) changes of contents (%) under S1, S2, S3 and S4 in relation to CK, Vertical bars indicate SE (*n = 3*).

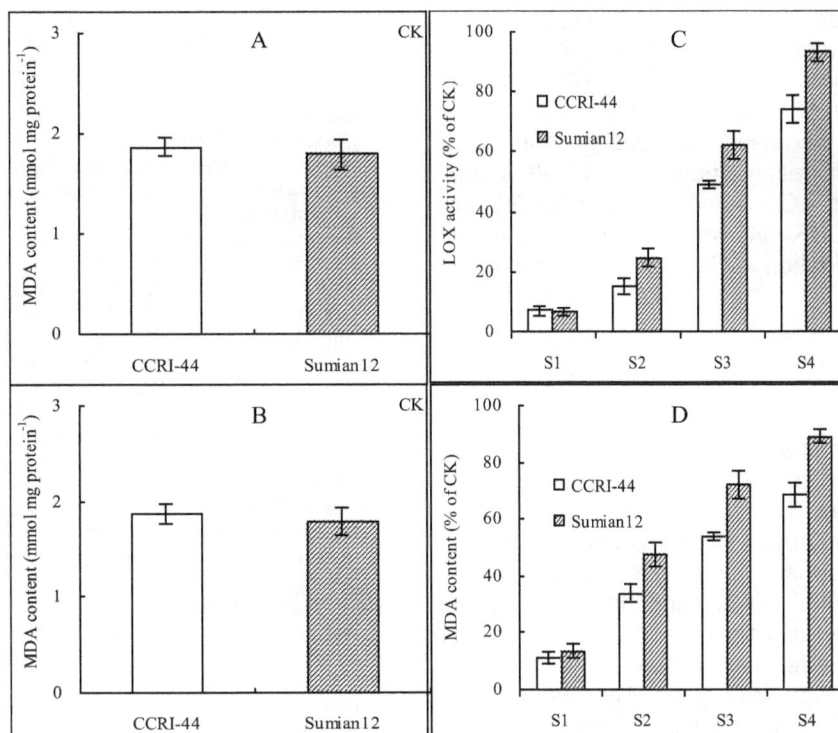

Figure 5. Malondialdehyde (MDA) content and lipoxygenase (LOX) activities in functional leaves of salt-tolerant CCRI-44 and salt-sensitive Sumian12 at flowering and boll-forming stage. (A, B) Plants grown under control conditions (CK), (*C, D*) changes of contents (%) under S1, S2, S3 and S4 in relation to CK, Vertical bars indicate SE (*n* = 3).

toxicity (Munns, 2005). In this study, Na^+ and Cl^- content was increased in both cultivars with increasing salinity levels, while K^+ content and K^+/Na^+ ratio were decreased (Table 2). These results were similar with that treated with NaCl (Ashraf and Ahmad, 2000) or mixed salts of NaCl and $CaCl_2$ (1:1 weight ratio) (Chen et al., 2010) in cotton. High Na^+ and Cl^- content may change composition and function of thylakoid membrane (Wang et al., 2008). Deficiency of K^+ may decrease activities of photosynthetic enzyme (Mahajan and Tuteja, 2005).

All these mentioned could decrease the stability of photosystem II reaction center. So maintaining a lower Na^+ and Cl^- content and higher K^+ content might be one of the reasons accounting for a high level of P_N in cotton under salinity stress. In addition, the K^+/Na^+ ratio has been used as a nutritional indicator to select salt-tolerant plants (Glenn et al., 1999). The reason for more significant decrease of K^+/Na^+ ratio in salt-sensitive cotton cultivar Sumian12 than that in salt-tolerant cotton cultivar CCRI-44 might be that, the K^+ selectivity of cell membrane in CCRI-44 was better than that in Sumian12, which was testified by always higher K^+ content in CCRI-44 than that in Sumian12, even in CK plants (Table 2). Soluble sugar and protein plays an important osmotic role in plants (Parida and Das, 2005). Higher soluble sugar and protein accumulation in CCRI-44 than those in

Sumian12 indicated that CCRI-44 had higher ability of osmotic adjustment under salinity. A direct consequence of higher organ osmolyte concentration in CCRI-44 is the maintenance of comparatively higher water and pigment content in leaf. However, Paul and Pellny (2003) observed that, higher soluble sugar, especially glucose, might repress the photosynthesis in most plants by disturbing the balance of carbon to nitrogen balance.

Salinity stress causes oxidative stress by inhibiting the CO_2 assimilation, exposing chloroplasts to excessive excitation energy, which in turn increases the generation of ROS from triplet chlorophyll (Tan et al., 2008). These ROS are all very reactive and cause severe damage to membranes, DNA and proteins (Demidchik et al., 2003). To alleviation of oxidative stress, plants detoxify ROS by up-regulating antioxidative enzymes (Mandhania et al., 2006). Higher activities of antioxidant enzymes in salt-tolerant cotton cultivar CCRI-44 than in salt-sensitive cotton cultivar Sumian12 indicated that CCRI-44 had higher ability to eliminate free active radicals than Sumian12 (Figure 5). A sharp decline observed in antioxidative enzymes activities of Sumian12 in S4 treatment could be due to formation of ROS beyond the critical limit which might pose serious threat to plant cell, causing higher membrane lipid peroxidation (Kumar et al., 2010). In addition, lipoxygenase can catalyzed polyunsaturated fatty acid oxygenation, producing MDA. Mixed salt stress activated LOX activity in cotton leaves, as observed by Elkahoui et al. (2005) in *Catharanthus roseus* suspension cells. With increasing LOX activity, MDA content also increased, indicating that LOX catalyzed PUFA peroxidation under mixed salt stress. The MDA content and LOX activity in Sumian12 were higher than those in CCRI-44, which provided evidence of a higher lipid peroxidation in Sumian12 in comparison to CCRI-44.

Conclusion

In conclusions, mixed salt stress significantly inhibited the growth of both cotton cultivars by reducing their gas exchange and leaf area. The salt-tolerant cotton cultivar CCRI-44 was better equipped than salt-sensitive cotton cultivar Sumian12 in maintaining gas exchange, SI, pigment content, K^+/Na^+ ratio, and in mechanisms resistant to secondary oxidative stress under salinity stress. CCRI-44 could effectively relieve the inhibition of salt stress and obtain high grain yield. The result of this study was similar with previous studies obtained from treatment with NaCl, indicating that, Na^+ and Cl^- were still the main factors causing salinity stress under the mixed salt treatment.

ACKNOWLEDGEMENT

This work was supported by a grant from the National High Technology Research and Development Program of China (863 Program) (No. 2007AA10Z206).

REFERENCES

Ashraf M, Ahmad S (2000). Influence of sodium chloride on ion accumulation, yield components and fibre characteristics in salt-tolerant and salt-sensitive lines of cotton (*Gossypium hirsutum* L.). Field Crop Res. 66:115–127.

Chen WP, Hou ZN, Wu LS, Liang YC, Wei CZ (2010). Effects of salinity and nitrogen on cotton growth in arid environment. Plant Soil 326:61–73.

Cuin TA, Shabala S (2007). Compatible solutes reduce ROS induced potassium efflux in Arabidopsis roots. Plant Cell Environ. 30:875–85.

Demidchik V, Shabala SN, Coutts KB, Tester MA, Davies JM (2003). Free oxygen radical regulate plasmam embrane Ca^{2+} and K^+-permeable channels in plant root cells. J. Cell. Sci. 116:81–88.

Didenko YT, Suslick KS (2002). The energy efficiency of formation of photons, radicals and ions during single-bubble cavitation. Nature 418:394–397.

Dong HZ, Li WJ, Tang W, Zhang DM (2009). Early plastic mulching increases stand establishment and lint yield of cotton in saline fields. Field Crop Res. 111:269–275.

Du Z, Bramlage WJ (1992). Modified thiobarbituric acid assay for measuring lipid oxidation in sugar-rich plant tissue extracts. J. Agric. Food Chem. 40:1566–1570.

Dubois M, Gilles KA, Hamilton JK, Rebers PA, Smith F (1956). Colorimetric method for determination of sugars and related substances. Anal. Chem. 38:350–356.

Elkahoui S, Hernández JA, Abdelly C, Ghrir R, Limam F (2005). Effects

of salt on lipid peroxidation and antioxidant enzyme activities of *Catharanthus roseus* suspension cells. Plant Sci. 168:607–613.

Evans JR, Jakobsen I, Ögren E (1993). Photosynthetic light response curves. Planta 189:191–200.

Foster JG, Hess JL (1980). Responses of superoxide dismutase and glutathione reductase activities in cotton leaf tissue exposed to an atmosphere enriched in oxygen. Plant Physiol. 66:482–487.

Garcia-Sanchez F, Jifon JL, Carvaial M, Syvertsen JP (2002). Gas exchange, chlorophyll and nutrient contents in relation to Na^+ and Cl^- accumulation in 'Sunburst' mandarin grafted on different rootstocks. Plant Sci. 162:705–712.

Giannopolitis CN, Ries SK (1977). Superoxide dismutases: I. Occurrence in higher plants. Plant Physiol. 59:309–314.

Glenn E, Brown JJ, Blumwald E (1999). Salt tolerance and crop potential of halophytes. Crit. Rev. Plant Sci. 18:227–255.

Kao WY, Tsai TT, Tsai HC, Shih CN (2006). Response of three *Glycine* species to salt stress. Environ. Exp. Bot. 56:120–125.

Koyro HW (2006). Effect of salinity on growth, photosynthesis,water relations and solute composition of the potential cash crop halophyte *Plantago coronopus* (L.). Environ. Exp. Bot. 56:136–146.

Kumar M, Kumari P, Gupta V, Reddy CRK, Jha B (2010). Biochemical responses of red alga Gracilaria corticata (*Gracilariales, Rhodophyta*) to salinity induced oxidative stress. J. Exp. Mar. Biol. Ecol. 391:27–34.

Liu J, Guo WQ, Shi DC (2010). Seed germination, seedling survival, and physiological response of sunflowers under saline and alkaline conditions. Photosynthetica 48:278–286.

Ma QQ, Wang W, Li YH, Li DQ, Zou Q (2006). Alleviation of photoinhibition in drought-stressed wheat (*Triticum aestivum* L.) by foliar-applied glycinebetaine. J. Plant Physiol. 163:165–175.

Mandhania S, Madan S, Sawhney V (2006). Antioxidant defense mechanism under salt stress in wheat seedlings. Biol. Plantarum. 50(2):227–231.

Mahajan S, Tuteja N (2005). Cold, salinity and drought stresses: An overview. Arch. Biochem. Biophys. 444:139–158.

Meloni DA, Oliva MA, Martinez CA, Cambraia J (2003). Photosynthesis and activity of superoxide dismutase, peroxidase and glutathione reductase in cotton under salt stress. Environ. Exp. Bot. 49:69–76.

Moradi F, Ismail AM (2007). Responses of photosynthesis, chlorophyll fluorescence and ROS-scavenging systems to salt stress during seedling and reproductive stages in rice. Annu. Bot. 99:1161–1173.

Munns R (2005). Genes and salt tolerance: Bringing them together. New Phytol. 167:645–663.

Parida AK, Das AB (2005). Salt tolerance and salinity effects on plants: A review. Ecotoxicol. Environ. Saf. 60:324–349.

Paul MJ, Pellny TK (2003). Carbon metabolite feedback regulation of leaf photosynthesis and development. J. Exp. Bot. 54:539–547.

Sairam RK, Rao KV, Srivastava GC (2002). Differential response of wheat genotypes to long-term salinity stress in relation to oxidantive stress, antioxidant activity and osmolyte concentration. Plant Sci. 163:1037–1046.

Shabala SN, Shabala SI, Martynenko AI, Babourina O, Newman IA (1998). Salinity effect on bioelectric activity, growth, Na^+ accumulation and chlorophyll fluorescence of maize leaves: a comparative survey and prospects for screening. Aust. J. Plant Physiol. 25:609–616.

Sultana N, Ikeda T, Itoh R (1999). Effect of NaCl salinity on photosynthesis and dry matter accumulation in developing rice grains. Environ. Exp. Bot. 42:211–220.

Tan W, Liu J, Dai T, Jing Q, Cao W, Jiang D (2008). Alterations in photosynthesis and antioxidant enzyme activity in winter wheat subjected to post-anthesis waterlogging. Photosynthetica 46:21–27.

Wang RL, Hua C, Zhou F, Zhou QC, Zhou BW (2008). Effects of salt stress on fatty acid composition of thylakoid membrane of two rice cultivars differing in salt tolerance. Agric. Sci. Tech. 9:8–13

Xie Z, Duan L, Tian X, Wang B, Egrinya-Eneji A, Li Z (2008). Coronatine alleviates salinity stress in cotton by improving the antioxidative defense system and radical-scavenging activity. J. Plant Physiol. 165:375–384.

Xu Q, Paulsen AQ, Guikema JA, Paulsen GM (1995). Functional and ultrastructural injury to photosynthesis in wheat by high temperature during maturation. Environ. Exp. Bot. 35:43–54.

Yang SP, Wei CZ, Liang YC (2010). Effects of NaCl stress on the characteristics of photosynthesis and chlorophyll fluorescence at seedlings stage in different sea island cotton genotypes. Scientia Agric. Sin. 43:1585–1593.

You SZ, Yang HQ, Zhang L, Shao XJ (2009). Effects of cadmium stress on fatty acid composition and lipid peroxidation of *Malus hupehensis*. Chin. J. Appl. Ecol. 20:2032–2037.

Zhang GW, Lu HL, Zhang L, Chen BL, Zhou ZG (2011). Salt tolerance evaluation of cotton (*Gossypium hirsutum* L.) at its germinating and seedling stages and selection of related indices. Chin. J. Appl. Ecol. 22:2045–2053.

Zheng Y, Jia A, Ning T, Xu J, Li Z.,Jiang G. (2008). Potassium nitrate application alleviates sodium chloride stress in winter wheat cultivars differing in salt tolerance. J. Plant Physiol. 165:1455–1465.

20

Association of seed coat colour with germination of three wild mustard species with agronomic potential

J. O. Ochuodho[1] and A. T. Modi [2]

[1]School of Agriculture and Biotechnology, Chepkoilel Campus, Moi University, P. O. Box 1125-30100, Eldoret, Kenya.
[2]School of Agricultural Sciences and Agribusiness (Crop Science), University of KwaZulu-Natal, Private Bag X01, Scottsville 3209, Pietermaritzburg, South Africa.

Seed dormancy and germination present significant challenges when wild species are domesticated for cultivation and economic exploitation. Wild plant species are generally characterized by dormant seeds with variable germination widely spread over time. The objective of this study was to evaluate the influence of colour selection on seed germination of wild mustard (Brassicaceae) species that have been identified as wild edible leafy vegetable in South Africa. Seed lots were separated by colour and germinated in a completely randomized design (CRD) after chilling and after-ripening for 6 months. The light seed lot of cultivar *Kwayimba* (K) showed higher germination percentage than the dark seed lot of the same cultivar but colour selection did not improve the germination in cultivars *Isaha* (I) and *Maslahlisane* (M). The dark seed lot of K recorded the lowest germination percentage and the slowest germination rate. Chilling improved the speed of germination in wild mustards, but after-ripening had no effect. Seed colour change in wild mustards intensifies after physiological maturity and may be accompanied with weight increase or not. The seed coat colour may not be a good indication of the physiological status of the seed but together with physiological tests (germination) can give insight on the quality of a seed lot.

Key words: Seed colour, image analysis, seed germination.

INTRODUCTION

Diversification through domestication of locally adapted wild species of known nutrient quality has been suggested as more appropriate and could improve food security (Jansen van Rensburg et al., 2004). However, wild plant species are generally characterized by dormancy and variable germination behaviours that hinder or slow down their cultivation. The seeds show wide morphological variations including colour, and colour selection has been recommended as a quick method of improving seed quality.

Seed colour has been reported to play a role in seed dormancy and germination, as seeds attain their specific colour at physiological maturity (Powell, 1989; Ochuodho,

2005). Seed coat pigmentation and structure have been shown to influence germination (Debeaujon and Koornneef, 2000; Debeaujon et al., 2000). In *Arabidopsis*, structural mutants lacking some seed coat layers and those that showed less pigment impregnation were lighter and germinated better than wild types (Debeaujon et al., 2000). While dark coloured testa accompanied with slow water uptake was attributed to the presence of phenolic compounds and tight adherence of the seed coat to the embryo in legumes, some dark soybean cultivars whose seed coats are loosely attached to the embryo showed greater rate of imbibition and fast germination (Chachalis and Smith, 2000). There were no differences detected in the content of phenolic material between permeable and impermeable seed coats of the accessions with dark coloured seeds (Chachalis and Smith, 2001).

However, the permeable seed coat was shown to have high density of deep and wide pores. Atanassova et al. (2004) showed that, seeds of 3 anthocyaninless mutants in tomatoes germinated faster than the wild type. The inner epidermal testa layer in these mutants did not have condensed tannins that contributed to the rigidity of cell structure, thereby reducing permeability. Light coloured seeds of radicchio were shown to have reduced germination and as seed colour became darker, seeds showed higher, faster and more uniform germination (Pimpini et al., 2002). The colour difference resulted from incomplete seed development as evidenced by lower seed weight and smaller embryos. Coste et al. (2005) found good correlation between pod colours measured in hue angles by spectrophotometer and seed water content. In this way, the authors were able to estimate physiological maturity and separate the pods and seeds. It is known that, seed germination of weedy species is highly variable and our observation was that, seeds of different colours and colour intensity showed varying germination percentages. The aim of this study was to evaluate the association of seed colour with the germination in 3 wild mustard cultivars.

MATERIALS AND METHODS

Plant materials

Seeds of wild mustard were produced under rainfed field conditions at Umbumbulu, 65 km South of Pietermaritzburg and mature dry seeds were harvested when the pods were brown. Three local vegetables known as *Isaha* (I), *Kwayimba* (K) and *Maslahlisane* (M), were used. M and I were originally collected from KwaZulu-Natal, while K was obtained from Eastern Cape Province. The seeds were visually selected on the basis of colour difference while shining white light on the seeds placed on a dark background and divided into light and dark components (seed lots). This selection produced 6 seed lots, which were each packaged in separate paper bags. The seeds were germinated directly or after prechilling at 5 to 10°C for 5 days or after storage in dry rooms with controlled temperatures of 5 and 20°C for 6 months.

Image analysis

The seed lots were analysed with AnalySIS(R), computer software for soft image analysis that determines the exact quality of colour by the wavelength (hue or colour), the saturation (colour purity) and the intensity (brightness). Using the software 50 seeds (positions) were randomly chosen and focused on for colour determination within a sample of 400 seeds. The program quantified the colour and the pixel values obtained were averaged.

Seed germination

Four replications of 100 wild mustard seeds were placed in Petri dishes (10 cm diameter) on 3 Whatman No.1 filter papers moistened

with 7 ml of distilled water. The Petri dishes were arranged in a completely randomized design (CRD) and were incubated for 10 days in growth chambers (Labcon, LTGC 20-40; Johannesburg, South Africa) at alternating temperature 30/20°C day/night (8/16 h) (Ochuodho and Modi, 2005). Seeds were considered germinated when the radicle protruded and germination counts were performed daily in order to calculate mean germination time (MGT) according to De Villiers et al. (2002). Seedling evaluation was performed on the tenth day based on ISTA (2004) guidelines for the *Brassicas*; normal seedlings have primary root, hypocotyl and cotyledons with terminal buds intact. Figures were developed from germination data for 7 days because the results were not different with those for 10 day. Germination tests were carried out before and after visual selection of the seed lots.

RESULTS

Image analysis

It was difficult to distinguish the cultivars visually on colour basis and soft image analysis showed that, there were no significant differences between the 3 cultivars. However, on visual selection coupled with image analysis, the colours of the dark seed lots were not significantly different but cultivars K and I were closer (Figure 1). While the colours of the light seed lots of cultivars K and I were not different, they were significantly different ($p < 0.001$) from M. This colour difference was brought out clearly by both the hue and saturation but not colour intensity. The light and dark seed lots of cultivar K appeared different visually and were almost significantly different ($p < 0.05$) in hue. Table 1 showed that, 70% of the seeds of cultivars K and I were dark and heavier. On the contrary, the same number of seeds of the light and dark seed lots of cultivar M had the same weight but the proportion (in number) of the dark component was higher than of the light.

Seed germination

The germination percentage of cultivar I and M were not different ($p > 0.05$) but were significantly higher than the germination of K. Cultivar M showed the fastest germination having attained > 70% germination after 1 day. Colour selection did not change the germination percentage of the cultivars M and I but the rate and percent germination improved in the light seed lot of K (Figure 2). The light seed lot of K showed a high germination percentage similar to those of cultivars M and I.

Dark M showed the fastest germination having a MGT of 1.5 days while dark K showed slow germination speed of 3.5 days (Figure 3). On the other hand, the light seed lots of all the cultivars showed similar MGT of 2 days. Pre-chilling the seeds at 5 to 10°C for 5 days showed a significantly increase in the speed of germination (Figure 4) as all the seed lots showed > 40% germination after day one. However, after-ripening at 5 and 20°C did not

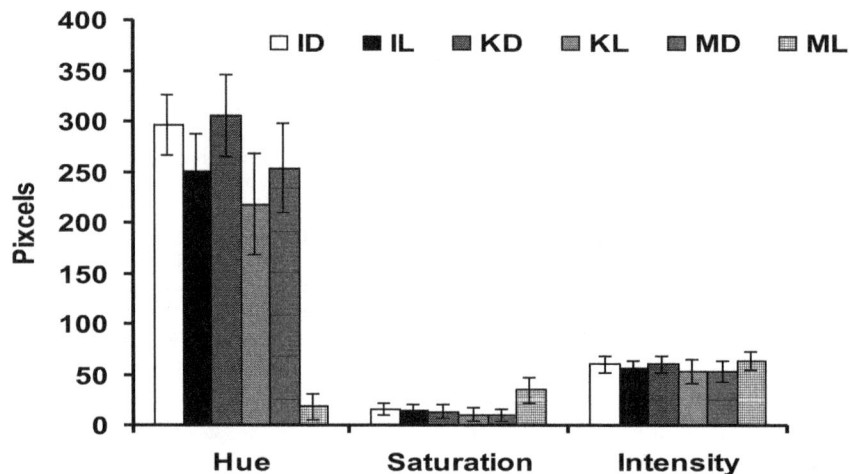

Figure 1. Colour analysis of seeds of wild mustard cultivars I, K and M by analysis after visual selection into light and dark coloured seed lots. ID, IL - dark and light I; KD, KL - dark and light K; MD, ML - dark and light M. The seed lot was spread out and between 50 points (seeds) analysed and the average obtained. The error bars represent SD, n = 50.

Table 1. Ten sub samples of 400 seeds were visually separated into light and dark colour components and seeds were counted [proportion (%)] and 100 seed weight obtained [Weight (g)].

Cultivar	Seed lots	Proportion (%)	Weight (g)
I	Light	28.3 ± 5.01	91.91 ± 5.78
	Dark	71.7 ± 5.32	103.19 ±0.60
K	Light	30.2 ± 4.96	85.85 ± 6.43
	Dark	69.8 ± 5.03	106.11±0.54
M	Light	44 ± 3.79	99.64 ± 2.92
	Dark	56 ± 3.44	100.29±0.87

The error values represent SD, n = 10.

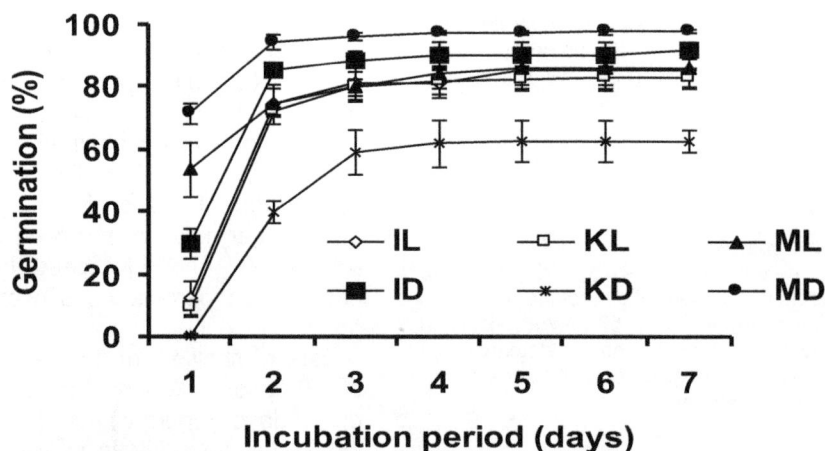

Figure 2. Germination percentage of the wild mustard cultivars I, K and M after visual selection into light (L) and dark (D) coloured seed lots. Seeds were incubated at alternating temperatures 20/30°C and 16/8 h, night/day, respectively. Seedling evaluation was done after 10 days. Error bars are standard deviation, n = 4.

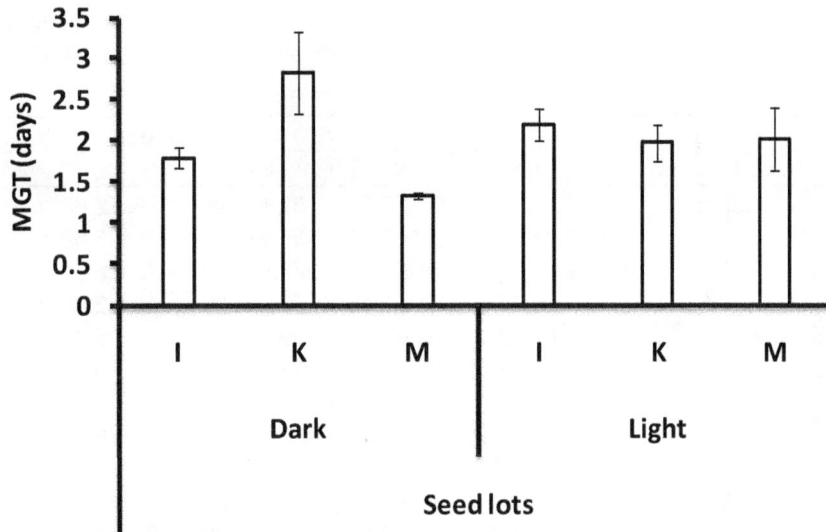

Figure 3. The speed of germination was computed as mean germination time (MGT) of the cultivars of wild mustard I, K and M before and after visual selection. The seeds were germinated at alternating temperatures 20/30°C and 16/8 h, night/day, respectively. Seeds that have germinated were counted every day for 10 days and MGT was calculated. Error bars represent SD, n = 4.

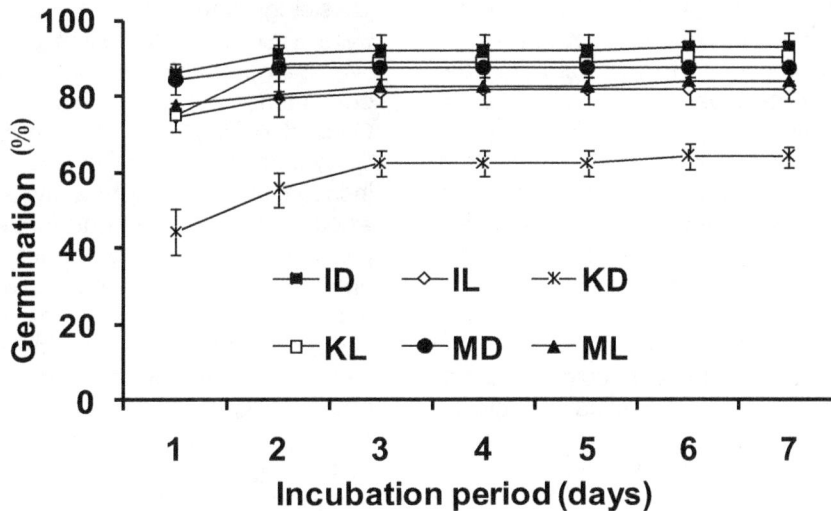

Figure 4. Germination of the seeds of cultivars of wild mustard I, K and M after visual selection into dark (D) and light (L) coloured seed lots. The seeds were germinated at alternating temperatures 20/30°C and 16/8 h, night/day, respectively after 5 days pre-chilling at 5 to 10°C in darkness. Seeds that have germinated were counted every day for 10 days.

improve the germination of cultivars M and I (Table 2) but by the fourth month there was marked decrease in the germination of cultivar K.

RESULTS AND DISCUSSION

It was not possible to separate the cultivars by colour with soft image analysis initially but after visual selection the hue and saturation were significantly different between the seed lots of cultivar M. While Coste et al. (2005) suggested that, hue angle measured on pods be used to monitor seed drying, hue (pixels) was used in this study to distinguish the seed lot colours. Hue is the measure of true colour (by wavelength) and this can be influenced by phenolics in *Arabidopsis* (Chachalis and Smith, 2000),

Table 2. Seed lots of wild mustard were stored in dry temperature controlled rooms at 5 and 20°C for six months and germinated at alternating temperatures of 20/30°C, 16/8 h night and day respectively. Germination percentage was scored daily for 10 days. Error values represent SD, n = 4.

Storage Temperature (°C)	Storage Period (month)	Seed lots					
		ID	IL	KD	KL	MD	ML
5	Control	93 ± 2.65	83 ± 5.57	57 ± 7.02	79 ± 2.65	95 ± 3.00	85 ± 2.52
	1	92 ± 3.42	81 ± 7.55	57 ± 6.19	79 ± 5.89	95 ± 3.30	83 ± 6.63
	2	96 ± 4.76	78 ± 7.83	40 ± 11.00	73 ± 5.74	97 ± 4.76	84 ± 8.16
	3	95 ± 2.58	88 ± 3.42	33 ± 5.03	66 ± 2.83	97 ± 2.52	88 ± 1.00
	4	90 ± 1.63	85 ± 8.70	32 ± 5.97	61 ± 5.74	92 ± 3.65	89 ± 3.42
	6	93 ± 1.91	82 ± 3.46	33 ± 7.72	63 ± 5.74	94 ± 1.03	89 ± 1.15
20	1	94 ± 3.65	78 ± 9.10	42 ± 14.84	71 ± 6.23	91 ± 3.00	87 ± 3.46
	2	93 ± 2.58	82 ± 1.91	45 ± 9.56	76 ± 2.58	96 ± 2.58	91 ± 4.76
	3	95 ± 2.52	85 ± 5.03	43 ± 4.16	73 ± 4.12	89 ± 3.00	90 ± 4.32
	4	95 ± 3.00	83 ± 3.83	37 ± 11.50	65 ± 9.59	94 ± 1.63	89 ± 4.76
	6	90 ± 3.27	79 ± 5.26	19 ± 3.46	58 ± 4.12	94 ± 2.83	87 ± 6.19

anthocyanins in soybean (Atanassova et al., 2004) or seed maturity stage in radicchio (Pimpini et al., 2002) and in *Cleome gynandra* (Ochuodho, 2005). On weight basis, the light and dark components of cultivar M were the same, while the dark component of cultivar K and I were heavier. These observations seem to suggest that, seed colour may not be a dependable determinant of the physiological state or maturity stage of these species.

The high germination percentage shown by the light and dark seed lots of M indicated that, the dark colouration did not negatively affect seed germination. These observations could imply that, the dark colour in cultivar M set in later after physiological maturity and the colour change did not affect seed mass. This observation is contrary to that by Debeaujon et al. (2000) and Chachalis and Smith (2001) that the dark coloured seeds showed poor germination. This was explained by the structural and morphological differences observed in the seeds studied then, but the present study did not go that far. The seed coat colouration in cultivar M did not influence germination as was observed in some accessions of soybean by Chachalis and Smith (2001). The light coloured seed lots of cultivars K and I weighed less compared to the dark seed lots, and yet they showed higher percent germinations. Pimpini et al. (2002) observed that, light coloured seeds of radicchio were immature and Ochuodho (2005) made the same conclusion with *C. gynandra*. It is possible that, the dark colouration intensified after physiological maturity or the climatic condition caused the darkening of wild mustard seeds.

However, in cultivar K, as the dark colour developed, seed mass increased and germination percentage decreased. This confirms further the observation made earlier that, the dark coloration intensified after physiological maturity, either through chemical action or

impregnation of compounds. These observations are in line with those by Hilhorst and Toorop (1997) that in some crucifers, maximum seed mass is attained after physiological maturity. The seeds that failed to germinate were fresh and tetrazolium test showed that, they were live. It is probable that secondary dormancy may have been induced in this cultivar. This finding is supported by Ochuodho (2005) who concluded that, prolonged stay of *C. gynandra* seeds in the field after physiological maturity induced secondary dormancy, a common phenomenon in weedy species. Furthermore the seed lots showed similar proteins at varying band intensity (data not shown), another factor supporting the earlier observation that all the seed lots had attained physiological maturity.

High imbibition and germination rates in dark soybean seeds were attributed to lose seed coat attachment to the embryo (Chachalis and Smith, 2000) and higher numbers of large pores (Chachalis and Smith, 2001). The morphology of seed (seed coat) was not examined in this study but the light seed lot of cultivar K (KL) showed faster imbibition and absorbed more water (Ochuodho and Modi, 2008). Pre-chilling improved percent germination slightly but significantly increased the rate of germination MGT in all the seed lots. This could be because pre-chilling allowed all the seed lots to imbibe enough water to facilitate germination. The high rate of germination could also be due to an increase of bioactive GA in the embryo during pre-chilling (Poljakoff-Mayber et al., 2002; Perez-Flores et al., 2003) and as GA increased, ABA decreased because of enhanced catabolism (Jacobsen et al., 2002; Gonai et al., 2004). When the seeds of cultivar K were treated with exogenous GA_1, they showed significant increase in germination speed and total germination percentage (Ochuodho and Modi, 2008). However, 6 months of after-ripening did not improve germination percentage but decreased the germination of cultivar K. This seems to support the argument that the seeds were harvested after physiological maturity and hence did not require after-ripening. The behaviour of cultivar K was found to be contrary to the statement by Copeland and McDonald (1995) that secondary dormancy was broken by low-temperature stratification and storage at 20°C, among other treatments.

Conclusion

In conclusion, colour selection resulted in increased seed germination in cultivar K but not in cultivars I and M. All the light coloured seed lots showed high germination percentage and one dark seed lot, which was heavier, showed poor germination. The seeds of the wild mustard cultivars turned black after physiological maturity but only cultivar K may have developed secondary dormancy as it persistently showed low germination. There was no clear relationship between seed colour and seed germination, and therefore colour selection may be an added cost to seed processing without adding value to seed quality.

ACKNOWLEDGEMENTS

Moi University, Eldoret (Kenya) for granting study leave and NRF- South Africa, through the University of KwaZulu-Natal for providing funds for this study.

REFERENCES

Atanassova B, Shtereva L, Georgieva Y, Balatcheva E (2004). Study on seed coat morphology and histochemistry in three anthocyaninless mutants in tomato (Lycopersicon esculentum Mill.) in relation to their enhanced germination. Seed Sci. and Technol. 32:79-90.

Chachalis D, Smith ML (2000). Imbibition behaviour of soybean (Glycine max (L.) Merrill) accessions with different testa characteristics. Seed Sci. Technol. 28:321-331.

Chachalis, D, Smith ML (2001). Seed coat regulation of water uptake during imbibition in soybean (Glycine max (L.) Merrill). Seed Sci. Technol. 29:401-412.

Copeland LO, McDonald MB (1995). Seed dormancy. Principles of seed science and technology. 3rd Ed. Chapman and Hall, New York.

Coste F, Raveneau MP, Crozat Y (2005). Spectrophotometrical pod colour measurement: a non-destructive method for monitoring seed drying. J. Agric. Sci. 143:183-192.

De Villiers AJ, van Rooyen MW, Theron GK (2002). Germination strategies of Strandveld succulent Karoo plant species for revegetation purposes: II. Dormancy-breaking treatments. Seed Sci. Technol. 30:35-49.

Debeaujon I, Koornneef M (2000). Gibberellin requirement for Arabidopsis seed germination is determined both by testa characteristics and embryonic abscisic acid. Plant Physiol. 122:415-424.

Debeaujon I, Leon-Kloosterziel KM, Koornneef M (2000). Influence of testa on seed dormancy, germination, and longevity in Arabidopsis. Plant Physiol. 122:403-413.

Gonai T, Kawahara S, Tougou M, Satoh S, Hashiba T, Hirai N, Kawaide H, Kamiya Y, Yoshioka T (2004). Abscisic acid in the thermoinhibition of lettuce seed germination and enhancement of its catabolism by gibberellin. J. Exp. Bot. 55:111-118.

Hilhorst HWM, Toorop PE (1997). Review on dormancy, germinability and germination in crops and weed seeds. Adv. Agron. 61:111-165.

International Seed Testing Association (ISTA). 2004. International Rules for Seed Testing Edition 2004. Zurich, Switzerland.

Jacobsen JV, Pearce DW, Poole AT, Pharis RP, Mander LN (2002). Abscisic acid, phaseic acid and gibberellin contents associated with dormancy and germination in barley. Physiol. Plant. 115:428-441.

Jansen van Rensburg WS, Venter SL, Netshiluvhi TR, Van den Heever E, Voster HJ, Ronde JA (2004). Role of indigenous leafy vegetables in combating hunger and malnutrition. S. Afri. J. Bot. 70:52-59.

Ochuodho JO (2005). Physiological basis for seed germination in Cleome gynandra L. PhD Thesis. University of KwaZulu-Natal, Pietermaritzburg, South Africa.

Ochuodho JO, Modi AT (2005). Temperature and light requirements for seed germination of Cleome gynandra L. S. Afr. J. Plant Soil 22:49-54.

Ochuodho JO, Modi AT (2008). Dormancy of wild mustard (Sisymbrium capense) seeds is related to seed coat colour. Seed Sci. Technol. 36(1):46-55.

Perez-Flores L, Carrari F, Osuna-Fernandez R, Rodriguez MV, Enciso S, Stanelloni R, Sanchez RA, Bottini R, Lusem RD, Benech-Arnold RL (2003). Expression analysis of a GA 20-oxidase in embryos from two sorghum lines with contrasting dormancy: possible participation of this gene in the hormonal control of germination. J. Exp. Bot. 54:2071-2079.

Pimpini F, Filippini MF, Sambo P, Gianquinto G (2002). The effect of seed quality (seed colour variation) on storability, germination temperature and field performance of radicchio. Seed Sci. Technol. 30:393-402.

Poljakoff-Mayber A, Popilevski I, Belausov E, Ben-Tal Y (2002). Involvement of phytohormones in germination of dormant and non-dormant oat (Avena sativa L) seeds. Plant Growth Regul. 37:7-16.

Powell AA (1989). The importance of genetically determined seed coat characteristics to seed quality in grain legumes. Ann. Bot. 63:169-175.

21

Lantana camara and Tithonia diversifolia leaf teas improve the growth and yield of Brassica napus

Chikuvire T. J., C. Karavina, C. Parwada and B. T. Maphosa

Department of Crop Science, Bindura University of Science Education, P. Bag 1020, Bindura, Zimbabwe.

Vegetable production is an economically viable enterprise in Zimbabwe. The most commonly grown leafy vegetable is *Brassica napus* (rape). The productivity of this vegetable is affected by poor soil fertility, with nitrogen being the most limiting nutrient. An experiment was carried out to evaluate the use of two invasive weeds *Lantana camara* and *Tithonia diversifolia* as leaf teas on harvestable biomass production of rape. The experiment was laid out as a randomized complete block design with six treatments that included different leaf tea concentrations. Leaf length was shortest (12.11 cm) on rape that did not receive any fertilizer application but was similar for all other treatments. Leaf widths of rape treated with leaf teas were not significantly different at three weeks after transplanting. At four weeks after transplanting, rape fertilized with ammonium nitrate and *T. diversifolia* (7.5 ℓ/week) had longest leaves. At five weeks after transplanting, rape that did not receive any fertilizer had the shortest leaves (12.79 cm) while ammonium nitrate gave the longest leaves. Highest rape fresh weight was from *T. diversifolia* (7.5 ℓ/week) and AN treatments. Treatments that did not receive any fertilizer consistently produced the lowest leaf length, width and fresh weight. The study showed that yield of rape from application of *T. diversifolia* (7.5 ℓ/week) was comparable to that of ammonium nitrate. Also, it is better to apply leaf teas than not to apply anything at all. Resource poor farmers who cannot afford to buy synthetic fertilizers could boost their rape production by using these leaf teas, and at the same time decelerate the spread of these weeds.

Key words: *Brassica napus*, leaf tea, *Tithonia diversifolia*, *Lantana camara*, smallholder farmer.

INTRODUCTION

In Zimbabwe, smallholder vegetable production is a fast growing enterprise due to increased vegetable demand from the rapidly growing urban population (Kuntashula et al., 2004; Chandiposha, 2007) and boarding schools. *Brassica napus* (rape) is one of the most commonly grown leaf vegetables (Turner and Chivinge, 1999). The vegetable is a cool season crop and its production is normally during the time of the year when temperature is low especially in autumn, winter and spring. However, poor soil fertility is one of the major biophysical constraints to increased productivity where nitrogen and phosphorus are the most severely depleted nutrients in many soils especially in sub-Saharan Africa (Sanchez et al., 1989; Sanchez, 1999; Jama et al., 2000). Compared to cereals, rape requires a higher amount of nutrients and available nitrogen (Rathke et al., 2005). Yield and nutrient uptake are highly dependent on nitrogen (N) fertility and peak yields occur with 120 to 180 kgN/ha (Jackson, 2000). As observed by Gachengo et al. (1999), the use of commercial fertilisers for vegetable production has generally been restricted to only a few farmers endowed with resources and high off-farm income. Inorganic fertilisers such as ammonium nitrate (AN) are at times inaccessible to smallholder farmers leading to reduced application rates or nothing being applied at all. Consequently, these farmers resort to use of traditional organic materials such as crop residues, livestock manure and tree leaf litter (Bradley and Dewees, 1993; Mafongoya et al., 2008). These resources are usually insufficient on most farms. Meanwhile unused, non-traditional organic resources are ubiquitous in smallholder farms. The commonest are the most unwanted *Lantana camara* and *Tithonia diversifolia* and utilization of these plants may help curb their spread.

L. camara, a fast growing shrub encroaching cultivated lands at an alarming rate, is labelled noxious in the Noxious Weeds Act of Zimbabwe (chapter 19:07). The plant is known to suppress the regeneration of neighbouring plants through allelopathic effects. The spread of *Lantana* is aided by the characteristic of its leaves which is somewhat poisonous to animals while its fruit is a delicacy for many birds which distribute the seeds (Fan et al., 2010). *T. diversifolia*, commonly known as Mexican sunflower, is an aggressive annual weed that grows to a height of about 2.5 m and is adapted to most soils (Sonke, 1997; Gachengo et al., 1999). It was probably introduced into Africa as an ornamental and is now widespread in Zimbabwe where it is found growing on waste lands, along major roads and waterways and on cultivated farmlands. The plant is very difficult to control especially during the rainy season because of its vigorous vegetative growth (Swift, 2000). Possible use of these plants for soil fertility improvement was generally explored. Both *L. camara* and *T. diversifolia* have the potential of being used as composts or green manures, especially supplying nitrogen as they both have high nitrogen content in their biomass (Mlangeni, 2010; Gachene and Kimaru, 2003).

Some studies reported that integrating *L. camara* trimmings with fertilizer, at 50% rates, resulted in higher maize yields than from either organic or inorganic resource (Kayuki and Wortmann, 2001). Similarly, green biomass of *T. diversifolia* applied either alone or in combination with fertilizer increased the yield of maize by 24 and 54%, respectively compared to plots which received no inputs (Party, 2010). However, there is a paucity of documented information on their use particularly as liquid fertilizers or leaf teas for vegetable production by the resource constrained smallholder farmers. Leaf teas are easier and quicker to prepare than composts. They have nutrients in mineral form and they can readily act as 'straight' fertilizers that have an important bearing on uptake by plants.

The objective of this study was therefore to assess harvestable biomass production of rape fertilized with leaf teas of *L. camara* and *T. diversifolia*.

MATERIALS AND METHODS

Study site

The study was carried out in January 2012 at Smaldeel Estate (20° 29' 33.35" S and 32° 39' 23.01" E), which is situated 47 km south east of Chipinge town in eastern Zimbabwe. Smaldeel Estate lies in

Agro-ecological Region 1, which receives 1100 mm of rainfall per annum. In the November to April rainfall season it received 1445 mm. It is at an altitude of 1130 m above sea level, with a mean temperature range between 15 and 21°C, average relative humidity of 68 and average wind speed of 8 km/h over the past 10 years starting from 2011. The vegetation resembles a disturbed natural forest. There is abundant *L. camara* and *T. diversifolia* plants at Smaldeel Estate.

Soil sampling and analysis of *Lantana camara* and *Tithonia diversifolia*

A total of six random sampling points were selected where samples were collected from the 0 to 15 cm depth. The samples were thoroughly mixed to form a composite sample. A kilogram of the composite sample was collected and passed through a 2 mm sieve. The soil pH (0.01 M CaCl$_2$) was measured using a pH meter and electrical conductivity (1:5) soil to deionised water using a conductivity meter. Total nitrogen was determined using the method where total nitrogen in digested samples was determined colorimetrically (Anderson and Ingram, 1993). Available P was determined colorimetrically (Watanabe and Olsen, 1965). Exchangeable K, Na, Ca and Mg were determined after extraction with 1 M ammonium acetate solution, adjusted to pH 7 (Anderson and Ingram, 1993). A kilogram sample of each of *L. camara* and *T. diversifolia* fresh leaves were dried in the oven at 60°C for 48 h. Samples were ground and analysed for percentage of total N, P, K, Na, Ca and Mg using protocols outlined earlier.

Experimental design

A Randomized Complete Block Design (RCBD) with six treatments replicated three times was used. The site slope was the blocking factor. The six treatments used were as follows: No fertilizer applied, ammonium nitrate (commonly used rate of 100 kg/ha), *T. diversifolia* (5 ℓ/week), *T. diversifolia* (7.5 ℓ/week), *L. camara* (5 ℓ/week) and *L. camara* (7.5 ℓ/week).

Land preparation and transplanting

The land was first disced and seedlings of the cultivar Hobson (improved English giant rape) raised in the nursery were transplanted two weeks post emergence. The nursery was watered to field capacity before transplanting. Plot sizes were 3.3 × 1.2 m with the longer side perpendicular to the direction of the slope. A plant spacing of 30 × 30 cm with 30 plants per plot in three rows was adopted as it is commonly used by smallholder farmers.

Leaf tea preparation and chemical composition

T. diversifolia and *L. camara* leaves were obtained from the farm and prepared according to recommended methods (KATC and SCC, 2007; Altierri, 2001). Fifteen kilograms (15 kg) of fresh and tender leaves of *T. diversifolia* and *L. camara* were put into separate sacks. Each sack was put into a 200 ℓ drum with about 150 ℓ of water to maintain the required leaf tea ratio of 1:10. The mixtures were covered with hessian sacks to reduce volatilization and were kept in a shade. Mixtures were agitated once after every 2 days for 3 weeks to allow for mineralisation until the water had turned dark brownish green, an indication that most of the nutrients had dissolved into the water. Serial dilutions of ratio 2:3 (leaf tea to water) were done to reduce the tea concentration. Aliquots of 50 ml leaf tea were taken and analysed for macronutrients N, P, K, Na,

Table 1. Chemical composition of soil before planting rape.

Soil pH (0.01 CaCl$_2$)	EC (1:5) (uS/cm)	N (%)	P (ppm)	K (cmol/kg)	Ca (cmol/kg)	Mg (cmol/kg)	Na (cmol/kg)
5.5	72	0.024	4.5	0.87	3.54	0.99	0.16

Table 2. Foliar macronutrient content of *L. camara* and *T. diversifolia*.

Organic resource	N (g/kg)	P (g/kg)	K (g/kg)	Ca (g/kg)	Mg (g/kg)	Na (g/kg)
L. camara	25	8	21.6	12.8	4.1	0.1
T. diversifolia	35	9	38.6	17.8	4.0	0.2

Table 3. Leaf tea macronutrient content of *L. camara* and *T. diversifolia*.

Organic resource	N (mg/L)	P (mg/L)	K (mg/L)	Ca (mg/L)	Mg (mg/L)	Na (mg/L)
L. camara	68.54	0.04	435.56	91.56	61.78	7.98
T. diversifolia	95.20	0.28	602.22	69.44	41.11	8.44

Ca and Mg (Anderson and Ingram, 1993; Watanabe and Olsen, 1965).

Crop management

The site received high amounts of rainfall during the research period in summer and watering was not frequent. During drought periods, watering was done once in two days. Ammonium nitrate and leaf teas were applied in respective plots a week after transplanting. The leaf teas were applied by pouring onto the soil around the plant at weekly intervals. Tea rates of 5 ℓ/week per plant row translated to a rate of 500 ml/plant/week or 15 ℓ/plot/week and that of 7.5 ℓ/week per plant row to 750 ml/plant/week or 22.5 ℓ/plot/week. Watering was done prior to treatment application. Weeds were rouged whenever they emerged. A swath of 3 m bordering the experimental area was sprayed with Tamaron (methamidophos) to discourage locust attack. Other pests like aphids, cutworms and white grubs were controlled by using Karate (lambda cyhalothrin). Shallow cultivation was done to prevent soil capping thus improving infiltration of water, leaf teas and aeration. Ridges were maintained at low heights to avoid waterlogging in the experimental site.

Data collection

Data was collected from the net plot made up of ten plants in the middle row. Leaf length and leaf width were averages of the ten plants in a bed. Harvesting was done at 3, 4 and 5 weeks after transplanting. Each harvest involved removing all but three apical leaves. Fresh leaf mass was measured immediately. Yield was calculated from the mass of vegetable leaves per plot and expressed in t/ha.

Data analysis

An analysis of variance (ANOVA, F test) was carried out on the data from the RCBD using Genstat (6th edition). Differences between means were tested using least significant differences

(LSD) at 5% significance level.

RESULTS

Chemical composition of soil, leaf and leaf tea

Soil chemical characteristics of the study area are presented in Table 1. Soil pH is medium acidity and soil electrical conductivity is generally low hinting at low levels of soluble salts. The total N is generally low indicating low N supplying power of the soil. According to Mashiringwani (1983), the available soil phosphorus is in the deficient range (7 to 15 ppm) while exchangeable K is in the rich range (>0.25 cmol/kg). Exchangeable Ca and Mg, though low, are relatively higher than in most communal areas of Zimbabwe (Shoko and Moyo, 2011; Chikuvire and Mpepereki, 2012). *T. diversifolia* had relatively higher nutrient content of N, P, K and Mg than *L. camara* (Table 2). *L. camara* had relatively higher levels of Mg and Na. The leaf teas displayed a disproportionate nutrient content pattern from that of the foliar. The leaf tea from *T. diversifolia* had higher levels of N, P, K and Na than that from *L. camara* (Table 3). However, the leaf tea from *L. camara* had relatively more Ca and Mg than that from *T. deversifolia*.

Effect of leaf tea on growth of *Brassica napus*

At 3 weeks after transplanting, leaf length was the same for all treatments except for the treatment where fertilizer was not applied that had the shortest leaf of 12.11 cm (Table 4). At 4 weeks after transplanting, leaf length was still significantly the shortest (17.59) in treatments that

Table 4. Effect of different fertilizers on leaf length (cm), leaf width (cm) and fresh weight (t/ha) of rape after 3, 4 and 5 weeks after transplanting.

Treatment	Leaf length (weeks)			Leaf width (weeks)			Fresh weight (weeks)		
	3	4	5	3	4	5	3	4	5
No fertilizer	12.11^a	17.59^a	22.21^a	6.26^a	10.41^a	12.79a	0.812^a	0.808^a	0.908^a
TD (5 L/W)	14.79^b	20.35^b	26.36^b	7.5^b	12.47^b	18.37b	1.14^b	1.41^{bc}	1.661^{ab}
TD (7.5 L/W)	17.07^b	24.13^{cd}	32.33^{cd}	8.36^b	15.48^c	21.79c	1.45^c	1.558^{cd}	2.122^{bc}
LC (5 L/W)	15.89^b	20.81^b	30.98^{cd}	7.51^b	13.41^b	17.49b	1.055^b	1.142^{ab}	1.247^a
LC (7.5 L/W)	14.95^b	22.69^{bc}	30.38^c	7.95^b	13.3^b	19.84bc	1.256^{bc}	1.167^{abc}	1.593^{ab}
AN	16.31^b	25.59^d	34.19^d	9.95^c	16.94^d	25.26d	1.432^c	1.864^d	2.724^c

Means followed by the same letter in a column are not significantly different (P > 0.05) using LSD test.
TD = *T. diversofolia*, LC = *L. camara*, AN = ammonium nitrate.

received no fertilizer. Whilst there was no significant difference between leaf length from treatments AN and *T. diversifolia* (7.5 ℓ/week), the former treatment had significantly longer leaves than those of treatment *L. camara* (7.5 ℓ/week). However, there was no significant difference between *T. diversifolia* (7.5 ℓ/week) and *L. camara* (7.5 ℓ/week). At 5 weeks, the leaf length from the treatment with no fertilizer continued to be the significantly shortest followed by that from *T. diversifolia* (5 ℓ/week). Leaf length from treatments AN, *T. diversifolia* (7.5 ℓ/week) and *L. camara* (5 ℓ/week) was similar. Conversely, that from AN treatment was significantly longer than from treatments with no fertilizer applied, *T. diversifolia* (5 ℓ/week) and *L. camara* (7.5 ℓ/week).

Widest leaves (9.95 cm) were recorded from the treatment AN while the treatment with no fertilizer had the narrowest leaves (6.26 cm) (Table 4). There were no significant differences in the leaf width of rape treated with leaf teas. At 4 weeks after transplanting, the leaf widths from treatments without fertilizer were significantly the narrowest (10.41 cm). Leaves from treatments AN were significantly the broadest of them all followed by those from treatment *T. diversifolia* (7.5 ℓ/week). Treatments *T. diversifolia* (5 ℓ/week), *L. camara* (5 ℓ/week) and *L. camara* (7.5 ℓ/week) had similar leaf width. At 5 weeks, significantly shortest (12.79 cm) leaves were consistently from the treatment that did not receive any fertilizer. The treatment with AN had the widest leaves (25.26 cm). Leaves were wider for *T. diversifolia* (7.5 ℓ/week) than for both treatments of *T. diversifolia* (5 ℓ/week) and *L. camara* (5.0 ℓ/week). The fresh leaf weight at 3 weeks after planting, from the treatment with no fertilizer, was significantly lowest (0.81 t/ha) (Table 4). Leaf weight from treatments AN and *T. diversifolia* (7.5 ℓ/week) were similar and significantly higher than those from *T. diversifolia* (5 ℓ/week) and *L. camara* (5.0 ℓ/week). At 4 weeks, there was no significant difference between fresh weight of leaves from the treatment with no fertilizer applied and *L. camara* treatments. Fresh leaf weight from treatments AN and *T. diversifolia* (7.5 ℓ/week) was similar but that from AN was significantly higher than weights from where no fertilizer was applied, *T. diversifolia*

(5 ℓ/week) and *L. camara* treatments.

The trend of fresh leaf weight from treatments displayed in week 5 was almost similar to that observed in week 4. As fresh and dry weights are closely related, the cumulative dry matter yield of rape after three consecutive weekly harvests revealed that the treatment of AN and *T. diversifolia* (7.5 ℓ/week) were almost similar (Figure 1). The treatment with no form of fertilization yielded the least.

DISCUSSION

The soil is generally suitable for cultivation of rape with supplemental nitrogen and phosphorus. Higher nutrient content of *T. diversifolia* than *L. camara* is consistent with literature (Gachengo et al., 1999; Jama et al., 2000; Kwabiah et al., 2001). Consequently, the leaf tea from *T. diversifolia* was relatively more superior to that of *L. camara* with respect to supplying the much needed nitrogen and phosphorus. Treatments with no applied fertilizer consistently produced rape of the lowest leaf size as indicated by leaf length, width and fresh mass over the study period. Generally, addition of fertilizers either as AN or leaf teas showed a more positive response in growth over the control. This might be as a result of addition of fertilizers that increase and ensure the availability of nitrogen which determines assimilates accumulation in the rape leaf. Probably, high concentration of N, similar to observation by Trapani et al. (1999) on sunflower, increased the cell number and size and in overall increased the leaf size. A study by Jama et al. (2000) on maize confirmed similar results. However, it could be noticed at an early age that treatments with *T. diversifolia* leaf tea at 7.5 ℓ/week and AN had the highest potential to influence growth. The crop is likely to show a positive response to higher N rates.

Small differences especially in the fresh weight in the second and third harvests may be attributed to heavy rains that could have led to leaching of N and hence became unavailable for plant uptake. The season

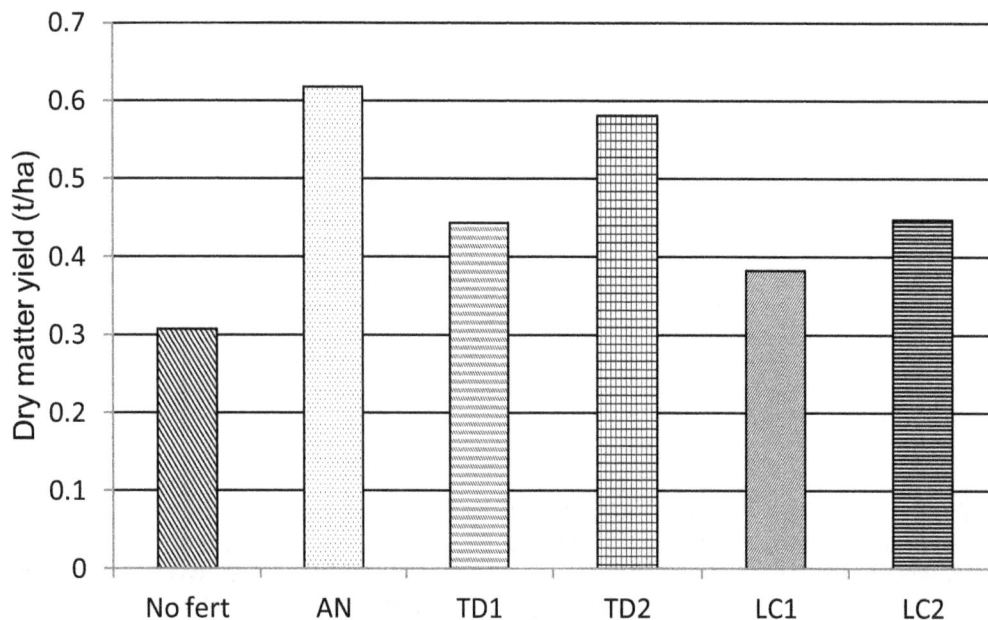

Figure 1. Cumulative dry matter yield (t/ha) of rape after three consecutive weekly harvests. Key: No fert = No fertilizer, AN = ammonium nitrate, TD1 = *T. diversifolia* (5 ℓ/week), TD2 = *T. diversifolia* (7.5 ℓ/week), LC1 = *L. camara* (5 ℓ/week), LC2 = *L. camara* (7.5 ℓ/week).

received higher normal rainfall amounts of 1445 mm. It might also be possible that volatilisation from the drums where the leaf tea was stored could have contributed to similarities of the leaf teas in terms of N content. The cumulative dry matter yield after three weekly harvests showed that the use of leaf teas is better than not fertilizing at all. *T. diversifolia* was apparently the better leaf tea than *L. camara* and the poor resourced farmers can alternatively utilize *T. diversifolia* at a higher concentration and attain similar results as those when ammonium nitrate is used. The application of leaf teas from these plants could help boost vegetable production while at the same time decelerates the spread of these invasive weeds. The trepidation of using *L. camara* due to presence of toxins such as putative hepatotoxin lantadene A is allayed by studies that revealed that *L. camara* biomass can be used, *inter alia*, as substrate for cultivation of edible mushroom, pesticide, antifungal agent to enhance crop yield and insecticide to control weevils in stored grains (Seema, 2011).

Conclusion

This study showed that generally, application of *T. diversifolia* at 7.5 ℓ/week was comparable to treatments with AN and there is scope for recommending it for use by smallholder farmers. *L. camara* at 7.5 ℓ/week though outperformed by *T. diversifolia,* can still be applied as it would be better than not applying at all. Spread of these invasive weeds can be curbed by exploiting them for soil

fertility improvement due to their high N content. More research needs to be carried out on leaf tea formulation, storage and application in different soil types.

REFERENCES

Altierri MA (2001). Sustainable agriculture extension manual for eastern and southern Africa. International Institute of Rural Reconstruction, Nairobi (Kenya).

Anderson JM, Ingram JSI (eds) (1993). Tropical Soil Biology and Fertility, A Handbook of Methods, CAB International, Wallingford.

Bradley PN, Dewees P (1993). Indigenous woodlands, agricultural production and households economy in the communal areas. In: Bradley PN and McNamara K (eds) Living with trees. World Bank Technical Paper No. 210, World Bank, Washington DC, pp. 63-137.

Chandiposha M (2007). Evaluation of moringa (*moringa oleifera* (lam) as a shade tree for rape production in the lowveld semi-arid region of Zimbabwe, Msc Thesis, University of Zimbabwe.

Chikuvire TJ, Mpepereki S (2012). Temporal variation of macronutrients in arable niches exploited by Zimbabwe's Semi Arid smallholder farmers. Bull. Environ. Pharmacol. Life Sci. 1(10):38-45.

Fan L, Chen Y, Yuan J, Yang Z (2010). The effect of *Lantana camara* invasion on soil chemical and microbiological properties and plant biomass accumulation in southern China. Geoderma 154:370-378.

Gachene P, Kimaru M (2003). Evaluation of agroforestry trees and indigenous shrubs as green manure based on maize performance at Kabete, Kenya.

Gachengo CN, Palm CA, Jama B, Othieno C (1999). *Tithonia* and Senna green manures and inorganic fertiliers as phosphorus sources for maize in western Kenya. Agrofor. Syst. 44:21-36.

Jackson GD (2000). Effects of nitrogen and sulphur on canola yield and nutrient uptake. Agron. J. 92:644-649.

Jama B, Palm CA, Buresh RJ, Niang A, Gachengo C, Nziguheba G, Amadalo B (2000). *Tithonia diversifolia* as a green manure for soil fertility improvement in western Kenya: A Rev. Agrofor. Syst. 49:201-221.

Kayuki KC, Wortmann CS (2001). Plant Materials for Soil Fertility Management in Subhumid Tropical Areas. Agron. J. 93:929-935.

KATC (Kasisi Agricultural Training Centre) and SCC (Swedish Cooperative Centre) (2007). Sustainable agriculture; study circle material. Aquila limited, Lusaka, Zambia.

Kuntashula E, Mafongoya PL, Sileshi G, Slungu H (2004). ICRAF project. Cambridge University.

Mafongoya PL, Chintu R, Sileshi G, Chirwa TS, Matibini J, Kuntashula E (2008). Sustainable maize production through leguminous tree and shrub fallows in eastern Zambia. In: Joint FAO/IAEA Division of Nuclear Techniques in Food and Agriculture (eds) Management of Agroforestry Systems for Enhancing Resource use Efficiency and Crop Productivity. AEA, Austria. pp. 203-220.

Mashiringwani NA (1983). The present nutrient status of the soils in the communal farming areas of Zimbabwe. Zimb. Agric. J. 80:73-75.

Mlangeni ANJT (2010). Improvements of total kjeldahl nitrogen and other quality related physico-chemical parameters of chimato composts using Tithonia diversifolia, MSC Thesis, University of Malawi.

Party ST (2010). The agronomic qualities of the Mexican sunflower (Tithonia diversifolia) for soil fertility improvement in Ghana: An exploratory study, DPhil Thesis, University of Science and Technology.

Rathke GW, Christen O, Diepenbrock W (2005). Effects of nitrogen source and rate on productivity and quality of winter oilseed rape (Brassica napus L) grown in different crop rotations. Field Crop. Res. 94:103-113.

Sanchez PA (1999). Improved fallows come of age in the tropics. New Delhi, India.

Sanchez PA, Palm CA, Szott LT, Cueva E, Lai R (1989). Organic Input Management in Tropical Agrosystems. In: DC Coleman, Oades JM, Uehara G (eds) Soil Organic Matter in Tropical Ecosystems, NifTal Project, University of Hawaii.

Seema P (2011). A weed with multiple utility: Lantana camara. Rev. Environ. Sci. Biotechnol. 10:341-351.

Shoko MD, Moyo S (2011). Soil characterization in contrasting cropping systems under the fast track land reform programme in Zimbabwe, Afr. J. Food, Agric. Nutr. Dev. 11(3):4801-4809.

Sonke D (1997). Tithonia weed: A potential green manure crop. Echo development notes.

Swift M (2000). Tropical soil biology and fertility programme, Kenya.

Trapani N, Hall AJ, Weber M (1999). Effects of Constant and Variable Nitrogen Supply on Sunflower (Helianthus annuus L.) Leaf Cell Number and Size. Ann. Bot. 84(5):599-606.

Turner A, Chivinge OA (1999). Production and Marketing of Horticultural Crops in Zimbabwe: A survey of smallholder farmers in the Mashonaland East Province. Cornell International Institute for Food, Agriculture and Development, New York, USA.

Watanabe FS Olsen SR (1965). Test of an ascorbic acid method for determining phoshorus in water and $NaHCO_3$ extracts from soil. SSSA Proc. 29:677-678.

22

Influence of temperature on germination performance of osmoprimed flue-cured tobacco (*Nicotiana tabacum* L.) seeds

T. H. Mukarati[1], D. Rukuni [1]and T. Madhanzi [2]

[1]Crop Productivity Services Division, Tobacco Research Board (Kutsaga Research Station), Airport Ring Road, Box 1909, Harare, Zimbabwe.
[2]Department of Agronomy, Faculty of Natural Resources Management, Midlands State University, P. Bag 9055, Gweru, Zimbabwe.

Seed enhancements such as seed priming can be used to improve germination uniformity and accelerate the rate of germination. The study focused on evaluating the interaction between light and temperature effects on germination percentages of osmoprimed tobacco seeds (*Nicotiana tabacum* L.) of eight flue-cured varieties namely K RK26, K RK28, T29, T64, T65, T66, T71 and T72. Seeds were osmoprimed for a maximum period of five days in different solutions of phytohormones (6-Benzylaminopurine (BA) and Gibberellic Acid (GA_3) and a salt (KNO_3). The lower water potential of these solutions was achieved by adding an osmoticum (Polyethylene Glycol 8000). The influence of osmotic priming was evaluated by incubating the treated seeds under light and dark conditions at optimal (20/30°C) and supra-optimal (33±2°C) temperatures. The experiment revealed that the eight tobacco varieties, gave lower germination percentages in total darkness at high temperatures (33±2°C) than when imbibed under light at 20/30°C. The non-primed seeds exhibited the least germination performance throughout the entire experiment. Osmopriming T64 and T29, in KNO_3+ PEG 8000 was superior in enhancing their germination percentages under dark conditions at 33±2°C. Response of the different varieties was different. K RK26, T65, T71 and K RK28 treated with BA+PEG 8000 recorded the highest germination performance for all the germination attributes used in this study whereas, T66 and T72 had a positive response to GA+KNO_3+PEG 8000 and BA+ KNO_3+PEG 8000 respectively. Osmotic priming resulted in increased germination percentages, rate of germination and improved germination uniformity of the different tobacco cultivars under supra-optimal conditions. Whereas under these conditions, the non-primed seeds of all the varieties exhibited poor germination performances. Therefore, osmotic priming can be used as a method of enhancing tobacco seed germination under stressful environmental conditions.

Key words: Osmotic priming, germination rate, germination uniformity, germination percentage, *Nicotiana tabacum* L., BA (6-Benzylaminopurine), GA_3 (Gibberellic acid), Potassium (KNO_3), Polyethylene Glycol (PEG 8000).

INTRODUCTION

Studying germination is difficult because a population of seeds does not complete the process synchronously (Still et al., 1997; Gallardo et al., 2001). In tobacco (*Nicotiana tabacum* L.) seeds, germination is constrained by the micropylar endosperm, which covers the radicle tip, and germination can proceed if this mechanical resistance decreases to such a level that radicle can protrude through the weakened tissues (El-Maarouf-Bouteau and Bailly, 2008). Tobacco seed does not germinate and emerge uniformly when there is poor seed to soil/media contact (Reed, 1997) and this affect uniformity in the growth of seedlings. Uniformity of growth and synchrony in development are highly desirable characters for mechanized cultural operations (Assefa, 2008) such as clipping in tobacco seedlings. Therefore, the search for improvement in the current productivity levels with reduction in production costs in tobacco culture have led to incorporation in the seedlings production system of new technologies, aiming to achieve seedlings in a shorter period of time, more vigorous and healthy, in order to improve their field performance (Almeida and Vieira, 2010). One method of improving seed germination performance both in the field and in the glass house has been through the use of presowing treatments such as priming (Assefa, 2008). Kester et al. (1997) and Abdolahi (2012) reported that this procedure can also ameliorate the detrimental effects of seed ageing. Osmopriming, refers to soaking seed in osmotic solutions such as polyethylene glycol (PEG), glycol, or mannitol (Kazemi and Eskandari, 2012). The low water potential of the treatment solution allows partial seed hydration so that pre-germination metabolic processes begin but germination is stalled (Bennett et al., 1992; McDonald, 2000; Pill and Necker, 2001). However, osmoprimed seeds may be dried back to their original moisture level (Kazemi and Eskandari, 2012) and stored for variable periods of time depending on the species (Assefa, 2008).

Priming stimulates many of the metabolic processes involved with the early phases of germination (Assefa, 2008). Gallardo et al. (2001), Harris et al. (1999) and McDonald (2000) reported that seed priming is one of the most important developments to help rapid and uniform germination and emergence of seeds, which have practical agronomic implications, notably under adverse conditions. According to Khan (1992) and Assefa (2008), osmotic conditioning in its modern sense, aims to reduce the time of seedling emergence, as well as synchronize and improve the germination percentage. Given that part of the germination processes has been initiated, the duration of the emergence period is decreased; as a result the seedlings grow faster, and more vigorously, leading to a more uniform plant stand (Assefa, 2008).

Priming helps to reduce the time course of germination (Steinmaus et al., 2000) and can also improve germination, especially when applied to poor quality seeds (Nerson, 2007). Gong Ping et al. (2000) and Finch-Savage et al. (2004) demonstrated that seed pretreatment with PEG-6000 increased seed germination and vigour index. Planting seeds treated in this way exhibit more rapid germination and reduced dormancy under adverse conditions (Cheng and Bradford, 1999).

The variable responses of different seed lots to seed priming are a continuing constraint on the commercial application of seed priming (Cheng and Bradford, 1999). Many factors, such as duration, temperature, osmotic concentrations, rehydration period, and the storage of the seed post-treatment, affect the results of the osmoconditioning treatment (Welbaum et al., 1998). The optimum condition required for osmoconditioning varies among the species as well as in relation to the osmotic condition (Assefa, 2008). The priming or osmoconditioning process, which serves to enhance seedling uniformity by improving germination and emergence, is delicate, and the stages of the process must be implemented with extreme care.

Gallardo et al. (2001) postulated that the optimization of priming treatments actually rests on carrying out subsequent germination assays, which only provide retrospective indications of the effectiveness of the priming conditions. Considerable research effort has been put into identifying how seed dormancy can best be broken to improve seed germination, but a definitive protocol is still far from having been achieved (Grundy, 2002). Primed tobacco seed is reported to germinate more uniformly, although the results are inconsistent (Hutchens, 1999; Clarke, 2001). Due to this variability in response, between seed lots, optimum priming conditions often need in practice to be determined on a case-by-case basis for many species (Halmer, 2004). Therefore this study involved laboratory components to determine seed priming effects on flue cured tobacco seed germination.

MATERIALS AND METHODS

Seed material

Pure seed lots of *N. tabacum* L., varieties T71, T72, T64, T66, T29, T65, KRK26 and KRK28 were used. These cultivars had varying degrees of dormancy ranging from very strong to moderate dormancy. The eight flue-cured varieties of tobacco seeds produced in 2009 and 2010 season were supplied by the Seed Production Division, Tobacco Research Board, and Harare, Zimbabwe.

Osmopriming solutions

Osmoticum – Polyethylene glycol 8000 (PEG 8000).

Polyethylene glycol (PEG 8000) solution was prepared according to Michel (1983) to give approximately -18.0 MPa osmotic potential at 25°C.

The priming solutions that were used are 100 mg 6-Benzylaminopurine (BA)/liter water, 100 mg Gibberellic acid (GA)/liter water, 100 mg (BA+GA), 0.2% Potassium Nitrate (KNO$_3$), distilled water and their various combinations thereof.

Osmopriming protocol

PEG priming of seeds was carried out by placing samples of seed (1 g) of known weight in test tubes to which 24 ml PEG 8000 (30%) mixed with salt solution or phytohormone had been added. Each test tube contained a constant amount of the priming solution (PEG,

BA, GA, KNO$_3$ and their various combinations thereof). The priming solution was continuously aerated by an aquarium pump. The maximum priming time in PEG 8000 mixed with different phytohormones and KNO$_3$ in our priming treatments was five days while those of distilled water were three days.

After the priming period, the seeds were rinsed thoroughly with distilled water to remove any PEG, surface dried on filter paper, then spread on dry blotters and left to be air dried for two days at 25°C on the lab bench. The seeds were transferred to paper packets and stored at room temperature before germination tests were conducted in water at 20/30°C or at 33±2°C.

Osmoprimed seeds were divided into two sub-samples, and one sub-sample was stored without any further treatment. The other sub-sample was subjected to heat shock (incubation at 35°C for 7 days). All the treated seeds were stored in paper envelopes at room temperature until required for sowing.

Germination assays

Response to priming was assessed by germination performance (rate, uniformity, total germination percentage, mean germination time and germination index) under dark and light conditions. Germination was expressed as the cumulative percentage of light and dark germinated seeds for both the indoors and greenhouse experiments.

Mean germination time (MGT) was calculated according to the equation of Ellis and Roberts (1981) and Afzal et al. (2005):

$$MGT = \frac{\sum Dn}{\sum n}$$

Where n is the number of seeds, which were germinated on day D and D is the number of days counted from the beginning of germination.

Rate of germination (R) was calculated following modified formula:

R= 1/MGT

Uniformity (GU) was calculated following modified formula:

$$GU = \frac{\sum n}{\sum [(Fn-t)^2 n]}$$

Where t is the time in days, starting from days 0, the day of germination and n is the number of seeds germinate t and F is equal to MGT (Abdolahi et al., 2012).

Germination index (GI) was calculated as described in the Association of Official Seed Analysts (1983) as the following formulae:

$$GI = \frac{\text{No. of germinated seeds}}{\text{Days of first count}} + ---------- + \frac{\text{No. of germinated seeds}}{\text{Days of final count}}$$

Light and dark germination

Effects of light on germination

Seeds were incubated in 4.7 cm diameter glass Petri dishes lined with cotton wool and two layers of Whatman No. 1 filter papers wetted with 16 ml of distilled water. Germination tests were conducted under constant temperature of 33±2°C and under alternate cycles of dark (16 h)/light (8 h) and low/high temperature of 20/30°C. Germination was defined as endosperm rupture, and protrusion of the radicle. Control tobacco seeds were not primed and they were put on germination tests as described above. Daily counts were conducted for 10 continuous days.

Effects of darkness on germination

For dark germination the dishes were wrapped in two layers of aluminium foil to ensure complete darkness. The Petri dishes were placed in a controlled-temperature growth room at 33±2°C and incubator at 20/30°C for 10 days. The percentage of germinated seeds was scored under a microscope. Control tobacco seeds were not primed and they were put on germination tests as described above.

Each dish contained 100 seeds and treatments were arranged in a split plot design (within each temperature regime) with six replicates per treatment.

Data collection

Germination counts were conducted on daily basis for light imbibed seeds and for the dark imbibed seeds the germination counts were conducted on day 10. Daily temperatures were recorded during the experimental period.

Statistical analysis

Data was analyzed using Genstat statistical package, 9th edition (Lawes Agricultural Trust, Rothamsted Experimental Station). Least significant difference (LSD) at 5% for discriminating treatment means was used. Data are means of 6 replicates ± SD.

RESULTS

Effects of temperature on germination uniformity (U), germination rate (R) and total germination percentage of dark imbibed tobacco seeds was evaluated for a continuous five days growth period following priming treatments. Seeds of eight tobacco cultivars T64 (Table 1), T29 (Table 2), K RK26 (Table 3), T71 (Table 4), T65 (Table 5), K RK28 (Table 6), T66 (Table 7) and T72 (Table 8) osmoprimed for five days were incubated for five days under two light conditions (light and dark) at an alternating temperature of 20/30°C and a constant temperature of 33±2°C.

Improved performance for T64 (Table 1) and T29 (Table 2) was obtained for seeds primed with KNO$_3$+PEG 8000. At a temperature of 33±2°C, KNO$_3$+PEG treated T64 seed germinability remained consistently high with the fastest germination rate, improved GU and a germination percentage of 69% under dark conditions which was statistically different from the non-primed seed (4%) (Table 1). T29 primed with the same treatment also exhibited improved performance for all the germination attributes under the same conditions (Table 2). Dark imbibed primed T29 seeds had germination percentage of 92% which was significantly different from most of the

Table 1. Effects of osmotic priming on germination parameters of T64 tobacco seeds. Data are means six replicates of the light (R and GU) and dark germinated seeds under two temperature regimes.

Treatment (s)	R		GU		Dark germination %	
	20/30°C	33±2°C	20/30°C	33±2°C	20/30°C	33±2°C
BA+PEG 8000	0.24971^c	0.26487^e	1.312^c	0.983^b	77.2^b	$49b^c$
GA+PEG 8000	0.2654^a	0.28384^a	$0.98a^b$	0.744^a	96.3^a	52.2^{bc}
KNO$_3$+PEG 8000	0.26617^a	0.28292^a	0.976^{ab}	0.783^a	95.7^a	69.3^a
BA+GA+PEG 8000	0.2566^b	0.27387^{cd}	1.148^b	0.889^{ab}	93.7^a	50.8^{bc}
BA+KNO$_3$+PEG	0.25039^c	0.26971^{de}	1.318^c	0.933^{ab}	91.3^{ab}	42.8^{bc}
GA+KNO$_3$+PEG	0.26645^a	0.28106^{ab}	0.954^{ab}	0.796^{ab}	95^a	49^{bc}
BA+GA+KNO$_3$	0.26628^a	0.27949^{ab}	0.945^a	0.811^{ab}	94.7^a	66^{ab}
Distilled H$_2$O	0.25395^{bc}	0.27205^d	1.244^c	0.91^{ab}	90.3^{ab}	38.2^c
PEG 8000	0.26392^a	0.2773^{bc}	1.004^{ab}	0.842^{ab}	95.2^a	48.5^{bc}
Non-primed	0.22556^d	0.24714^f	3.033^d	1.502^c	65^{bc}	3.5^d
l.s.d. treatment	0.00521		0.1983		14.56	
l.s.d. temperature	0.00233		0.0887		6.51	
F Pr.	≤0.001		≤0.001		≤0.001	

R = Germination rate; GU = Germination uniformity. Data with the same letters are not significantly different ($P \le 0.05$).

Table 2. Effects of osmotic priming on germination parameters of T29 tobacco seeds. Data are means six replicates of the light (R and GU) and dark germinated seeds under two temperature regimes.

Treatment (s)	R		GU		Dark germination %	
	20/30°C	33±2°C	20/30°C	33±2°C	20/30°C	33±2°C
BA+PEG 8000	0.27194^{bc}	0.28878^b	1.111^a	0.702^a	85^{ab}	64.5^{bc}
GA+PEG 8000	0.26956^{cd}	0.28169^{cd}	0.917^a	0.786^{ab}	91^a	72.7^{bc}
KNO$_3$+PEG 8000	0.2794^a	0.29771^a	0.824^a	0.647^a	78.5^{ab}	92^a
BA+GA+PEG 8000	0.24431^g	0.26593^f	1.507^d	1.014^b	72^b	54.7^c
BA+KNO$_3$+PEG	0.26296^{de}	0.27268^{ef}	1.004^{ab}	0.911^{ab}	93.7^a	72.8^{bc}
GA+KNO$_3$+PEG	0.27717^{ab}	0.28926^b	0.834^a	0.723^{ab}	81.3^{ab}	65.3^{bc}
BA+GA+KNO$_3$+PEG	0.27725^{ab}	0.28782^{bc}	0.817^a	0.723^{ab}	96.5^a	72.7^{bc}
Distilled H$_2$O	0.25243^f	0.271^f	1.307^{bd}	0.933^{ab}	79.3^{ab}	23.8^d
PEG 8000	0.26216^e	0.27903^{de}	1.441^{cd}	0.816^{ab}	95.3^a	73.7^{ab}
Non-primed	0.23199^h	0.25348^g	2.307^e	1.346^c	31.7^c	0.3^e
l.s.d. treatment	0.00694		0.3036		18.61	
l.s.d.temperature	0.0031		0.1358		8.32	
F Pr.	<0.001		<0.001		<0.001	

R = Germination rate; GU = Germination uniformity. Data with the same letters are not significantly different ($P \le 0.05$).

treatments including the control which failed to germinate five days after planting (Table 2).

The highest germination performance for K RK26 (Table 3), T71 (Table 4) and T65 (Table 5) was obtained from seeds primed with BA+PEG 8000 and this was significantly different from their respective non-primed seeds. Priming K RK26 (Table 3) with BA+PEG 8000 resulted in improved GU and increased R under both temperature conditions. The final germination percentage of the dark imbibed seeds was high at 93±4% averaged over all treatments except for the seeds primed with distilled water and the non-primed seeds which had germination percentages of 77 and 56% respectively (Table 3) at a constant temperature of 33±2°C. The same trend was also observed for T65 seeds primed with the same solution (Table 4). The final germination percentage (96±2%) of the dark imbibed T65 seeds was significantly high at 20/30°C for all treatments compared to seeds primed in distilled water (77%) and the non-primed seeds (64%) (Table 4). Incubation at low

Table 3. Effects of osmotic priming on germination parameters of K RK26 tobacco seeds. Data are means six replicates of the light (R and GU) and dark germinated seeds under two temperature regimes.

Treatment (s)	R		GU		Dark germination %	
	20/30°C	33±2°C	20/30°C	33±2°C	20/30°C	33±2°C
BA+PEG 8000	0.29648[a]	0.30989[a]	0.64[a]	0.571[a]	93.5[a]	89.5[a]
GA+PEG 8000	0.28954[abc]	0.30891[abcd]	0.725[a]	0.594[a]	97[a]	95.33[a]
KNO$_3$+PEG 8000	0.28596[bcd]	0.30304[bcd]	0.759[a]	0.635[a]	96.33[a]	95.17[a]
BA+GA+PEG 8000	0.29301[ab]	0.30861[abcd]	0.684[a]	0.595a	96.17[a]	95.83[a]
BA+KNO$_3$+PEG	0.27647[e]	0.3056[abcd]	1.087[b]	0.613[a]	96.5[a]	94.67[a]
GA+KNO$_3$+PEG	0.29054[ab]	0.30933[abc]	0.712[a]	0.588[a]	96.5[a]	95.5[a]
BA+GA+KNO$_3$+PEG	0.28294[cde]	0.3023[cd]	0.781[a]	0.642[a]	92.67[a]	95.17[a]
Distilled H$_2$O	0.28934[abc]	0.31064[a]	0.727[a]	0.581[a]	94.83[a]	76.83[b]
PEG 8000	0.28149[de]	0.30165[d]	0.788[a]	0.636[a]	97.17[a]	94.5[a]
Non-primed	0.24599[f]	0.26709[e]	1.553[c]	1.065[b]	94.17[a]	55.67[c]
l.s.d. treatment	0.00748		0.1668		7.882[NS]	
l.s.d. temperature	0.00334		0.0746		3.525	
F Pr.	<0.001		<0.001		<0.001	

R = Germination rate; GU = Germination uniformity. Data with the same letters are not significantly different ($P \leq 0.05$).

Table 4. Effects of osmotic priming on germination parameters of T65 tobacco seeds. Data are means of six replicates of the light (R and GU) and dark germinated seeds under two temperature regimes.

Treatment (s)	R		GU		Dark germination %	
	20/30°C	33±2°C	20/30°C	33±2°C	20/30°C	33±2°C
BA+PEG 8000	0.29917[a]	0.29791[a]	0.6118[a]	0.6335[a]	94.5[a]	92.7[a]
GA+PEG 8000	0.28948[bc]	0.29045[bcd]	0.6891[b]	0.6841[b]	95[a]	80.3[b]
KNO$_3$+PEG 8000	0.28923[bcd]	0.29143[bcd]	0.6915[b]	0.6963[b]	96.3[a]	95.2[a]
BA+GA+PEG 8000	0.29906[a]	0.29508[ab]	0.607[a]	0.6495[ab]	97.5[a]	89.8[ab]
BA+KNO$_3$+PEG	0.29104[b]	0.29251[bc]	0.6732[b]	0.6818[b]	96.7[a]	93.5[a]
GA+KNO$_3$+PEG	0.28469[de]	0.2892[cd]	0.7517[cd]	0.7008[c]	97.3[a]	86.5[ab]
BA+GA+KNO$_3$+PEG	0.28507[cde]	0.29087[bd]	0.7051[b]	0.6958[b]	94.5[a]	86.3[ab]
Distilled H$_2$O	0.28434[e]	0.28774[d]	0.7088[bc]	0.6956[b]	77.2[b]	23[d]
PEG 8000	0.27814[f]	0.27825[e]	0.7669[d]	0.7783[d]	94.2[a]	64.7[c]
Non-primed	0.2407[g]	0.24409[f]	1.6669[e]	1.5684[e]	63.8[c]	17.5[d]
l.s.d. treatment	0.00471		0.04826		12.38	
l.s.d. temperature	0.0021		0.02158		5.53	
F Pr. treatment	<0.001		<0.001		<0.001	
F Pr. temperature	0.120		0.421		<0.001	

R = Germination rate; GU = Germination uniformity. Data with the same letters are not significantly different ($P \leq 0.05$).

temperatures resulted in a reduced germination percentages of the T65 seeds. Under these conditions BA+PEG 8000 treated seeds had a germination percentage of 93% which was significantly different from the seeds primed with distilled water, PEG 8000 and the non-primed seeds which had germination percentages of 23, 65% and 18% respectively (Table 4).

Priming T71 seeds in distilled water resulted in increased germination speed and improved germination uniformity compared to all other treatments (Table 5).

Although priming with distilled water improved germination under light conditions, the germination percentages (83% at 20/30°C and 40% at 33±2°C) of this seed lot was significantly lowered under dark conditions for both temperature levels. BA+PEG 8000 treated seeds were consistently high for the same germination parameters and this treatment resulted in high germination percentages (95% at 20/30°C and 73% at 33±2°C) for the dark imbibed seeds which were significantly different from the seeds primed with distilled

Table 5. Effects of osmotic priming on germination parameters of T71 tobacco seeds. Data are means of six replicates of the light (R and GU) and dark germinated seeds under two temperature regimes.

Treatment (s)	R		GU		Dark germination %	
	20/30°C	33±2°C	20/30°C	33±2°C	20/30°C	33±2°C
BA+PEG 8000	0.283^b	0.28374^b	0.695^b	0.697^b	94.5^a	72.67^a
GA+PEG 8000	0.25747^{de}	0.26318^d	1.079^{ef}	0.99^d	95.67^a	55.17^{bc}
KNO$_3$+PEG	0.25468^e	0.25767^e	1.169^f	1.127^e	95.83^a	56.83^{bc}
BA+GA+PEG	0.27599^c	0.27715^c	0.765^{bc}	0.747^{bc}	97^a	59.83^b
BA+KNO$_3$+PEG	0.27206^c	0.2748^c	0.783^{bcd}	0.766^{bc}	96.67^a	49.33^c
GA+KNO$_3$+PEG	0.26204^d	0.26442^d	0.99^e	0.945^d	95.67^a	49^c
BA+GA+KNO$_3$+PEG	0.27155^c	0.2731^c	0.828^{cd}	0.821^c	96.5^a	70.17^a
Distilled H$_2$O	0.2909^a	0.29424^a	0.601^a	0.597^a	82.67^b	39.5^d
PEG 8000	0.24713^f	0.25544^e	1.316^g	1.13^e	92.17^a	26^e
Non-primed	0.23892^g	0.23949^f	1.704^h	1.751^f	93.67^a	23.33^e
l.s.d. treatment	0.0048		0.0908		8.532^{NS}	
l.s.d. temperature	0.00215		0.0406		3.816	
F Pr. treatment	<0.001		<0.001		<0.001	
F Pr. temperature	0.008		0.084		<0.001	

R = Germination rate; GU = Germination uniformity. Data with the same letters are not significantly different ($P\leq0.05$).

Table 6. Effects of osmotic priming on germination parameters of K RK 28 tobacco seeds. Data are means of six replicates of the light (R and GU) and dark germinated seeds under two temperature regimes.

Treatment (s)	R		GU		Dark germination %	
	20/30°C	33±2°C	20/30°C	33±2°C	20/30°C	33±2°C
BA+PEG 8000	0.30013^{ab}	0.30307^a	0.589^a	0.595^a	94.3^a	86.8^{abc}
GA+PEG 8000	0.29563^c	0.29718^b	0.638^{ab}	0.626^a	93.2^{ab}	81.3^{bc}
KNO$_3$+PEG 8000	0.29085^d	0.29273^c	0.696^{abc}	0.678^{ab}	96.5^a	93.3^a
BA+GA+PEG 8000	0.30401^a	0.30449^a	0.586^a	0.593^a	98.5^a	79.8^c
BA+KNO$_3$+PEG	0.29579^c	0.30331^a	0.626^{ab}	0.599^a	98^a	94.7^a
GA+KNO$_3$+PEG	0.28994^d	0.29761^b	0.691^{ab}	0.639^{ab}	98.3^a	87.3^{abc}
BA+GA+KNO$_3$+PEG	0.29651^{bc}	0.30074^{ab}	0.638^{ab}	0.609^a	96.3^a	89.5^{ab}
Distilled H$_2$O	0.26709^f	0.2704^e	0.936^c	0.869^b	84.3^b	14.2^e
PEG 8000	0.27341^e	0.27629^d	0.861^b	0.83^{ab}	96.8^a	62.7^d
Non-primed	0.22075^g	0.22714^f	3.497^d	3.021^c	26.8^c	0.2^f
l.s.d. treatment	0.00425		0.2425		9.23	
l.s.d. temperature	0.0019		0.1084		4.13	
F Pr. treatment	<0.001		<0.001		<0.001	
F Pr. temperature	<0.001		0.203		<0.001	

R = Germination rate; GU = Germination uniformity. Data with the same letters are not significantly different ($P\leq0.05$).

water and the non-primed seeds (23% at 33±2°C) (Table 5).

Mixing BA and GA in PEG 8000 promoted germination speed and uniformity of primed K RK28 seeds, but this trait did not significantly differ from BA+PEG treated seeds (Table 6). Just like K RK26 (Table 3), T71 (Table 4) and T65 (Table 5), K RK28 seeds treated with BA+PEG resulted in improved GU and increased R under both temperature conditions (Table 6). Priming K RK28

seeds with BA+GA and in BA only all in combination with PEG resulted in increased germination percentages under dark conditions at high temperature (33±2°C) which were significantly different from the seeds primed with distilled water (14%), PEG (63%) and the non-primed seeds (0%) (Table 6). There was, however, no statistical difference ($P\leq0.05$) between seeds primed with BA+GA and the BA treated seeds (Table 6).

Although priming T66 tobacco seeds with distilled

Table 7. Effects of osmotic priming on germination parameters of T66 tobacco seeds. Data are means six replicates of the light (R and GU) and dark germinated seeds under two temperature regimes.

Treatment (s)	R		GU		Dark germination %	
	20/30°C	33±2°C	20/30°C	33±2°C	20/30°C	33±2°C
BA+PEG 8000	0.25728[d]	0.27332[f]	1.133[c]	0.878[c]	93.33[b]	78.17[c]
GA+PEG 8000	0.28369[bc]	0.30125[bc]	0.769[b]	0.631[ab]	96.5[ab]	88.83[b]
KNO$_3$ +PEG	0.2848[bc]	0.30327[bc]	0.769[b]	0.62[ab]	95.67[ab]	95.33[a]
BA+GA+PEG	0.28719[bc]	0.30634[b]	0.734[b]	0.596[ab]	97.17[ab]	96.83[a]
BA+KNO$_3$+PEG	0.28749[bc]	0.29998[cd]	0.74[b]	0.63[ab]	95.5[ab]	94.83[a]
GA+KNO$_3$+PEG	0.28932[b]	0.30622[b]	0.725[b]	0.598[ab]	98.83[a]	95.67[a]
BA+GA+KNO$_3$+PEG	0.27692[d]	0.29505[de]	0.833[b]	0.666[b]	97.33[ab]	97.5[a]
Distilled H$_2$O	0.32471[a]	0.32426[a]	0.518[a]	0.519[a]	96.83[ab]	85.83[b]
PEG 8000	0.28295[c]	0.29179[e]	0.764[b]	0.701[b]	97.83[ab]	85.17[b]
Non-primed	0.23957[e]	0.26319[g]	1.899[d]	1.121[d]	88[c]	10.17[d]
l.s.d. treatment	0.00592		0.131		5.09	
l.s.d. temperature	0.00265		0.0586		2.276	
F Pr.	<0.001		<0.001		<0.001	

R = Germination rate; GU = Germination uniformity. Data with the same letters are not significantly different ($P \leq 0.05$).

Table 8. Effects of osmotic priming on germination parameters of T72 tobacco seeds. Data are means six replicates of the light (R and GU) and dark germinated seeds under two temperature regimes.

Treatment (s)	R		GU		Dark germination %	
	20/30°C	33±2°C	20/30°C	33±2°C	20/30°C	33±2°C
BA+PEG 8000	0.28562[d]	0.2901[de]	0.7272[cd]	0.6834[bc]	93.7[abc]	85.7[a]
GA+PEG 8000	0.27513[e]	0.27889[f]	0.8195[e]	0.7777[d]	97.7[a]	58[c]
KNO$_3$ +PEG 8000	0.27904[e]	0.28575[e]	0.7929[de]	0.7309[cd]	97[ab]	67.8[bc]
BA+GA+PEG 8000	0.29279[bc]	0.29887[bc]	0.6454[abc]	0.6089[ab]	97.3[ab]	81.7[a]
BA+KNO$_3$+PEG	0.30647[a]	0.31281[a]	0.5785[a]	0.5573[a]	97.2[ab]	91.2[a]
GA+KNO$_3$+PEG	0.28734[cd]	0.28783[e]	0.703[bc]	0.708[cd]	96.8[ab]	71[b]
BA+GA+ KNO$_3$+PEG	0.29558[b]	0.30131[b]	0.632[ab]	0.608[ab]	97[ab]	86.8[a]
Distilled H$_2$O	0.29128[bcd]	0.29534[cd]	0.6426[ab]	0.6164[ab]	86.5[c]	38.3[d]
PEG 8000	0.26592[f]	0.26869[g]	0.9583[f]	0.9377[e]	94.7[abc]	45.3[d]
Non-primed	0.24133[g]	0.25021[h]	1.6577[g]	1.3959[f]	87.5[bc]	10.5[e]
l.s.d. treatment	0.00573		0.08216		9.92	
l.s.d. temperature	0.00256		0.03674		4.43	
F Pr. treatment	<0.001		<0.001		<0.001	
F Pr. temperature	<0.001		0.005		<0.001	

R = Germination rate; GU = Germination uniformity. Data with the same letters are not significantly different ($P \leq 0.05$).

water resulted in improved R and GU the dark imbibed seeds of the same treatment exhibited reduced germination percentage (86%) at 33±2°C (Table 7). This was significantly lower than that of the seeds primed with GA+KNO$_3$ (96%). Under the same conditions non-primed seeds had a total germination percentage of 10% which was significantly lower than all other treatments (Table 7).

Priming T72 seeds with BA+KNO$_3$+PEG resulted in improved GU (0.5785 at 20/30°C and 0.5773 at 33±2°C), increased germination speed (0.306479 at 20/30°C and 0.31281 at 33±2°C) and percentage (92%) under dark conditions at high temperature (33±2°C) which were significantly different from the seeds primed with distilled water (14%), PEG (63%) and the non-primed seeds (0%) (Table 6).

DISCUSSION

Germination of osmoconditioned seeds was determined to assess its relationship with germination rate, uniformity and germination percentage under both light and dark conditions at two temperature levels (Tables 1 to 8). Primed seeds of eight tobacco cultivars (T71, T72, T64, T66, T29, T65, KRK26 and KRK28) and their controls were incubated for five days at an alternating temperature of 20/30°C and a constant temperature of 33±2°C. A definite trend toward increased germination performance by combining/mixing the phytohormones and or the inorganic salt KNO_3, with PEG 8000 was evident compared to PEG 8000 alone. Bakht et al. (2011), Bonvissuto and Busso (2007) and Farashah et al. (2011), in their studies found out that PEG 6000 and 8000 only, improved germination performance of various crop species. Unlike from the findings of this study it is evident that the influence of PEG 8000 on germination can be further improved by mixing with phytohormones. To further confirm this, in some studies by Khan (1992) and Assefa (2008) they observed that osmoconditioning had an incremental effect on the germination rate, uniformity of emergence, and the capacity of seeds to withstand adverse environmental conditions. Ghassemi-Golezanik et al. (2008) reported that osmopriming contributes to significant improvement in seed germination and seedling growth in different plant species. The response of the different cultivars in this study differed with respect to the increase in germination rate, uniformity and the total germination percentage under dark conditions (Tables 1 to 8). Similar results were obtained by Lanteri et al. (1994) who observed that the effect of a given osmotic treatment differed between seed lots of the same pepper cultivar. Hartley et al. (2001a) also reported different responses to priming in the flue-cured cultivars NC 71 and NC 72.

High temperature (33±2°C) had a promotory effect on the germination, performance of light imbibed seeds with germination proceeding more rapidly for the primed seeds compared to the non-primed seeds (Tables 1 to 8). The germination speed at both temperatures (20/30 and 33± 2°C) following priming was dependent on the priming treatment. However, seed germination rate was significantly affected by seed priming for all the varieties. For all treatments, germination proceeded most rapidly in the full-light treatment except for the non-primed which showed slow germination speed under both temperatures (Tables 1 to 8). It is surprising that germination was slowest in the incubator (20/30°C), where we had the optimum temperatures for germination, but it should be kept in mind that this condition comparably performed better than the 33±2°C condition in darkness (Tables 1 to 8). At a temperature of 20/30°C under dark conditions seed germinability was consistently high at around 91% averaged over all species (Tables 1 to 8). This further confirms Hartley et al. (2002) suggestion that, the light requirement for the germination of some tobacco cultivars and other photodormant species can be bypassed by alternating the day/night temperature during germination to mimic a typical diurnal fluctuation. Thus, it remains to be determined whether the main factor responsible for

the reduced germination under high temperature in dark was thermoinhibition or the small range of temperature fluctuation. The fact that non-primed seed actually performed well in dark under 20/30°C than in darkness under high temperature implies that inhibition was involved. Considerable increases in germination rates were evident in primed seeds. Seeds primed in BA + osmotica (PEG 8000) and then incubated at 33±2°C exhibited fast germination speed at this supra-optimal temperature (Tables 3, 4, 5). This confirms Hutchens (1999) and Clarkes' (2001) findings, that primed tobacco seed germinate faster and maintain higher germination rates at a wider temperature range than non-primed seed. KNO_3 was effective in improving all the germination parameters of T64 (Table 1) and T29 (Table 2), similar to results from previous studies in tomato seeds by Kester et al. (1997).

Basra et al. (2007) reported that primed seeds usually exhibit increased germination rate, greater germination uniformity, and sometimes greater total germination percentage. This was further strengthened by the findings of this study where priming significantly improved germination uniformity of all the varieties compared to the non-primed seeds (Tables 1 to 8). Hartley et al. (2001a, b) demonstrated that tobacco plants that emerged two days apart were significantly different in size, and therefore homogeneity at 55 days of age. Hartley et al. (2001a) and Clarke (2001) also reported that the survivability of later germinating plants is lowered. Germination tests document seed viability, but not non-uniform emergence, which influences seedling vigor (Clarke, 2001). Uniformity is weighted more heavily than size because a uniform small population can be managed more effectively than a large non-uniform population (Clarke, 2001) (Table 3a, b, c, d, e).

The germination percentages of non-primed seeds were lower as compared to seeds from primed seeds (Tables 1 to 8). Similar germination difficulties were observed in seeds treated with distilled water and PEG 8000 (Tables 1 to 8). The effect of seed priming on germination percentage of tobacco seeds was significant. While some priming treatments like BA+KNO_3+PEG 8000 (Table 5) and BA+KNO_3+PEG 8000 and KNO_3+PEG 8000 (Table 6) resulted in large germination percentages they had no significant effect on germination rate. Exclusion of light at a temperature of 33±2°C reduced the germination percentages of both the primed seeds and the non-primed seeds, but the total germination of the primed seeds exceeded that of the controls (non-primed and PEG 8000 treated seeds) (Tables 1 to 8). This indicates that seed priming can alleviate the negative effects of supraoptimal temperatures. Orzeszko-Rywka and Podlaski (2003) found out that priming alone and or in combination with rubbing improved seed tolerance to unfavourable environmental conditions. Rapid seed germination and stand establishment are critical factors for crop production under stress conditions (Rouh et al., 2011).

Although, there was germination for all the treatments and there were significant differences in total germination percentages of most of the treatments. Germination at 20/30°C under dark conditions was statistically similar for almost all the treatments for each respective variety (Tables 1 to 8). Non-primed seed at 20/30°C had a mean germination percentage of 69 averaged over all varieties and this was significantly different to that at 33±2°C which was below 17%. Priming with distilled water alone had significant effect on R and GU (Tables 3, 5, 7) but had minimal effect on germination percentage throughout the entire osmotic priming experiment. Limited information on BA as germination enhancing treatment suggests that its potential has not been fully explored in seed priming.

Conclusion

Seed priming in this experiment appears to be a useful technique to break dormancy and improve seed germinability even under suboptimal conditions. Increasing knowledge of the molecular mechanisms of dormancy will reveal further ways to improve seed technologies and seed quality. Priming in this experiment resulted in a significant increase in germination performance. Therefore the future looks bright for innovative research and practice in this field, and the opportunity for success appears to be great. An important advantage is the availability of research and measurement instrumentation as a basis for immediate progress for those entering the program.

ACKNOWLEDGEMENTS

Technical assistance for this study was provided by the seed technologist Mrs. Mudarikwa and the biometrician Mr. S. Banana. This research was supported jointly by Midlands State University and Tobacco Research Board through a graduate fellowship to Takura Mukarati.

REFERENCES

Abdolahi M, Andelibi B, Zangani E, Shekari F, Jamaati-e-Somarin S (2012). Effect of accelerated aging and priming on seed germination of rapeseed (*Brassica napus* L.) cultivars. Intl. Res. J. Appl. Basic Sci. 3(3):499-508.

Afzal I, Basra SMA, Iqbal A (2005). The effects of seed soaking with plant growth regulators on seedling vigor of wheat under salinity stress. J. Stress Physiol. Biochem. 1(1):6-14.

Almeida AQ de, Vieira EL (2010). Gibberellin action on growth, development and production of tobacco. Sci. Agrar. Paranaensis 9:45-57.

Assefa MK (2008). Effect of seed priming on storability, seed yield and quality of soybean (*Glycine max* (L.) Merill). Department of Seed Science and Technology College of agriculture, Dharwad, University of Agricultural Sciences, Dharwad 580:005.

Bakht J, Shafi M, Shah R, Raziuddin, Munir I (2011). Response of maize cultivars to various priming sources. Pak. J. Bot. 43(1): 205-212.

Basra SMA, Farooq M, Rehman H, Saleem BA (2007). Improving the germination and early seedling growth in melon (*Cucumis melo* L.) by pre-sowing salicylicate treatments. Int. J. Agric. Biol.

Bennett M, Fritz VA, Callan NW (1992). Impact of seed treatments on crop stand establishment. Hort. Technol. 2:345-349.

Bonvissuto GL, Busso CA (2007). Germination of grasses and shrubs under various water stress and temperature conditions. Int. J. Exp. Bot. 76:119-131.

Cheng Z, Bradford KJ (1999). Hydrothermal time analysis of tomato seed germination responses to priming treatments. Department of Vegetable Crops, One Shields Avenue, University of California, Davis, 50(330):89-99.

Clarke JJ (2001). Development of a Greenhouse Tobacco Seedling Performance Index. Blacksburg, Virginia.

Ellis RA, Roberts EH (1981). The quantification of ageing and survival in orthodox seeds. Seed Sci. Technol. 9:373-409.

El-Maarouf-Bouteau H, Bailly C (2008). Oxidative signaling in seed germination and dormancy. Plant Signal. Behav. 3(3):175-182.

Farashah DH, Afshari RT, Sharifzadeh F, Chavoshinasab S (2011). Germination improvemtn and α-amylase and β-1,3-glucanase activity in dormant and non-dormant seeds of Oregano (*Origanum vulgare*). Aust. J. Crop Sci. 5(4):421-427.

Finch-Savage WE, Dent KC, Clark LJ (2004). Soak conditions and temperature following sowing influence the response of maize (*Zea mays* L.) seeds to on-farm priming (pre-sowing seed soak). Field Crops Res. 90:361-374.

Gallardo K, Job C, Groot SPC, Puype M, Demol H, Vandekerckhove J, Job D (2001). Proteomic Analysis of Arabidopsis Seed Germination and Priming[1]. Plant Physiol.126:835-848.

Ghassemi-Golezanik K, Aliloo AA, Valizadeh M, Moghaddam M (2008). Effects of Hydro and Osmo-Priming on Seed Germination and Field Emergence of Lentil (*Lens culinaris* Medik.). Not. Bot. Hort. Agrobot. Cluj 36(1):29-33.

Gong Ping GU, GuoRong WU, ChangMei L, ChangFang L (2000). Effects of PEG priming on vigour index and activated oxygen metabolism in soaybean seedlings. Chin. J. Oil Crop Sci. 22:26-30.

Grundy AC (2002). Predicting weed emergence: A review of approaches and future challenges. Weed Res. 43:1-11.

Halmer P (2004). Methods to improve seed performance in the field: In: Benech-Arnold RL, Sanchez RA eds. Handbook of seed physiology, New York, Food Products Press, The Harworth Press, inc., pp. 125-166.

Harris D, Joshi A, Khan PA, Gothkar P, Sodhi PS (1999). On-farm seed priming in semi-arid agriculture: Development and evaluation in maize, rice and chickpea in India using participatory methods. Exp. Agric. 35:15-29.

Hartley MD, Smith WD, Spears JF, Fisher LR, Schultheis JR (2001a). Response of flue-cured tobacco cultivars NC 71 and NC 72 to seed priming: II. Influence on transplant production under variable float-system environments. Tob. Sci. doi: 10.3381/0082-4623-45.1.11. pp. 11-14.

Hartley MD, Smith WD, Spears JF, Fisher LR, Schultheis JR (2002). Effect of uniformity of seedling emergence on the percentage of usable transplants produced in the greenhouse float system. Tob. Sci. 45:1-5.

Hutchens TW (1999). Tobacco seed. In Davis DL, Nielsen MT (eds.) Tobacco: Production, chemistry, and technology. Blackwell Science, London, UK. pp. 66-69.

Kazemi K, Eskandari H (2012). Does primingimprove seed performance under salt and drought stress? J. Basic and Appl. Sci. Res. 2(4):3503-3507.

Kester ST, Geneve RL, Houtz RL (1997). Priming and accelerated ageing affect L-isoaspartylmethyltransferase activity in tomato (*Lycopersicon esculentum* Mill.) seed. J. Exp. Bot. 48(309):94-949.

Khan AA (1992). Preplant physiological seed conditioning. Hortic. Rev. 13:131-181.

Lanteri S, Saracco F, Kraak HL, Bino RJ (1994). The effect of priming on nuclear replication activity and germination of pepper (*Capsicum annum* L.) and tomato (*Lycopersicon esculentum* L.) seeds. Seed Sci. Res. 4:81-87.

McDonald MB (2000). Seed Priming. In: "Seed Technology and its Biological Basis" (Black M, Bewley JD Eds.), Sheffield Academic

Press Ltd., Sheffield, USA, pp. 287-325.

Nerson H (2007). Seed Production and Germinability of Cucurbit Crops, Seed Sci. Biotechnol. Israel.

Orzeszko-Rywka A, Podlaski S (2003). The effect of sugar beet seed treatments on their vigour. Plant Soil Environ. 49(6):249-254.

Pill WG, Necker AD (2001). The effects of seed treatments on germination and establishment of Kentucky bluegrass (*Poa pratense* L.). Seed Sci. Technol. 29:65-72.

Reed TD (1997). Float greenhouse tobacco transplant production guide. VA Agriculture. Experimental Station., Blacksburg, A. Publ. pp. 436-051, 189-193.

Rouh HR, Aboutalebian AM, Sharif-zadeh F (2011). Seed priming improves the germination traits of tall Fescue (*Festuca arundinacea*). Not. Sci. Biol. 3(2):57-63.

Steinmaus SJ, Prather TS, Holt JS (2000). Estimation of base temperatures for nine weed species. J. Exp. Bot. 51(343):275-286.

Still DW, Dahal P, Bradford KJ (1997). A single-seed assay for endo-β-mannanase activity from tomato endosperm and radicle tissues. Plant Physiol. 113:13-20.

Welbaum GE, Shen ZX, Oluoch MO, Jett LW (1998). The evolution and effects of priming vegetable seeds. A Symposium on Vegetable and Flower Seed Quality, Boise, Idaho, USA, Seed Technol. 20:209-235.

23

Effects of chilling stress on membrane lipid peroxidation and antioxidant system of *Nicotiana tabacum* L. Seedling

C. Cui, Q. Y. Zhou, C. B. Zhang, L. J. Wang and Z. F. Tan

College of Agronomy and Biotechnology, Southwest University, 400715 Chongqing, P. R. China.

Chilling stress is one of main constraint factors for tobacco production in many regions of the world. In order to study the lipid peroxidation and antioxidant system of flue-cured tobacco seedling after chilling stress, the experiment was conducted with different stressful period which was 2, 4 and 6 days under 5 to 7°C by the material named Yunyan87, Msk326 and Yunyan85, respectively. The results indicated that the rate of O_2^- production in leaves of three cultivars significantly declined at short-term (2 to 4 days) chilling stress. The malondialdehyde (MDA) contents and electric conductivities presented increasing at 2 to 4 days chilling treatments for YY87 and YY85 while those were similar to that of control seedling for K326. Superoxide dismutase (SOD) activities of YY87 remained similar to that of control plants while SOD activities of K326 and YY85 took reverse change. Catalase (CAT) and peroxidase (POD) activities of YY87 and YY85 enhanced under chilling stress, and reached remarkable difference under 6 days chilling stress; As for K326, CAT activities were the maximum under 4 days chilling stress, and POD activities had a prominent decline under 2 days cold stress and increased significantly after that. Except glutathione (GSH) contents of YY87 declined at 6 days chilling, the contents of ascorbic acid (ASA) and glutathione (GSH) were significantly higher in leaves of others treatments comparing with that of controls.

Key words: Chilling stress, flue-cured tobacco, lipid peroxidation, antioxidant enzymes, ascorbic acid, glutathione.

INTRODUCTION

Temperature is one of important factors which affect yield and quality of plant finally. Low temperature is a major factor limiting the productivity and geographical distribution of chilling-sensitive plant species (Zhang et al., 2008). According to previous studies, stress can induce a great deal of physiological alterations to ameliorate oxidative stress, such as reducing accumulation of ROS and malondialdehyde (Chiang et al., 2006; Parvaiz and Prasad, 2012). Chilling stress

affects growth of plant by accumulating reactive oxygen species (ROS). These cytotoxic active oxygen species (AOS), which are also generated during metabolic processes in the mitochondria and peroxisomes, can destroy normal metabolism through oxidative damage of lipids, proteins, and nucleic acids (McCord, 2000). Lipid peroxidation, induced by free radicals is also important in membrane deterioration (McCord, 2000). Chilling stress could impair membrane permeability by the transition of membrane lipids from a liquid–crystalline phase to a gel phase (Lukatkin 2003;Parvaiz; Prasad, 2012). To scavenge AOS, higher plants have developed several strategies to cope with oxidative stress(Hasan et al., 2009; Zhou et al., 2012). These defense tactics are involved of both enzymatic and non-enzymatic antioxidant mechanisms, which include superoxide dismutase(SOD), peroxidase(POD), and catalase (CAT) (Dazy et al., 2009), and non-enzyme antioxidant metabolites such as ascorbic acid (ASA) and glutathione (GSH) (Jebara et al., 2005). Activities of oxygen-scavenging enzymes under chilling stress are correlated with tolerance to the stress. Numerous experiments suggested that chilling tolerance was related to the composition and structure of plant membrane lipids (Guo et al., 2006; Morsy et al., 2007).

Tobacco is an economical leave crop and important multiplex mode organism which is cultured widely in the world. Chilling stress is one of primary constraints to tobacco production in many parts of the world (Xu et al., 2011). The growth of tobacco seedlings would be inhibited when exposed to nonfreezing temperature below approximately 5°C (Gechev et al., 2003). Chilling in its sowing of early spring is the most important hindrance which affects poor and erratic germination and culture of healthy and strong flue-cured seedling. These are not the only result in prolongs periods of seedling-raising, which does not help for regrowth after transplant. Previous reports have described fatty acid desaturation (Kodama et al., 1995), antioxidant enzymes (Gechev et al., 2003), chloroplastic NAD(P)H dehydrogenase (Li et al., 2004) of tobacco seedling under different chilling stress, respectively. In this paper, we simulated tobacco seedlings often suffer low temperature and its lasting period in practice to analyze lipid peroxidation and antioxidant system of tobacco seedling of three genotypes and increased their understanding of those reactive oxygen free radicals. And this would help researchers define how tobacco seedling adapts to chilling and make a right choice of varieties.

MATERIALS AND METHODS

Plant material and experiment design

Three varieties of flue-cured tobacco Msk326 (K326), Yunyan87 (YY87) and Yunyan85 (YY85), were used for experiment materials which was cultivated widely in Chinese tobacco region. After seeds of similar size germinated for 14 days in a Petri dish containing 2 layer of filter paper and distilled water at 25°C, 3 young seedlings were then planted in a plastic containers filled with commercial soil, and reared in a growth chamber, and supplementary lighting (12 h photoperiod), and irrigated water with 1/2MS solution. Tobacco seedling with 5 to 6 true leaves were divided in two groups for each cultivar, one group of flue-cured seedlings were put in a growth chamber with low temperature of 5 to 7°C for 2, 4 and 6 days for chilling stress treatments, and the other group remained at the initial temperature and illumination conditions, which was controlled and grew in temperature at 23 to 25°C with other similar conditions.

90 replicate plants (30 plastic containers) in each treatment. At the end of each treatment, 36 randomly selected replicate plants (12 containers) of each treatment were examined for antioxidant enzyme activities (SOD, CAT, POD) and antioxidant compounds (ASA, GSH). All treatments were done in four replicates.

Extraction and estimation of lipids peroxidation

The rate of O_2^- production was measured as described in Ke et al. (2002) by monitoring the nitrite formation from hydroxylamine in the presence of O_2^- with 6 ml of 65 mM sodium phosphate (pH 7.8), 0.5 g samples were grinded and centrifuged at 10,000 ×g for 15 min. Then, 0.5 ml of the supernatant was incubated with 0.5 mL of 50 mM phosphate buffer (pH 7.8) and 1 mL of 1 mM hydroxylamine hydrochloride at 25°C for 1 h. After incubation, 1 mL of 17 mM sulfanilamide and 1 mL of 7 mM a-naphthylamine were added to the incubation mixture. After reaction at 25°C for 20 min, the absorbance in the aqueous solution was read at 530 nm. A standard curve with NO_2^- was used to calculate the production rate of O_2^- from the chemical reaction of O_2^- and hydroxylamine.

Malondialdehyde (MDA) is a decomposition product of lipid peroxidation. The MDA content was determined by the reaction of thiobarbituric acid (TBA). Briefly, 0.5 g fresh leaves from each treatment were homogenized with 4 mL of 20% (w/v) trichloroacetic acid (TCA) and the homogenate was centrifuged at 3500 g for 20 min. Then 1 ml of the aliquot of the supernatant was mixed with 2 mL of 20% TCA containing 0.5% (w/v) TBA and 100 µl 4% (w/v) butylated hydroxytouluene in ethanol. The mixture was heated at 95°C for 30 min and then cooled down to room temperature and centrifuged at 10 000 g for 10 min. The optical density value of the supernatant was measured at 532 and 600 nm. The MDA content was calculated according to the extinction coefficient of the MDA (156 mM^{-1} cm^{-1}).

Electric conductivity was measured according to Yao et al. (2009), 0.5 g of leaf discs with diameter of 0.35 cm were rinsed 3 times by deionized water, and then were dipped into deionized water in closed tube for 30 min vacuum pumping. The primary conductivity was measured using a DDS-11A (Shanghai, China) conductivity meter after 3 h oscillation, and CK conductivity was also measured. The tubes containing samples were treated with boiling water for 30 min and the final conductivity was measured when the samples were allowed to cool to room temperature. The results were expressed as: relative conductivity [%] = (primary conductivity − CK) × 100/(final conductivity − CK).

Antioxidant enzymes extraction and assay

1.0 g fresh leaves of flue-cured tobacco from each treatment were homogenized in a pestle and mortar with 0.05 M sodium phosphate buffer (pH 7.8) at the end of treated days. The homogenate was centrifuged at 10,000×g for 20 min, and the supernatant was used for analyzing SOD, POD and CAT. The above steps were carried out at 4°C.

The SOD activity was detected according to the modified method of Zhang et al. (2005). The reaction mixture was made of 1.5 mL phosphate buffer (pH 7.8), 0.3 mL 130 mmol/L methionine, 0.3 mL 750 µmol/L nitroblue tetrazolium chloride (NBT), 0.3 mL 100 µmol/L EDTA-Na_2, and 0.3 mL 20 µmol/L riboflavin. Appropriate quantity of enzyme extract was added to the reaction mixture. The reaction started by placing tubes below two 15 W fluorescent lamps for 15 min. Reaction stopped by keeping the tubes in dark for 10 min. Absorbance was recorded at 560 nm. One unit of SOD activity was defined as the quantity of SOD enzyme required to produce a 50% inhibition of reduction of NBT under the experimental conditions, and the specific enzyme activity was expressed as units per

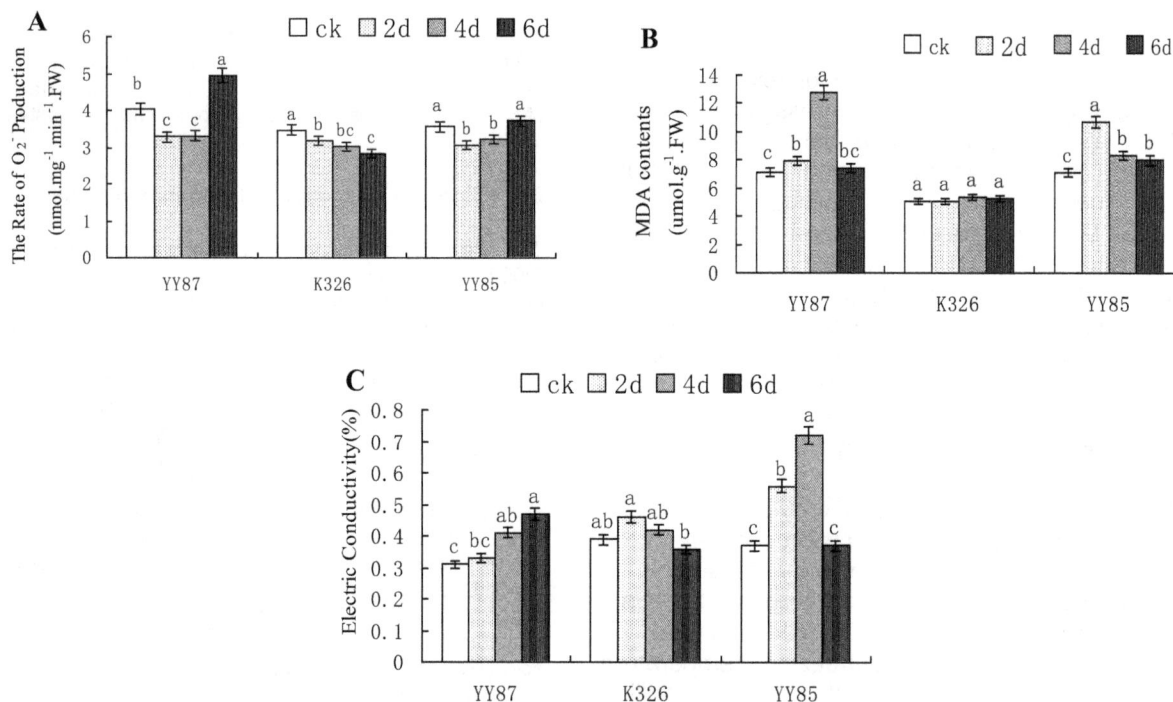

Figure 1. Effect of chilling on oxidant of leaves of tobacco seedling A. the rate of O_2^- production; B. malondialdehyde (MDA); C. electric conductivity. The bars with different letters are significant different (p < 0.05, *t* test).

gram fresh weight (FW) of leaves.

The POD activity was examined according to the modified method of Zhang et al. (2005). The reaction mixture in a total volume of 6.9 mL 0.1 M of sodium phosphate buffer (pH 5.5) containing 1 mL H_2O_2 (30%), 2 mL deionized H_2O, and 1 mL 0.05 M guaiacol was prepared immediately before use. Then, 0.2 ml enzyme extract was added to reaction mixture. Increase in absorbance was measured at 470 nm at 1 min intervals up to 4 min using a UV-Vis spectrophotometer. Enzyme specific activity is defined as units (one POD activity unit defined as absorbance at 470 nm changes per minute) per gram of fresh weight of leaves.

The CAT activity was assayed according to the method of Zhang et al. (2005). CAT activity was determined at 25°C in 2.7 ml reaction mixture containing 2.25 ml 0.05 M sodium phosphate buffer (pH 7.8), 1.5 ml deionized water, and 0.45 mL 0.1 M H_2O_2 prepared immediately before use, and then 0.3 ml enzyme extract was added. The CAT activity was measured by monitoring the decrease in absorbance at 240 nm as a consequence of H_2O_2 consumption. Activity was expressed as units (one CAT activity unit defined as absorbance at 240 nm changes per minute) per gram of fresh weight of leaves.

Measure of ascorbic acid and glutathione

Ascorbic acid (ASA) was determined as described in Hodges et al. (1996). 0.5 g tobacco leaves were homogenized in 7 ml of cold 5% (w/v) *m*-phosphoric acid and centrifuged at 3,000 g for 15 min. About 0.3 ml of supernatant was incubated for 5 min in a 0.7 ml total volume of 100 mM KH_2PO_4 and 3.6 mM EDTA. Color was developed with 0.4 ml of 44% *o*-phosphoric acid, 0.4 mL of 65 mM *a,a'*–dipyridyl in 70% ethanol, and 0.2 ml of 110 mM $FeCl_3$. The reaction mixtures were then incubated at 40°C for 1 h and quantified at 525 nm.

GSH contents were determined according to Israr et al. (2006). Over ice, 1 g of rozen tissue was grounded with sterilized sand and 5 mL ice-cold 6% (v/v) phosphoric acid (pH 2.8) containing 0.5 mM EDTA in a mortar and pestle. The homogenate was centrifuged at 12000 g for 15 min and the supernatant was removed and used for estimation of glutathione. 0.5 ml of reaction buffer including 0.1 M phosphate buffer (pH 7) and 0.5 mM EDTA was added in 0.5 ml of above aliquot, and 0.05 ml of 3 mM 5-dithio-bis-(2-nitrobenzoic acid) was added. After 5 min, absorbance was taken at 412 nm.

Statistical analysis

Analysis of variance (ANOVA) was used to detect the effects of chilling. Multiple comparisons were also performed to permit separation of effect means using the least significant difference test at significant level of P = 0.05. All statistical analyses were done using the software statistical package (DPS) version 3.01.

RESULTS

Cold periods induced lipid peroxidation of tobacco seedling

The rate of O_2^- production of YY87 and YY85 decreased significantly below those of controls under 2 and 4 days chilling stress, but it jumped quickly and prominently higher than that of control for YY87 and was similar level to the control for YY85 with 6 days chilling stress (Figure 1A). The rate of O_2^- production for K326 declined gradually with the development of chilling periods.

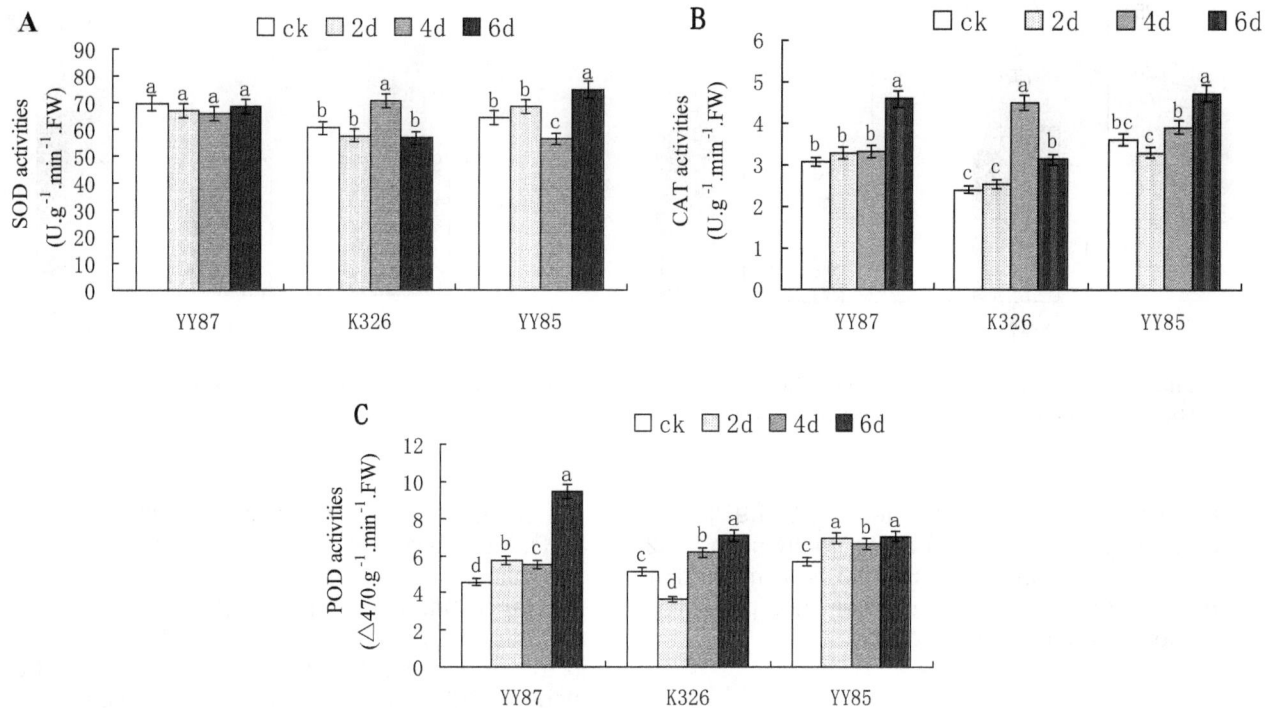

Figure 2. Effect of chilling on antioxidant enzyme activities of leaves of tobacco seedling A. SOD activities; B. CAT activities; C. POD activities. The bars with different letters are significant different (p < 0.05, t test); d, days.

The MDA contents in YY85 and YY87 increased quickly and significantly under 2 to 4 days chilling stress, and that in K326 remained at control's level after treatment by chilling, while that of YY87 increased gradually with treatment (Figure 1B). K326 showed little change of electric conductivity compared with its control plants. And electric conductivity of YY87 was insignificant different under 2 days chilling stress, but obviously increased after 4 days chilling (Figure 1C). While electric conductivity of YY85 rise up firstly and dropped later, which ascend by 94.59% than control seedling under 4 days chilling stress. Taken together, it showed that YY85 is sensitive and K326 is tolerance to 5 to 7°C low temperature, and YY87 between them.

Cold periods induced antioxidant enzymes systems of tobacco seedling

To determine the role of ROS scavenging systems in combating the oxidative stress, antioxidant enzymes were characterized in leaves of three genotype tobacco cultivars (Figure 2). Under 2 to 6 days chilling conditions, SOD activities of YY87 remained similar to that of control plants while SOD activities of K326 and YY85 were the maximum and the minimum for 4 days chilling stress, respectively (Figure 2A), which took reverse change.

Catalase (CAT) activities of YY87 and YY85 enhanced with the development of the chilling stress, which went beyond 49.51 and 30.66%, respectively, with remarkable difference under 6 days chilling stress (Figure 2B). CAT activities of K326 were the maximum under 4 days chilling stress, which ascend by 86.31% comparing with that of control. POD activities of YY87 and YY85 were significant higher than those of their control under chilling stress for 2 to 6 days (Figure 2C). POD activities of K326 had a prominent decline under 2 days cold stress and increased significantly after that, which was the maximum under 6 days cold stress. Change of POD activities under chilling stress for 2 to 6 days showed that it was one of very important enzyme of oxygen-scavenging and defense system.

Response of ascorbic acid and glutathione of tobacco seedling on cold periods

After chilling stress for 2 to 6 days concentrations of ASA and GSH of three genotype flue-cured tobacco seedling significantly increased in comparison with control seedlings (Figure 3A and B). Under 4 days chilling stress, ASA contents of YY87 and YY85 were the maximum and increased by 95.77 and 105.36% and GSH contents were times of 1.97 and 2.19 above those of control, respectively.

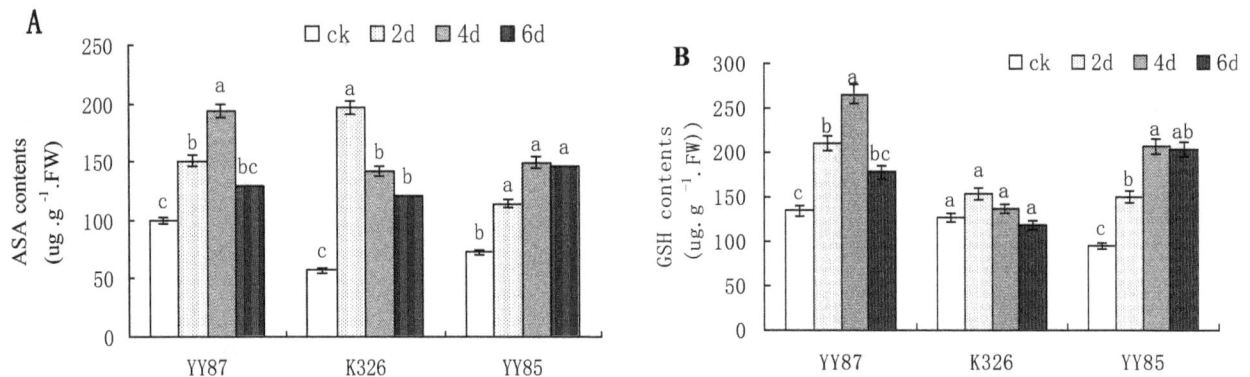

Figure 3. Effect of chilling on antioxidant enzyme activities of leaves of tobacco seedling. A. ASA contents; B. GSH contents; the bars with different letters are significant different (p< 0.05, *t* test); d, days.

While ASA contents of K326 elevated by 244.99% and GSH contents were times of 3.77 of controls under 2 days chilling stress. Therefore, ASA and GSH might play critical role in oxygen-scavenging system and protection of cell membrane for tobacco seedling.

DISCUSSION

ROS can cause lipid peroxidation resulted in breakdown of functional and structural integrity of biological membranes, in turn damage cell membrane increase the permeability of plasma membrane, leakage of K^+ ions, and eventually cause cell death (Tewari et al., 2008). MDA formation is used as the general indicator of the extent of lipid peroxidation resulting from oxidative stress (Yong et al., 2013). Electric conductivity reflected the damage of stresses to the plasmalemma. In our study, enhancement of MDA and electric conductivity for YY87 and YY85 suggested that membrane stability had been damage and happened ion leakage, which might mean the protective mechanisms against oxidative stress was not enough to keep balance and control of ROS. This showed that YY87 and YY85 were sensitive to 5 to 7°C chilling, and K326 could resist 5 to 7°C chilling stress during periods of seedling judged from small changes of MDA contents and electric conductivity. Antioxidant enzymes, such as SOD, CAT, and POD are endogenous factors that protect cells from oxidative damage caused by ROS. SOD catalyzes the dismutation of the O_2^- to molecular oxygen and H_2O_2.

However, H_2O_2 is still toxic to plants, and it is metabolized to harmless water and oxygen by CAT (Chiang et al., 2006). Under short-term (2 days) chilling stress, the rates of O_2^- production of three flue-cued varieties evidently decreased while activities of SOD and CAT remained similar to controls. This might be explained that SOD and CAT were efficient O_2^- scavenger in that SOD catalyzed the conversion of O_2^- into H_2O_2 and CAT converted it into harmless water and oxygen. With

the increasing of natural scavengers such as CAT and SOD, MDA in tissues reduced to low levels.

A positive relationship between peroxidase activity and resistance has been reported (Edreva et al., 1989). Cell wall-bound peroxidases were probably involved not only in the oxidative polymerization of hydroxylated cinnamyl alcohols but also in the generation of hydrogen peroxide necessary for lignification (Coffey and Cassidy, 1984; Edreva et al., 1989; Goldberg et al., 1985). Strong localization and high intensity of the peroxidase response around the sites of infection were characteristic of systemic resistance induced by Mg^{2+} in tobacco seedlings (Edreva et al., 1989). In our experiments, the activities of POD were a few higher in seedling treated with chilling in most treatments for three varieties, this showed long-term (4 to 6 days) chilling might catalysed aged and lignification of tobacco seedling, also exerted influences on some pathogens induced, for peroxidase could not only participate in the biosynthesis of antimicrobial compounds and lignin but also serve as a regulator for the entire metabolic process (Peng and Kuc, 1992). On the other hand, K326 could resist in infection under short-term (2 days) chilling stress because of low POD contents to some extent. The result disclosed that K326 is tolerant to chilling and easy to raise strong seedling while YY85 and YY87 is much susceptible to chilling.

Ascorbic acid and glutathione are important ROS scavenging metabolites. ASA was a non-enzymatic antioxidant acting as a substrate for extracellular enzymes for attenuation ROS levels (Burkey et al., 2006). Glutathione is crucial for biotic and abiotic stress management. Oxidative stress enhances ascorbate and glutathione levels. In the present study, enhancement of ASA and GSH contents showed that the chilling-treated flue-cured seedling accumulated greater amount of toxic oxygen compounds than the control, and long-term chilling stress resulted in attenuating of physiological metabolism and enhancement of antioxidant compound contents. ASA and GSH had the positive effects on protecting tobacco seedling from chilling injure.

Antioxidant system took complicated changes and had been affected by chilling periods and genotypes of flue-cured. Chilling periods influenced physiological difference of seedling and triggering of gene of antioxidant enzyme expression. In this study, with the development of chilling stress, the rate of O_2^- production, MDA contents, electric conductivity and non-enzymatic antioxidant had different changes among three varieties, and three types antioxidant enzymes activities were the maximum for YY87 and YY85 under 6 days chilling stress while SOD and CAT activities reached to be the maximum for K326 under 4 days chilling stress. It is possible that triggering of different antioxidant enzyme protection mechanism is different among three genotype flue-cured tobacco, this need to be further study.

Conclusion

The rate of O_2^- production in tobacco leaves of three cultivars significantly declined at short-term (2 to 4 days) chilling, but happened different change at 6 days chilling. The MDA contents and electric conductivities presented increasing at 2 to 4 days chilling treatments for YY87 and YY85 while that remain similar to level of control for K326. SOD activities of YY87 were similar to that of control plants while SOD activities of K326 and YY85 took reverse change. CAT and POD activities of YY87 and YY85 enhanced under the chilling stress, and reached to remarkable difference under 6 days chilling stress; CAT activities of K326 were the maximum with 5 to 7°C for 4 days and POD activities of K326 had a prominent decline under 2 days cold stress and increased significantly after that. Except GSH of YY87 at 6 days chilling, the contents of ASA and GSH were significantly higher in leaves of others treatments comparing with that of controls. On the whole, K326 is tolerant to chilling and easy to raise strong seedling while YY85 and YY87 is much susceptible to chilling.

ACKNOWLEDGEMENTS

This study was supported by the Science and Technique Foundation of CQ (CSTC, 2011jjA0326), "the fundamental research funds for the Central University (XDJK2011C002)".

REFERENCES

Burkey KO, Neufeld HS, Souza L, Chappelka AH, Davison AH (2006). Seasonal Profiles of Leaf Ascorbic Acid Content and Redox State in Ozone-Sensitive Wildflowers, Environ. Pollut. 143:427-434.

Chiang AN, Wu HL, Yeh HI, Chu CS, Lin HC, Lee WC (2006). Antioxidant Effects of Black Rice Extract through the Induction of Superoxide Dismutase and Catalase Activities. Lipids 41:797-803.

Coffey M, Cassidy DSM (1984). Peroxidase activity and induced

lignification in rusted flax interactions varying in their degree of in compatibility. Can. J. Bot. 62:134-141.

Dazy M, Masfaraud JF, Férard JF (2009). Induction of Oxidative Stress Biomarkers Associated With Heavy Metal Stress in Fontinalis Antipyretica. Hedw. Chemosphere 75:297-302.

Edreva A, Georgieva J, Cholakova N (1989). Pathogenic and nonpathogenic stress effects on peroxidase in leaves of tobacco. Environ. Exp. Bot. 29:365-377.

Gechev T, Willekens H, Montagu MV, Inzé D, Van Camp W, Toneva V, Minkov I (2003). Different responses of tobacco antioxidant enzymes to light and chilling stress. J. Plant Physiol. 160(5):509–515.

Goldberg R, Le T, Catesson AM (1985). Localization and properties of cell wall enzyme activities related to the final stage of lignin biosynthesis. J. Exp. Bot. 38:503-510.

Hodges DM, Andrews CJ, Johnson DA, Hamilton RI (1996).Antioxidant compound responses to chilling stress in differentially sensitive inbred maize lines. Physiol. Plant. 98:685–692.

Guo Z, Ou W, Lu S, Zhong Q (2006). Differential responses of anti-oxidative system to chilling and drought in four rice cultivars differing in sensitivity. Plant Physiol. Biochem. 44:828-836.

Hasan SA, Fariduddin Q, Ali B, Hayat S, Ahmad A (2009). Cadmium: Toxicity And Tolerance In Plants. J. Environ. Biol. 30:165-174.

Hodges DM, Andrews CJ, Johnson DA, Haniton RI (1996). Antioxidant Compound Responses to Chilling Stress in Differentially Sensitive Inbred Maize Lines. Physiol. Plant 98:685-692.

Israr M, Sahi SV, Jain J (2006). Cadmium Accumulation and Antioxidant Responses in the Sesbania Drummondii Callus. Arch. Environ. Contam. Toxicol. 50:121-127.

Jebara S, Jebara M, Limam F, Aouani ME (2005). Changes in Ascorbate Peroxidase, Catalase, Guaiacol Peroxidase and Superoxide Dismutase Activities in Common Bean (Phaseolus Vulgaris) Nodules Under Salt Stress. J. Plant. Physiol.162:929-936.

Ke D, Wang A, Sun G, Dong L (2002). The Effect of Active Oxygen on the Activity of ACC Synthase Induced By Exogenous IAA. Act. Bot.Sin. 44:551-556 (in Chinese).

Kodama H, Horiguchi G, Nishiuchi T, Nishimuraand M, Iba K (1995). Fatty Acid Desaturation during Chilling Acclimation Is One of the Factors Involved in Conferring Low-Temperature Tolerance to Young Tobacco Leaves. Plant Physiol. 4:1177-1185 .

Li XG, Duan W, Meng QW, Zou Q, Zhao SJ (2004). The Function of Chloroplastic NAD(P)H Dehydrogenase in Tobacco during Chilling Stress under Low Irradiance. Plant Cell Physiol. 45(1):103-108.

Lukatkin S (2003). Contribution of Oxidative Stress to the Development of Cold-Induced Damage to Leaves of Chilling-Sensitive Plants: 3. Injury of Cell Membranes by Chilling Temperatures. Russian J. Plant Physiol. 50: 243-246.

McCord JM (2000) . The evolution of free radicals and oxidative stress. Am. J. Med.108:652-659

Morsy MR, Jouve L, Hausman JF, Hoffmann L, Stewart JM (2007). Alteration of oxidative and carbohydrate metabolism under abiotic stress in two rice (Oryza sativa L.) genotypes contrasting in chilling tolerance. J. Plant Physiol. 164(2):157-167.

Parvaiz A, Prasad MNV (2012). Environmental Adaptations and Stress Tolerance Of Plants In The Era Of Climate Change. In Zoldan et al. Understanding chilling tolerance traits using arabidopsis chilling-sensitive mutants. Springer. Pp.159-171.

Peng M, Kuc J (1992). Perxodidase-generated hydrogen peroxide as a source of antifungal activity in vitro and on tobacco leaf disks. Phytopathology 82:696-699.

Tewari A, Singh R, Singh NK, Rai UN (2008). Amelioration of Municipal Sludge By Pistia Stratiotes L.: Role of Antioxidant Enzymes in Detoxification of Metals. Bioresource Technol. 99:8715-8721.

Xu SC, Hu J, Li YP, Ma WG, Zheng YY, Zhu SJ (2011). Chilling tolerance in Nicotiana tabacum induced by seed priming with putrescine. Plant Growth Regul. 63:279-290.

Yao AA, Bernard W, Philippe T (2009). Effect of Protective Compounds on the Survival, Electrolyte Leakage, and Lipid Degradation of Freeze-Dried Weissella Paramesenteroides LC11 During Storage. J. Microbiol. Biotechnol. 19:810-817.

Yong L, Zhang SS, Jiang WS, Liu DH (2013). Cadmium Accumulation, Activities of Antioxidant Enzymes, and Malondialdehyde (MDA)

Content in *Pistia stratiotes* L. Environ. Sci Pollut. Res. 20(2):1117-1123.

Zhang HY, Jiang YN, He ZY, Ma M (2005). Cadmium Accumulation and Oxidative Burst In Garlic (Allium sativum). J. Plant Physiol. 162:977-984.

Zhang YP, Qiao YX, Zhang YL, Zhou YH, Yu JQ (2008). Effects of root temperature on leaf gas exchange and xylem sap abscisic acid concentrations in six Cucurbitaceae species. *Photosynthetica* 46(3):356-362.

Zhou J, Wang J, Shi K, Xia XJ, Zhou YH, Yu JQ (2012). Hydrogen peroxide is involved in the cold acclimation-induced chilling tolerance of tomato plants. Plant Physiol. Biochem. 60:141-149.

24

Different expression of *S*-locus cysteine-rich protein *(SCR)* alleles in self-incompatible and self-compatible *Brassica napus* breeding lines and cultivars: Can be *SCR/SP11* used as a selectable marker in breeding?

Jana Žaludová, Božena Kukolíková, Lenka Havlíčková* and Vladislav Čurn

Biotechnological centre, Faculty of Agriculture, University of South Bohemia, Studentská 13, České Budějovice, Czech Republic.

There are several approaches available for hybrid breeding in oil seed rape, *Brassica napus*, as cytoplasmic male sterility (CMS), genic male sterility (GMS), self-incompatibility (SI), and chemical hybridizing agens (CHA). In comparison with others, SI is regarded as one of the most valuable strategies in hybrid breeding. Unlike self-incompatible (SI) *B. rapa* and *B. oleracea*, two ancestor species, *B. napus* is naturally self-compatible (SC). However, occasionally SI also occurs in rapeseed cultivars. SI in *Brassicaceae* plants is sporophytically controlled by a single multi-allelic locus (S-locus), which contains at least three highly polymorphic genes expressed in the stigma (S locus glycoprotein, *SLG and* S receptor kinase, *SRK*) and in the pollen (*SCR/SP11*). In segregating population derived from crosses between DH SI lines and 00-quality donors we found two recessive alleles of a *SCR class II* gene. We developed new primers for detection of unique cv. Tandem derived allele and this allele was successfully amplified in SI donor plants and SI plants after first cycle of crossing. Analyses of other accessions (SI donor different from cv. Tandem) and varieties did not show so clear pattern of segregation and different expression of both alleles does not correspond to phenotypic manifestation of self-incompatibility and we can assume that it is caused by the presence of repressor gene that does not lie on the *S*-locus.

Key words: Self-incompatibility, *Brassica napus*, S receptor kinase (*SRK*), S locus glycoprotein (*SLG)*, S locus cysteine-rich protein (*SCR)*.

INTRODUCTION

Brassica napus is one of the most important source of edible oil (Tan et al., 2011). At present, hybrid cultivars have higher productivity than conventional ones and their seed quality (contents of erucic acid and glucosinolates) has also been greatly improved (Li et al., 2011; El-Beltagi and Mohamed, 2010; Ahmad et al., 2013). Therefore, the breeders are interested in development of commercial F_1 hybrids of *Brassica* species. There are several ways available for hybrid breeding in *Brassica* species, including cytoplasmic male sterility (CMS), genic male sterility (GMS), self-incompatibility (SI), and chemical hybridizing agents (CHA) (Li et al., 2011). In comparison with other ways for hybrid breeding, SI is regarded as one of the most valuable strategies for the following reasons: (1) it does not have any of the negative effects that exist in male sterility system (Dong et al., 2013); (2) it can be easy overcome; and (3) it can be selected easily in a breeding programme. Self-incompatibility is a widespread mechanism used by flowering plants to prevent inbreeding depression and helps to create and maintain genetic diversity within a species (Goring and Indriolo, 2010). Oil seed rape (*B. napus* L.) is an optionally cross-pollinating species, but self-pollination prevails. In an effort to increase the production, breeding is focused on the F1 hybrid breeding programmes and self-incompatibility here could find its application.

After genetic background investigation of SI in *Brassica* species was found out that SI is under control of *S*-locus (Bateman, 1955). Three genes of the *Brassica S*-locus (*SP11/SCR, SRK,* and *SLG*) are inherited together as a unit, and each comes in many versions, or *S*-alleles; different *S*-allele combination of these genes is referred to as *S* haplotypes (Goring and Indriolo, 2010). For SI to be maintained, alleles from each *S*-haplotype must remain as a tightly linked genetic union. Single alleles of *S*-locus were divided according their phenotype effects into two classes, Class I and II supposed to be dominant and recessive, respectively (Nasrallah et al., 1991). Recently many review articles focused on molecular mechanism of sporophytic SI in *Brassica* species were published (Takayama and Isogai, 2003; Isogai and Hinata, 2002; Kachroo et al., 2002; Takayama and Sakagami, 2002; Watanabe et al., 2003, Ivanov and Gaude, 2009, Haasen and Goring, 2010, Chapman and Goring, 2010, Charlesworth, 2010). Physical maps have been constructed of the *S*-locus region of several *S*-haplotypes from *B. oleracea* (Boyes et al., 1997), *B. rapa* (Boyes et al., 1997; Suzuki et al., 1999), and *B. napus* (Cui et al., 1999, Casselman et al., 2000). Using different techniques (direct sequencing, cDNA selection techniques, and RNA differential display), a number of genes have been identified in the *S*-locus region (Cui et al., 1999; Suzuki et al., 1999; Casselman et al., 2000).

Even though the female determinant was known, the pollen determinant was always elusive. Candidate for the pollen determinant was predicted to have several characteristics. As well as S receptor kinase (*SRK)* and S locus glycoprotein (*SLG*) gene it should be located at the *S*-locus. The pollen grain would have showed doubled haploid character in *S*-haplotype of male determinant as it is required in sporophytic form of SI. To fulfil this condition pollen determinant should be expressed before meiosis in pollen mother cells or expressed later in the tapetum cells (Watanabe et al., 2003). Characterization of the coating has revealed the presence of several families of gametophytically expressed small cysteine-rich proteins pollen coat proteins (PCPs). PCP-A class proteins have specific affinities for stigmatic and S- and S-related proteins (Doughty et al., 2000).

Two independent studies have revealed a gene for pollen determinant at the same time. *S*-locus cysteine-rich protein *(SCR)* (Schopfer et al., 1999), syn. *SP11* (Suzuki et al., 1999) is expressed in anthers only. Anther tapetum cells produce SCR/SP11 protein that bind with a high affinity to its receptor, the pistil-specific SRK protein (Goring and Indriolo, 2010). Putative pollen determinant fulfilled all requirements such as highly polymorphic character, anther-specific expression, physical linkage with SLG and SRK. *SCR/SP11* is invariably located between *SLG* and *SRK* in upstream orientation to *SLG*, closer to *SRK* than to *SLG* (Takayama et al., 2000). *SCR/SP11* encodes a small (<8 kDa) hydrophobic and positively charged peptide that exists as a monomer (Takayama et al., 2001). Based on immunostaining is suggested that S_8 SCR/SP11 was secreted from the tapetal cell into anther locule as a cluster and translocated to the pollen surface at the early developmental stage of the anther. During the pollination process, SCR/SP11 was translocated from the pollen surface to the papilla cell, and then penetrated the cuticle layer of the papilla cell to diffuse across the pectine cellulose layer (Iwano et al., 2003). Eight conserved cysteine residues, a glycine residue, and an aromatic amino acid residue are characteristic for amino acid arrangement of *SCR/SP11* gene product. *SCR/SP11* gene consist of two ORF, the first ORF for putative signal protein is highly conservative among *Brassica* species, while the second is highly polymorphic. L1 loop, designated as the hypervariable (HV) region is the most variable region of the SCR, suggesting to be responsible for imparting the allele-specific interaction with receptors. It folds into an α/β sandwich that resembles those of plant defensins (Mishima et al., 2003). As well as *S*-haplotypes the *SCR/SP11* gene is divided into two classes. *SCR/SP11* alleles are highly divergent (Watanabe et al., 2000). In four *B. rapa* class-II *S*-haplotypes, linear dominance relationship were observed. Using RNA gel blot analysis, linear dominance relationship regulation by expression of *SCR/SP11* was found (Kakizaki et al., 2003). The mRNA of class II SCR/SP11 was detected predominantly in the anther tapetum in homozygotes and was not detected in the heterozygotes of class I and class II *S*-haplotypes, suggesting that the dominant/recessive relationships of pollen are regulated at the mRNA level of SCR/SP11 (Shiba et al., 2002). The 522 bp 5' upstream region support the correct function of SCR gene (Shiba et al., 2001). It was demonstrated that SCR/SP11 alleles determined *S*-haplotype (Shiba et al., 2001). Duplicated SCR found in *B. oleracea* S^{15} haplotype produced two different sizes of transcripts (Shiba et al., 2004). The linear dominance relationship of SI phenotype on pollen side in class II SP11 is regulated by the expression of *SP11* (Kakizaki et al., 2003). In A self-compatible *B. rapa* of class I *S*-haplotype insertion of retrotransposon-like sequence in first introne of *SRK* and 89-bp deletion in the SP11 promoter was revealed (Fujimoto et al., 2006). Transcription of functional *SP11-60* allele of *B. rapa* was suppressed in heterozygotes with S-f2 allele originating from class I self-compatible *B. rapa* (Fujimoto et al., 2006).

The present research work was therefore designed to: (1) study the genetic variability of the *SCR/SP11* gene in SI donors, self-compatible cultivars and segregating population derived from crosses between these DH SI lines and 00-quality donors respectively; (2) explore a possible way to identify specific marker associated with SI trait (recessive alleles of a *SCR* class II gene); and (3) verify this molecular marker in segregating population.

MATERIALS AND METHODS

Plant material

A segregating doubled haploid (DH) populations of 118 oilseed rape plants was derived from four crosses between self-compatible (SC) cultivar 'Lisek' and self-incompatible (SI) line 'AIK 6', SC cultivar 'Rasmus' and SI line 'AIK 6', SC cultivar 'Rasmus' and SI line 'AIK 3', and finally SC line 'OP BN-03' and SI line 'AIK 3'. 'AIK 3' and 'AIK 6' SI lines were derived from SI line 'Tandem' with recessive type of self-incompatibility. DH lines were produced in the Crop Research Institute Prague – Ruzyne, CZ.

Analysed DH lines:
AIK-6 x Rasmus   DH lines No. 1-33
AIK 6 x OP-BN-03   DH lines No. 34-66
AIK-6 x Lisek   DH lines No. 67-92
AIK-3 x Rasmus   DH lines No. 93-118

Analysed SI donors:
'Start (86/1)', 'WRG 15', 'Tandem 6/85', 'Tandem 1/85', 'AIK 3', 'AIK 6'.

Analysed cultivars:
Rasmus, Lisek, Jesper, Regent, Lirajet, Solida (genotypes used in specified breeding programme).

Isolation of genomic DNA and PCR-RFLP of SCR gene

Genomic DNA of *B. napus* cultivars and DH lines was extracted from young leaves of 2-week-old seedlings by the DNeasy Plant Mini kit (QIAGEN). *SCR* gene was amplified with Class II *SCR*-specific oligonucleotide primers for genomic DNA (Shiba et al., 2002). Allele specific amplifications were performed with sets of primers (allele 1: 5'-TTTGATTTTGACATATGTTC-3' and 5'-CCCCTCAACTTCATAGTGTT-3'; allele 2: 5'-TTGGACTTTGACATATGTTC-3' and 5'-CTCTGAAGTGGGTTTTACAG-3') designed according to highly variable segments between two analysed *SCR* alleles. Plant genomic DNA approximately 50 ng was mixed with 10 pmoles primers, 10x buffer (10 mM Tris HCl pH 8.3, 50 mM KCl, 3 mM MgCl$_2$, 1% Triton X-10), 100 µM dNTP, 1U *Taq* polymerase (TaKaRa) in a final volume of 25 µl. The PCR conditions were 45 cycles of 30 s at 94°C, 30 s at 55°C and 1 min at 72°C. Polymerase Chain Reaction (PCR) fragments were analysed using agarose and polyacrylamide gel electrophoresis and stained with ethidium bromide. PCR products were then cleaved by specific restriction endonucleases (*Mnl*I, *Hinf* I, *Hha*I). The resulting fragments were then divided into 10% PAGE or 1.5% agarose.

mRNA isolation and *cDNA* synthesis

Anthers of the DH SI lines and rapeseed cultivars were collected from buds at 2 to 3 days before opening. Total RNA was isolated using RNasy Plant Mini Kit (Qiagen). Isolation included DNA degradation step using DNase. The mRNA was isolated from total RNA using Oligotex mRNA Kit (Qiagen). Approximately 20 µg of RNA was subjected first-strand cDNA synthesis using Omniscript (Qiagen) with an oligo(dT)$_{18}$ primer. Second strand was amplified with a set of SCR II specific primers (5'-GCGAAAATCTTATATACTCATAAG-3' and 5'-TTCGTTGATCAATTATGATT-3' Shiba et al., 2002). Reverse transcription-polymerase chain reaction (RT-PCR) was performed under these conditions: 42 cycles of 30 s at 93°C, 30 s at 45°C, 1 min at 72°C and one cycle of 72°C for 5 min.

DNA cloning and sequencing

For determination of nucleotide sequences, PCR fragments were extracted from gel with QIAquick Gel extraction kit QIAGEN and ligated with TOPO TA Cloning kit (invitrogen). The insert of the expected size was analysed using polymerase chain reaction–restriction fragment length polymorphism (PCR–RFLP) (*Mnl* I, *Hha* I) and individual clones were sequenced. Sequencing reaction was prepared with cycle sequencing ready reaction kit (Applied Biosystem). Sequence analysis was performed on the ABI PRISM 3130 sequencer.

Seed test

On one branch with buds from each plant was isolated and number of seed were counted in open-pollinating and isolated flowers/siliquae. Phenotypic expression of SI was evaluated by the number of seeds in the silique.

RESULTS

With Class II specific *SCR/SP11* primers a 450 bp band from genomic DNA was amplified. Subsequently this product was cloned and sequenced. Resulting sequence had all characteristic of SCRs such as eight conserved cysteines, one glycine, and one tyrosine in assumed positions. After analysis of genomic DNA only one single allele was found in analysed plants from segregating population. To detect other expected allele, cDNA synthesis of Class II *SCR* from anthers was performed and 350 bp long fragment was obtained and subsequently again cloned and sequenced. In order to find allelic polymorphism, screening of clones was performed by PCR-RFLP method. Two different PCR-RFLP pattern were revealed corresponding to two Class II *SCR* alleles (Figure 1). The two Class II *SCR/SP11* gene alleles were sequenced and marked as allele 1 and 2. The sequence similarity between these two alleles was 85% (Figure 2). Comparison with database NBCI (http://www.ncbi.nlm.nih.gov) showed that allele 1 was identical with S[15] allele from *B. oleracea* and allele 2 was unique and typical for cv. Tandem as SI donor. Using PCR-RFLP of single cloned cDNAs we have found out that both alleles were present in SC plants as well as SI plants (Figure 1). It seemed that allele 2 expression is often much higher in both phenotypes. Occurrence of single nucleotide mutations in both alleles was quite

AIK6 x Lisek   AIK6 x Rasmus

| 86SI F | 86SI G | 86SI H | 86SI I | 74SI A | 74SI B | 84SC A | 84SC B | 84SC C | 84SC D | 91SC B | 91SC D | 6SI 5 | 6SI 6 | 10SI A | 10SI B | 10SI C | 4SI A | 4SI C | 20SC A | 20SC B | 20SC C | 20SC D | M |

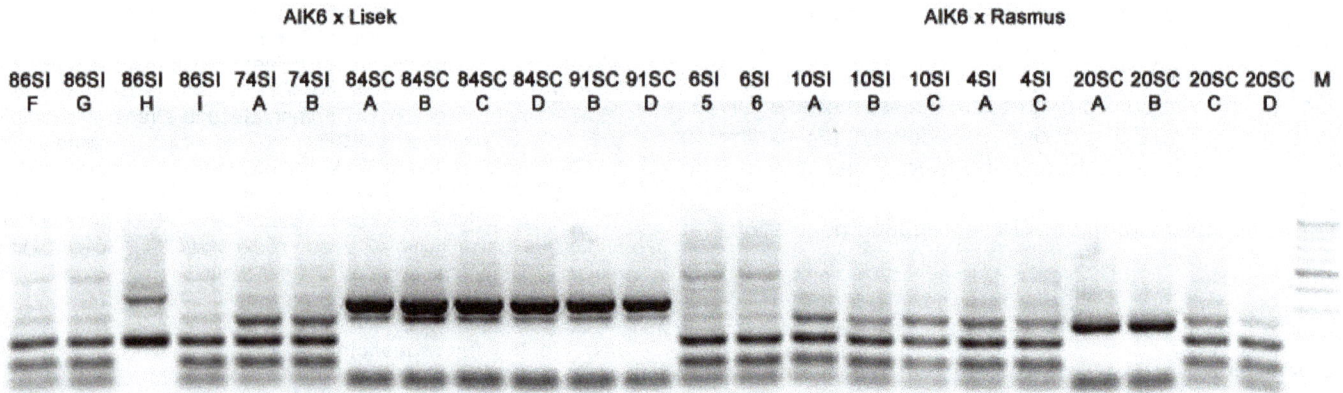

Figure 1. PCR-RFLP of *SCR* class II cDNA clones. Cloned cDNAs were amplified by PCR and then restricted with *MnI* II. Allele 1 had three restriction sites (sample 84SC A to 91SC D, 20SC A to 20SC B). Allele 2 had two restriction sites (sample 86SI F to 74SI B, 6 SI 5 to 4SI C, 20SC C to 20SC D). Clone 86SI H had 1 bp mutation in one recognition site. Putative phenotype was marked with letter SI or SC.

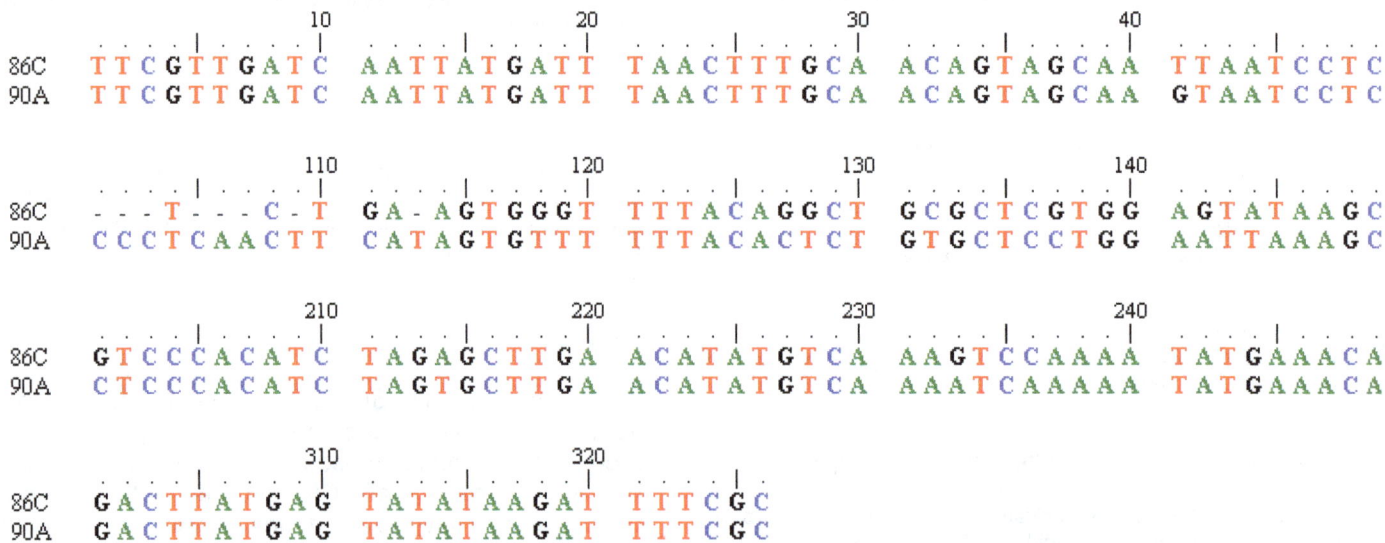

Figure 2. Comparison of cDNA sequences of two revealed *SCR* class II alleles in *Brassica napus*. The sequence similarity between these two alleles was 85%.

frequent. Furthermore, we found that S^{15} allele was duplicated. Similarly we observed that two forms of transcripts exist. We performed screening of segregating doublehaploid population with allele specific primers. A single approximately 280 bp band was present in SI DH lines only (Figure 3). Original donor of SI in this set of DH lines was Tandem 6/85 and screening of other accessions and varieties did not show so clear segregation. Amplification of allele 1 carried out at temperature of annealing 55°C but did not carry out at temperature of annealing 60°C. Also presence of allele 1 and 2 in wide group of cultivars and in several SI lines was tested. Amplification of allele 1 in cultivars resulted in about 700 bp long fragment in cultivar 'Jesper' and about 550 bp long fragment in cultivar 'Regent' despite these

specific primers are designed to amplify approximately 300 bp long fragment. Allele 2 specific primers amplified 280 bp long fragment in cultivar 'Lirajet' and cultivar 'Solida'. In SI lines two fragments about 480 and 520 bp long of allele 1 in SI line 'WRG 15' and 'AIK 6' and one single fragment about 500 bp long in SI line 'AIK 3' were amplified. Allele 2 was detected in SI lines 'Start (86/1)' (1), 'Tandem 6/85' (3), 'AIK 3' (5) and 'AIK 6' (6).

DISCUSSION

SCR/SP11 is sole pollen determinant of self-incompatibility (Schopfer et al., 1999; Suzuki et al., 1999). In *B. oleracea* S^{15} haplotype Class II duplicated *SP11*

86SI 74SI 84SC 91SC 90SC 6SI 10SI 4SI 20SC 102SI 107SC 99SI 60SI 66SC M

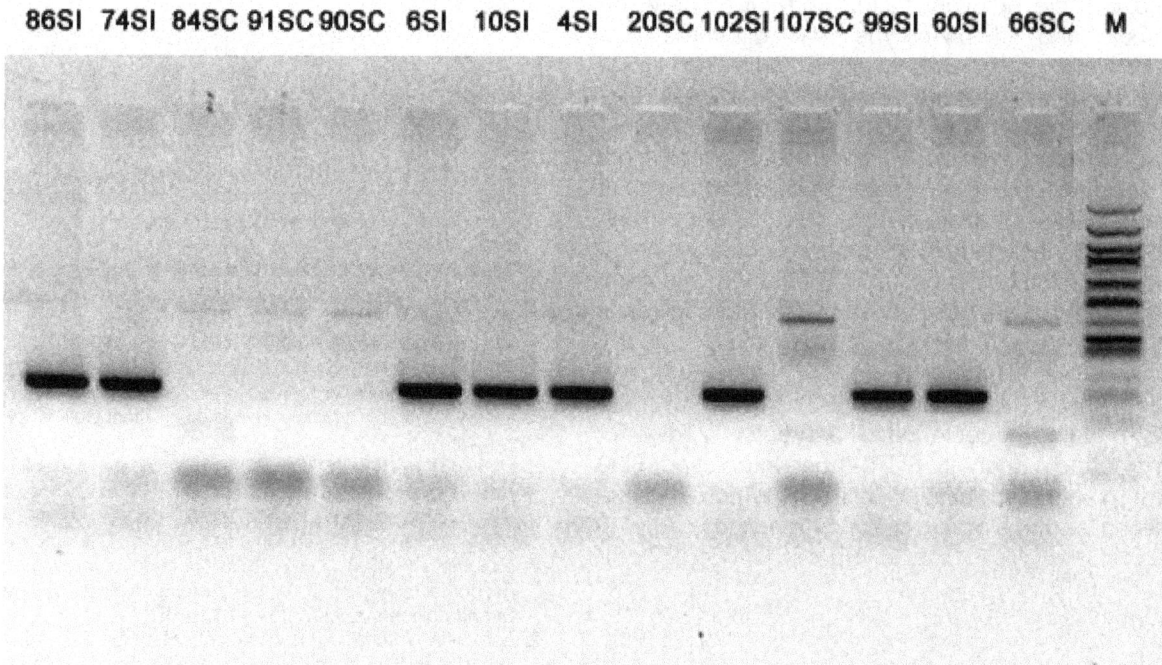

Figure 3. Screening of segregating doubled haploid population with allele 2 specific primers. A single approximately 280 bp fragment was amplified in putative SI DH lines.

gene was found - *SP11 – SP11a*, *SP11b* a *SP11b´*, and these duplicated genes were connected by integeneric spacer. The difference between *SP11a* a *SP11b* were only in one deletion and were oriented in the same direction. As result of this duplication also transcrips of different length were found (Shiba et al., 2004). In *B. napus* we also found duplicated *SCR II* gene but without presence of different transcripts. Cabrillac et al. (1999) found duplicated *SLG* genes in *B. oleracea*, line S^{15}. Allele 1 is probably derived from *B. oleracea*, because it is also duplicated and has 100% similarity to S^{15} alele from *B. oleracea*. This could suggest that this allele originates from *B. oleracea*. The duplication could lead to the producing of incorrect lengths of transcripts responsible for dysfunction. The second allele was specifically occurred in SI lines. Two types of transcripts, AL+ and AL- were observed in *B. oleracea* and it is supposed that AL- type of transcript is the later form (Shiba et al., 2002). This finding is in consistent with our results. Both types of transcripts were expressed in SI and SC DH plants. Considering that the allele specific primers were designed on the basis of sequences obtained from specific plant material and the *SCR* polymorphic nature, it is evident that they cannot easily amplify single alleles in different cultivars. Moreover, there is a big sequence similarity among genes at the S-locus and for instance SCR binding domain area of *SRK* gene can be amplified as well. Previously we do not succeed in amplification of allele 2 with Class II specific *SCR* primers from genomic DNA. The difference between allele 2 amplification from cDNA and genomic DNA may be caused by different numbers of copies of which the expression might be affected in a certain way. The level of expression of S-alleles from *B. oleracea* in interspecific crosses was dependent on the different backgrounds of *B. napus*, that is, the presence of specific S-alleles (Ripley and Beversdorf, 2003). Probably there is another gene present and this gene affects the expression of SCR II alleles and response to SI. We can assume that this unknown gene have multiple alleles with different relations of dominance and recessivity, because of character of suppression of particular S-alleles and different effects on SI reaction. Ekuere et al. (2004) believes that self-incompatibility is controlled by two loci, the S-locus and supressor *sp* locus, which also has multiple alleles. Latent S-alleles were discovered in DH population derived from F_1 hybrids after crossing of resynthetised rape and different varieties. Ekuere et al. (2004) supposed that these latent alleles are hidden due to presence of common supressor system and this suppressor system is not associated with S-locus. SCR/SP11 protein is quite small and for interaction with eSRK just a few specific amino acids it is necessary, but the SI reaction is much more complicated and any specific protein is needed (Chookajorn, 2004). Hypothetical suppressor should be specifically expressed in tapetal cells and is located outside of S-locus. We found that SC phenotype correlates well with the presence of class I *SLG* gene. Correlation with *SCR* gene was unambiguous only in specific crosses when AIK3 and AIK 6 DH lines were used as SI donors and only after first cycle of hybridization with 00 quality donors.

After second cycle of hybridization correlation between SCR marker and SI phenotype was disrupted and specific correlation was only between SLG marker and SI phenotype (data not shown). Assuming that there is a suppressor gene at another locus, the question arises, how it can correlate with *SLG I* gene. *SLG* gene is not strictly necessary, but it supported full performance of self-incompatibility (Takasaki et al., 2000). Self-incompatibility is associated with normal expression of *SRK*, but with low expression *SLG* in naturally occurring variants of *B. rapa* (Nasrallah et al., 1992). Also in many SC *Brassica* lines A10 allele was found (*SLG I* gene), which was at the same locus as *SRK* gene with one nucleotide deletion (Goring et al., 1993). We can assume that SLG protein (I) could have a greater affinity for suppressor protein than to SCR. After interaction SLG protein with suppressor binding to eSRK is possible and this reactions lead to blocking of the SI reaction.

Conclusion

The results of this study showed that rapeseeds SI donors and SC cultivars differed in their SCR/SP11 transcripts. Two different transcripts of *SCR/SP11* gene observed in cv. Tandem revealed unique recessive allele 2 and new primers amplifying 280 bp band only in SI plant derived from cv. Tandem (as SI donor) were designed. It is important to note that that the detected allele 2 (and its related sequence) was unique and typical for cv. Tandem and that therefore the allele specific primers designed for specific plant material here cannot easily amplify single alleles in different SI material derived from different SI donors. Thus, our results could be used also as a guide describing how to develop new specific primers for particular population.

ACKNOWLEDGEMENTS

We gratefully acknowledge the financial support of the National Agency for Agricultural Research grant No. NAZV QI111A075, University of South Bohemia grants No. GAJU 064/2010/Z, GAJU 063/2013/Z and the project Postdoc No. CZ.1.07/2.3.00/30.0006 realized through EU Education for Competitiveness Operational Programme. We thank Dr. Keith Edwards for their critical reading of the manuscript.

REFERENCES

Ahmad B, Mohammad S, Azam F, Ali I, Ali J, Rehman S (2013). Studies of genetic variability, heritability and phenotypic correlations of some qualitative traits in advance mutant lines of winter rapeseed (Brassica napus L.). Am-Eurasian J. Agric. Environ. Sci 13:531-538.

Bateman AJ (1955). Self-incompatibility systems in angiosperms: III. Cruciferae. Heredity 9:52-58.

Boyes DC, Nasrallah ME, Vrebalov J, Nasrallah JB (1997). The self-incompatibility (S) haplotypess of Brassica contain highly divergent and rearranged sequences of ancient orgin. Plant Cell. 9:237-247.

Cabrillac D, Delorme V, Garin J, Ruffio-Chable V, Giranton JL, Dumas C, Gaude T, Cock JM (1999). The S15 self-incompatibility haplotype in Brassica oleracea includes three S gene family members expressed in stigmas. Plant Cell. 11:971-986.

Casselman AL, Vrebalov J, Conner JA, Singal, A, Giovanni J, Nasrallah ME, Nasrallah JB (2000). Determining the physical limits of the Brassica S locus by recombinational analysis. Plant Cell 12:23-33.

Chapman LA, Goring DR (2010). Pollen-pistil interactions regulating successful fertilization in the Brassicaceae. J. Exp. Bot. 61(7):1987-1999.

Charlesworth D (2010). Self-incompatibility. Biol. Rep. 2:68.

Chookajorn T, Kachroo A, Ripoll DR, Clark AG, Nasrallah JB (2004). Specificity determinants and divesification of the Brassica self-incompatibility pollen ligand. PNAS 4:911-917.

Cui Y, Brugiere N, Jackman L, Bi Y-M, Rothstein, SJ (1999). Structural and transcriptional comparative analysis of the S locus regions in two self-incompatible Brassica napus lines. Plant Cell 11:2217-1131.

Dong X, Feng, H, Xu M, Lee J, Kim YK, Lim YP, Piao Z, Park YD, Ma H, Hur Y (2013). Comprehensive analysis of genic male sterility-related genes in Brassica rapa using a newly developed Br300K oligomeric chip. Plos One 8:e72178.

Doughty J, Wong HY, Dickinson,G (2000). Cysteine-rich pollen coat proteins (PCPs) and their interactions with stigmatic S (incompatibility) and S-related proteins in Brassica: Putative roles in SI and pollination. Ann. Bot. 85:161-169.

Ekuere UU, Parkin IA, Bowman C, Marshall D, Lydiate DJ (2004). Latent S alelles are widwspread in cultivated self-compatible Brassica napus. Genome 47:257-65.

El-Beltagi H.E-D.S, Mohamed A (2010). Variations in fatty acid composition, glucosinolate profile and some phytochemical contents in selected oil seed rape (Brassica napus L.) cultivars. Grasas Y Aceites 61:143-150.

Fujimoto R, Okazaki K, Fukai E, Kusaba M, Nishio T (2006). Comparison of the genome structure of the self-incompatibility (s)locus in interspecific pairs of s haplotypes. Genetics 173:1157-1167.

Goring D, Indriolo E (2010). How plants avoid incest. Nature 466:296-928.

Goring DR, Glavin TL, Schafer U, Rothstein SJ (1993). An S receptor kinase gene in self-compatible Brassica napus has a 1-bp deletion. Plant Cell 5:531-539.

Haasen E, goring DR (2010). The recognition and rejection of self-incompatible pollen in the Brassicaceae. Botanical Stud. 51:1-6.

Isogai A, Hinata K (2002). Molecular mechanism for recognition reaction in self-incompatibilityof Brassica species. Proc. Jpn. Acad. 78B:241-249.

Ivanov R, Gaude T (2009). Brassica self-incompatibility. Plant signalling. Behavior 4:996-998.

Iwano M, Shiba H, Funato M, Shimosato H, Takayama S, Isogai A (2003). Immunohistochemical studies on translocation of pollen S-haplotype determinant in self-incompatibility of Brassica rapa. Plant Cell Physiol. 44:428-436.

Kachroo A, Nasrallah ME, Nasrallah JB (2002). Self-incompatibility in the Brassicaceae: receptor-ligand signaling and cell-to-cell communication. Plant Cell Sup: S227-S238.

Kakizaki T, Takada Y, Ito A, Suzuki G, Shiba H, Takayama S, Isogai A, Watanabe M (2003): Linear Dominance Relationship among Four Class-II S Haplotypes in Pollen is Determined by the Expression of SP11 in Brassica Self-Incompatibility. Plant Cell Physiol. 44:70-75.

Li Y, Cai Q, Yang G, He Q, Liu P (2011). Identification of a microsatellite marker linked to the fertility-restoring gene for a polima cytoplasmic male-sterile line in Brassica napus L. Afr. J. Biotechnol. 10:9563-9569.

Mishima M, Takayama S, Sasaki K-i, Jee J-G, Kojima C, Isogai A, Shirakawa M (2003). Structure of the Male Determinant Factor for Brassica Self-incompatibility. J. Biol. Chem. 278:36389-36395.

Nasrallah JB, Nishio T, Nasrallah ME (1991). The self-incompatibility genes of Brassica: Expression and use in genetic ablation of floral tissues. Annu. Rev. Plant Physiol. Plant Mol. Biol. 42:393-422.

Nasrallah ME., Kandasamy MK, Nassrallah JB (1992). A genetically defined trans-acting locus regulates S Locus function in

Brassica. Plant. J. 2:497-506.

Ripley VL, Beversdorf WD (2003). Development of self-incompatible Brassica napus: (III) B. napus genotype effects on S-allele expression. Plant Breed. 122:12-18.

Schopfer CR, Nasrallah ME, Nasrallah JB (1999). The male determinant of self-incompatibility in Brassica. Science 286:1697-1700.

Shiba H, Iwano M, Entani T, Ishimto K, Shimohato H, Che F-S, Satta Y, Ito A, Takada, Y, Watanabe M, Isogai A, Takayama, S (2002). The Dominance of Alleles Controlling Self-Incompatibility in Brassica Pollen Is Regulated at the RNA Level. Plant Cell 14:491-504.

Shiba H, Park J-I, Suzuki G, Matsushita M, Nou I-S, Isogai A, Takayama S, Watanabe M (2004). Duplicated SP11genes produce alternative transkripts in the S15 haplotype of Brassica oleracea. Genes Gent Syst. 79:87-93.

Shiba H, Takayama S, Iwano M, Shimohato, H., Funato M, Nakagawa T, Che F-S, Suzuki G., Watanabe M, Hinata K, Isogai A (2001). A Pollen Coat Protein, SP11/SCR, Determines the Pollen S-Specificity in the Self-Incompatibility of Brassica Species. Plant Physiol. 125:2095-2103.

Suzuki G, Kai N, Hirose T, Fukui K, Nishio T, Takayama S, Isogai A, Watanabe M, Hinata K (1999). Genomic organization of the S locus: identification and characterization of genes in SLG/SRK region of S9 haplotype of Brassica campestris (syn. rapa). Genetics 153:391-400.

Takasaki T, Hatakeyama K, Suzuki G, Watanabe M, Isogai A, Hinata, K (2000). The S receptor kinase determines self-incompatibility in Brassica stigma. Nature 403:913-916.

Takayama S, Isogai A (2003). Molecular mechanism of self-recognition in Brassica self-incompatibility. J. Exp. Bot. 54:149-156.

Takayama S, Sakagami Y (2002). Peptide signaling in plants. Cur. Opin. Plant Biol. 5:382-387.

Tan H, Yang X, Zhang F, Zheng X, Qu C, Mu J, Fu F, Li J, Guan R, Zhang H, Wang G., Zuo J (2011). Enhanced seed oil production in canola by conditional expression of Brassica napus LEAFY COTYLEDON1 and LEC1-LIKE in developing seeds. Plant Physiol. 156:1577-1588.

Watanabe M, Ito A, Takada Y, Ninomiya C, Kakizaki T, Takahata Y, Hatakeyama K, Hinata K, Suzuki G, Takasaki T, Satta Y, Shiba H, Takayama S, Isogai A (2000). Highly divergent sequence of the pollen self-incompatibility (S) gene in class-I S haplotypes of Brassica campestris (syn. rapa). L. FEBS Lett. 12:139-144.

Watanabe M, Takayama S, Isogai A, Hinata K (2003). Recent Progresses on self-incompatibility research in Brassica species. Breed. Sci. 53:199-208.

Permissions

All chapters in this book were first published in AJAR, by Academic Journals; hereby published with permission under the Creative Commons Attribution License or equivalent. Every chapter published in this book has been scrutinized by our experts. Their significance has been extensively debated. The topics covered herein carry significant findings which will fuel the growth of the discipline. They may even be implemented as practical applications or may be referred to as a beginning point for another development.

The contributors of this book come from diverse backgrounds, making this book a truly international effort. This book will bring forth new frontiers with its revolutionizing research information and detailed analysis of the nascent developments around the world.

We would like to thank all the contributing authors for lending their expertise to make the book truly unique. They have played a crucial role in the development of this book. Without their invaluable contributions this book wouldn't have been possible. They have made vital efforts to compile up to date information on the varied aspects of this subject to make this book a valuable addition to the collection of many professionals and students.

This book was conceptualized with the vision of imparting up-to-date information and advanced data in this field. To ensure the same, a matchless editorial board was set up. Every individual on the board went through rigorous rounds of assessment to prove their worth. After which they invested a large part of their time researching and compiling the most relevant data for our readers.

The editorial board has been involved in producing this book since its inception. They have spent rigorous hours researching and exploring the diverse topics which have resulted in the successful publishing of this book. They have passed on their knowledge of decades through this book. To expedite this challenging task, the publisher supported the team at every step. A small team of assistant editors was also appointed to further simplify the editing procedure and attain best results for the readers.

Apart from the editorial board, the designing team has also invested a significant amount of their time in understanding the subject and creating the most relevant covers. They scrutinized every image to scout for the most suitable representation of the subject and create an appropriate cover for the book.

The publishing team has been an ardent support to the editorial, designing and production team. Their endless efforts to recruit the best for this project, has resulted in the accomplishment of this book. They are a veteran in the field of academics and their pool of knowledge is as vast as their experience in printing. Their expertise and guidance has proved useful at every step. Their uncompromising quality standards have made this book an exceptional effort. Their encouragement from time to time has been an inspiration for everyone.

The publisher and the editorial board hope that this book will prove to be a valuable piece of knowledge for researchers, students, practitioners and scholars across the globe.

List of Contributors

Lara Ramaekers
Centre of Microbial and Plant Genetics, Department of Microbial and Molecular Systems, KU Leuven, Leuven, Belgium

Alfred Micheni
Kenyan Agricultural Research Institute (KARI), Nairobi, Kenya

Paul Mbogo
Kenyan Agricultural Research Institute (KARI), Nairobi, Kenya

Jozef Vanderleyden
Centre of Microbial and Plant Genetics, Department of Microbial and Molecular Systems, KU Leuven, Leuven, Belgium

Miet Maertens
Division of Bio-economics, Department of Earth and Environmental Sciences, KU Leuven, Leuven, Belgium

Muhammad Ahsan
Department of Plant Breeding and Genetics, University of Agriculture Faisalabad, Pakistan

Amjad Farooq
Department of Plant Breeding and Genetics, University of Agriculture Faisalabad, Pakistan

Ihsan Khaliq
Department of Plant Breeding and Genetics, University of Agriculture Faisalabad, Pakistan

Qurban Ali
Department of Plant Breeding and Genetics, University of Agriculture Faisalabad, Pakistan

Muhammad Aslam
Department of Plant Breeding and Genetics, University of Agriculture Faisalabad, Pakistan

Muhammad Kashif
Department of Plant Breeding and Genetics, University of Agriculture Faisalabad, Pakistan

Sokona Dagnoko
Rural Polytechnic Institute for Training and Applied Research (IPR/IFRA), Katibougou, Koulikoro, BP 06, Mali Republic Seneso Limited, BPE 5459, Bamako, Mali Republic

Niamoye Yaro-Diarisso
Institute of Rural Economy (IER), Rue Mohamed V, BP 258, Bamako, Mali Republic

Paul Nadou Sanogo
Rural Polytechnic Institute for Training and Applied Research (IPR/IFRA), Katibougou, Koulikoro, BP 06, Mali Republic

Olagorite Adetula
National Horticultural Research Institute (NIHORT), Jericho Reservation Area - Idi-Ishin, PMB 5432, Ibadan, Oyo State, Nigeria

Aminata Dolo-Nantoumé
Institute of Rural Economy (IER), Rue Mohamed V, BP 258, Bamako, Mali Republic

Kadidiatou Gamby-Touré
Institute of Rural Economy (IER), Rue Mohamed V, BP 258, Bamako, Mali Republic

Aissata Traoré-Théra
Institute of Rural Economy (IER), Rue Mohamed V, BP 258, Bamako, Mali Republic

Sériba Katilé
Institute of Rural Economy (IER), Rue Mohamed V, BP 258, Bamako, Mali Republic

Daoulé Diallo-Ba
Seneso Limited, BPE 5459, Bamako, Mali Republic

E. TASGIN
Primary Department, Bayburt University, 69000 Bayburt, Turkey

H. NADAROGLU
Department of Food Technology, Erzurum Vocational Training School, Atatürk University, Erzurum, Turkey

Aniruddha Maity
Seed Technology Division, Indian Grassland and Fodder Research Institute, Jhansi, UP- 284003, India

S. K. Chakrabarty
National Fund for Basic, Strategic and Frontier Application Research in Agriculture, 707, Krishi Anusadhan Bhavan-I, Pusa, New Delhi- 110012, India

S. Sathish
Department of Seed Science and Technology, Seed Centre, Tamil Nadu Agricultural University, Coimbatore, Tamil Nadu, India

M. Bhaskaran
Department of Seed Science and Technology, Seed Centre, Tamil Nadu Agricultural University, Coimbatore, Tamil Nadu, India

Hillal Ahmad
Faculty of Forestry, SK University of Agricultural Sciences and Technology of Kashmir, Shalimar, Srinagar, J and K, 19112, India

T. H. Masoodi
Faculty of Forestry, SK University of Agricultural Sciences and Technology of Kashmir, Shalimar, Srinagar, J and K, 19112, India

Altamash Bashir
Faculty of Forestry, SK University of Agricultural Sciences and Technology of Kashmir, Shalimar, Srinagar, J and K, 19112, India

G. I. Hassan
Division of Pomology, SK University of Agricultural Sciences and Technology of Kashmir, Shalimar, Srinagar, J and K, 19112, India

S. A. Mir
Division of Statistics, SK University of Agricultural Sciences and Technology of Kashmir, Shalimar, Srinagar, J and K, 19112, India

P. A. Sofi
Faculty of Forestry, SK University of Agricultural Sciences and Technology of Kashmir, Shalimar, Srinagar, J and K, 19112, India

G. M. Bhat
Faculty of Forestry, SK University of Agricultural Sciences and Technology of Kashmir, Shalimar, Srinagar, J and K, 19112, India

M. Maqsood
Department of Botany, University of Kashmir, Srinagar, India

O. S. Udengwu
Department of Plant Science and Biotechnology, Faculty of Biological Sciences, University of Nigeria Nsukka, Enugu State, Nigeria

Andre Luiz Piva
Universidade Estadual do Oeste do Paraná/ Centro de Ciências Agrárias/ Programa de Pós Graduação em Agronomia. Rua Pernambuco, 1777, CEP: 85960-000. Marechal Cândido Rondon, Paraná, Brazil

Eder Junior Mezzalira
Universidade Estadual do Oeste do Paraná/ Centro de Ciências Agrárias/ Programa de Pós Graduação em Agronomia. Rua Pernambuco, 1777, CEP: 85960-000. Marechal Cândido Rondon, Paraná, Brazil

Anderson Santin
Universidade Estadual do Oeste do Paraná/ Centro de Ciências Agrárias/ Programa de Pós Graduação em Agronomia. Rua Pernambuco, 1777, CEP: 85960-000. Marechal Cândido Rondon, Paraná, Brazil

Daniel Sschwantes
Universidade Estadual do Oeste do Paraná/ Centro de Ciências Agrárias/ Programa de Pós Graduação em Agronomia. Rua Pernambuco, 1777, CEP: 85960-000. Marechal Cândido Rondon, Paraná, Brazil

Jeferson Klein
Universidade Estadual do Oeste do Paraná/ Centro de Ciências Agrárias/ Programa de Pós Graduação em Agronomia. Rua Pernambuco, 1777, CEP: 85960-000. Marechal Cândido Rondon, Paraná, Brazil

Leandro Rampim
Universidade Estadual do Oeste do Paraná/ Centro de Ciências Agrárias/ Programa de Pós Graduação em Agronomia. Rua Pernambuco, 1777, CEP: 85960-000. Marechal Cândido Rondon, Paraná, Brazil

Fabíola Villa
Universidade Estadual do Oeste do Paraná/ Centro de Ciências Agrárias/ Programa de Pós Graduação em Agronomia. Rua Pernambuco, 1777, CEP: 85960-000. Marechal Cândido Rondon, Paraná, Brazil

Claudio Yuji Tsutsumi
Universidade Estadual do Oeste do Paraná/ Centro de Ciências Agrárias/ Programa de Pós Graduação em Agronomia. Rua Pernambuco, 1777, CEP: 85960-000. Marechal Cândido Rondon, Paraná, Brazil

Gilmar Antônio Nava
Universidade Tecnológica Federal do Paraná, Campus Dois Vizinhos, Brazil

C. C. Onyeonagu
Department of Crop Science University of Nigeria, Nsukka, Enugu State, Nigeria

J. E. Asiegbu
Department of Crop Science University of Nigeria, Nsukka, Enugu State, Nigeria

A. M. Parmar
Department of Vegetable crops, Punjab Agricultural University, Ludhiana-141004, Punjab, India

A. P. Singh
Department of Vegetable crops, Punjab Agricultural University, Ludhiana-141004, Punjab, India

N. P. S. Dhillon
Department of Vegetable crops, Punjab Agricultural University, Ludhiana-141004, Punjab, India

M. Jamwal
Division of Fruit Science, Sher-e- Kashmir University of Agricultural Sciences and Technology of Jammu, Chatha, Jammu-180 009, (J&K), India

M. Jayanthi
Department of Seed Science and Technology, Tamil Nadu Agricultural University, Coimbatore-641 003, India

R. Umarani
Department of Seed Science and Technology, Tamil Nadu Agricultural University, Coimbatore-641 003, India

V. Vijayalakshmi
Department of Seed Science and Technology, Tamil Nadu Agricultural University, Coimbatore-641 003, India

M. Yarnia
Department of Crop Production and Plant Breeding, Tabriz Branch, Islamic Azad University, Tabriz, Iran

E. Khalilvand Behrouzyar
Department of Crop Production and Plant Breeding, Tabriz Branch, Islamic Azad University, Tabriz, Iran

F.R. Khoii
Department of Crop Production and Plant Breeding, Tabriz Branch, Islamic Azad University, Tabriz, Iran

M. Mogaddam
Department of Crop Production and Plant Breeding, Tabriz University. Iran

M. N. Safarzadeh Vishkaii
Department of Crop Production and Plant Breeding, Rasht Branch, Islamic Azad University, Rasht, Iran

Casper Nyaradzai Kamutando
Department of Crop Science,University of Zimbabwe, P.O.Box MP167, Mt Pleasant, Harare, Zimbabwe
Agriseeds (PvtLtd,5 Wimbledon Drive, P.O.Box, 6766, Eastlea, Harare, Zimbabwe

Dean Muungani
Agriseeds (PvtLtd,5 Wimbledon Drive,P.O.Box,6766,Eastlea,Harare,Zimbabwe

Doreen Rudo Masvodza
Department of Biosciences, Bindura University of Science Education, P. Bag 1020, Bindura, Zimbabwe

Edmore Gasura
Department of Crop Science,University of Zimbabwe,P.O.Box MP167,Mt Pleasant,Harare,Zimbabwe

Abebe Hinkossa
Bule Hora University, P. O. Box 144 Bule Hora, Ethiopia

Setegn Gebeyehu
National Bean Research Program, Melkassa Agricultural Research Center, P. O. Box, 436, Adama, Ethiopia

Habtamu Zeleke
Haramaya University, P. O. Box 76, Haramaya, Ethiopia

C. Cui
College of Agronomy and Biotechnology, Southwest University, 400715 Chongqing, P. R. China

Q. Y. Zhou
College of Agronomy and Biotechnology, Southwest University, 400715 Chongqing, P. R. China

R. E. Brevedan
Department of Agronomía, Universidad Nacional del Sur (UNS), 8000 Bahía Blanca, Argentina
CERZOS (CONICET), 8000 Bahía Blanca, Argentina

C. A. Busso
Department of Agronomía, Universidad Nacional del Sur (UNS), 8000 Bahía Blanca, Argentina
CERZOS (CONICET), 8000 Bahía Blanca, Argentina

M. N. Fioretti
Department of Agronomía, Universidad Nacional del Sur (UNS), 8000 Bahía Blanca, Argentina

M. B. Toribio
Department of Agronomía, Universidad Nacional del Sur (UNS), 8000 Bahía Blanca, Argentina

S. S. Baioni
Department of Agronomía, Universidad Nacional del Sur (UNS), 8000 Bahía Blanca, Argentina

Y. A. Torres
Department of Agronomía, Universidad Nacional del Sur (UNS), 8000 Bahía Blanca, Argentina
CERZOS (CONICET), 8000 Bahía Blanca, Argentina

O. A. Fernández
Department of Agronomía, Universidad Nacional del Sur (UNS), 8000 Bahía Blanca, Argentina
CERZOS (CONICET), 8000 Bahía Blanca, Argentina

H. D. Giorgetti
Chacra Experimental de Patagones, Ministerio de Asuntos Agrarios, (8504) Carmen de Patagones, Argentina

D. Bentivegna
CERZOS (CONICET), 8000 Bahía Blanca, Argentina

J. Entío
Facultad de Ciencias Agrarias, Universidad Nacional de La Plata, (1900) La Plata, Argentina

L. Ithurrart
Department of Agronomía, Universidad Nacional del Sur (UNS), 8000 Bahía Blanca, Argentina.
CERZOS (CONICET), 8000 Bahía Blanca, Argentina

O. Montenegro
Chacra Experimental de Patagones, Ministerio de Asuntos Agrarios, (8504) Carmen de Patagones, Argentina

Mujica M. de las M.
Facultad de Ciencias Agrarias, Universidad Nacional de La Plata, (1900) La Plata, Argentina

G. Rodríguez
Chacra Experimental de Patagones, Ministerio de Asuntos Agrarios, (8504) Carmen de Patagones, Argentina

G. Tucat
Department of Agronomía, Universidad Nacional del Sur (UNS), 8000 Bahía Blanca, Argentina
CERZOS (CONICET), 8000 Bahía Blanca, Argentina

Ana Dolores S. de Freitas
Soil Microbiology Laboratory, Agronomy Department, University Federal Rural of Pernambuco (UFRPE). Rua Dom Manoel de Medeiros, s/n, Dois Irmãos 52171-900 - Recife. PE - Brazil

Maria Luiza R. B. da Silva
Genomics Laboratory, Agronomical Institute of Pernambuco (IPA). Av. Gal San Martin 1371 Bonji 50761-000 - Recife. PE - Brazil

Adália C. E. S. Mergulhão
Genomics Laboratory, Agronomical Institute of Pernambuco (IPA). Av. Gal San Martin 1371 Bonji 50761-000 - Recife. PE - Brazil

Maria do Carmo C. P. de Lyra
Genomics Laboratory, Agronomical Institute of Pernambuco (IPA). Av. Gal San Martin 1371 Bonji 50761-000 - Recife. PE - Brazil

Guowei Zhang
Key Laboratory of Crop Physiology and Ecology in Southern China of Ministry of Agriculture, Nanjing Agricultural University, Nanjing, Jiangsu Province, 210095, P. R. China

Lei Zhang
Key Laboratory of Crop Physiology and Ecology in Southern China of Ministry of Agriculture, Nanjing Agricultural University, Nanjing, Jiangsu Province, 210095, P. R. China

Binglin Chen
Key Laboratory of Crop Physiology and Ecology in Southern China of Ministry of Agriculture, Nanjing Agricultural University, Nanjing, Jiangsu Province, 210095, P. R. China

Zhiguo Zhou
Key Laboratory of Crop Physiology and Ecology in Southern China of Ministry of Agriculture, Nanjing Agricultural University, Nanjing, Jiangsu Province, 210095, P. R. China

J. O. Ochuodho
School of Agriculture and Biotechnology, Chepkoilel Campus, Moi University, P. O. Box 1125-30100, Eldoret, Kenya

A. T. Modi
School of Agricultural Sciences and Agribusiness (Crop Science), University of KwaZulu-Natal, Private Bag X01, Scottsville 3209, Pietermaritzburg, South Africa

T. J. Chikuvire
Department of Crop Science, Bindura University of Science Education, P. Bag 1020, Bindura, Zimbabwe

C. Karavina
Department of Crop Science, Bindura University of Science Education, P. Bag 1020, Bindura, Zimbabwe

C. Parwada
Department of Crop Science, Bindura University of Science Education, P. Bag 1020, Bindura, Zimbabwe

B. T. Maphosa
Department of Crop Science, Bindura University of Science Education, P. Bag 1020, Bindura, Zimbabwe

T. H. Mukarati
Crop Productivity Services Division, Tobacco Research Board (Kutsaga Research Station), Airport Ring Road, Box 1909, Harare, Zimbabwe

D. Rukuni
Crop Productivity Services Division, Tobacco Research Board (Kutsaga Research Station), Airport Ring Road, Box 1909, Harare, Zimbabwe

T. Madhanzi
Department of Agronomy, Faculty of Natural Resources Management, Midlands State University, P. Bag 9055, Gweru, Zimbabwe

C. Cui
College of Agronomy and Biotechnology, Southwest University, 400715 Chongqing, P. R. China

Q. Y. Zhou
College of Agronomy and Biotechnology, Southwest University, 400715 Chongqing, P. R. China

C. B. Zhang
College of Agronomy and Biotechnology, Southwest University, 400715 Chongqing, P. R. China

L. J. Wang
College of Agronomy and Biotechnology, Southwest University, 400715 Chongqing, P. R. China

Z. F. Tan
College of Agronomy and Biotechnology, Southwest University, 400715 Chongqing, P. R. China

Jana Žaludová
Biotechnological centre, Faculty of Agriculture, University of South Bohemia, Studentská 13, České Budějovice, Czech Republic

Božena Kukolíková
Biotechnological centre, Faculty of Agriculture, University of South Bohemia, Studentská 13, České Budějovice, Czech Republic

Lenka Havlíčková
Biotechnological centre, Faculty of Agriculture, University of South Bohemia, Studentská 13, České Budějovice, Czech Republic

Vladislav Čurn
Biotechnological centre, Faculty of Agriculture, University of South Bohemia, Studentská 13, České Budějovice, Czech Republic

www.ingramcontent.com/pod-product-compliance
Lightning Source LLC
Chambersburg PA
CBHW080644200326
41458CB00013B/4731